ENNIS AND NANCY HAM LIBRARY
ROCHESTER COLLEGE
800 WEST AVON ROAD
ROCHESTER HILLS, MI 48307

EXPLORING THE WORLD OF pLASTICS

EXPLORING THE WORLD OF plastics

GERALD L. STEELE
Professor of Industrial Education
and Technology
Ball State University

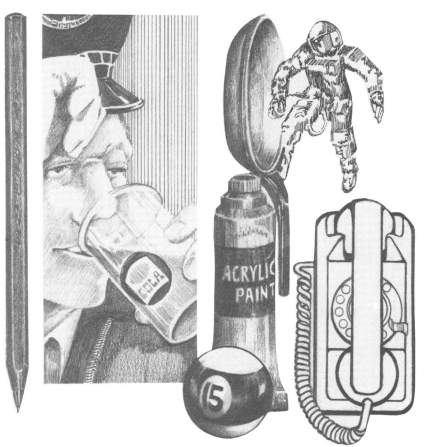

McKNIGHT PUBLISHING COMPANY
BLOOMINGTON, ILLINOIS

FIRST EDITION

Lithographed in U.S.A.

**Copyright © 1977 by McKnight Publishing Company
Bloomington, Illinois**

All rights reserved. No part of this book may be reproduced or utilized in any form or by any means, electronic or mechanical, including photocopying or recording, or by any information storage or retrieval system without permission in writing from the publisher.

**Library of Congress
Card Catalog Number: 75-42964**

SBN: 87345-411-1

To my family for their patience

and

Gary, a close friend lost

Preface

Plastics of all kinds touch our lives everyday. From the time we get up in the morning until we go to bed we come in constant contact with plastics products. The clothes we wear, the dishes we use, and even the bed we sleep in may be made all or partly of plastics. United States Government statisticians have predicted that by 1983 we will enter the "plastics age" when the volume of plastics will exceed that of metals and by 2000 A.D. plastics consumption will reach three and one-half times that of all metals. Because of these facts, it seems necessary that synthetic materials, its processes, products, and uses are understood as well as more traditional materials. This understanding is needed both from an employment and a consumer standpoint.

A recent study by The Society of the Plastics Industry, Inc. has indicated a shortage of trained personnel available to the plastics industry. It showed the need for more mold and die makers, supervisors and foremen, laboratory technicians, cycle analysts, setup personnel, quality control technicians, plastics engineers, and other employees. This survey also showed that nearly three-fourths of the plastics processors responding would pay higher wages for personnel who have pre-employment training in plastics. The basic information for an introduction to and exploration of the exciting field of plastics is contained in this book. Study of its contents may lead many to a more in-depth study of plastics — the material of the future.

This book, then, has been written as an introductory text for units or courses in plastics. The amount of time spent on specific processes will depend somewhat on the amount of time, space, and equipment available. Process selection will also depend upon the age level and prior experience of the students. Each of the basic processes of the plastics industry has been included so that a complete selection is available. It is hoped that the student will be exposed to as many of the processes and materials described as possible.

The book is divided into six sections. Each section contains information on materials or processes which are related to each other. Section I is an introduction to plastics including their testing, identification, and chemistry. Section II contains the molding processes; Section III, forming and lamination processes; Section IV, casting processes; and Section V, the finishing operations. Section VI is devoted to the development of a simulated industry for the manufacture and sale of products made partly or wholly of plastics.

The introductory section is divided into three chapters. All three should be studied as part of a comprehensive unit or course in plastics. If time is limited, either Chapters 2 or 3 or both may be reserved for later study.

Processes to be included in a plastics course or unit should be carefully selected according to their relative importance to the industry if that industry is to be truly represented or interpreted in the classroom. Processes, then, should be selected according to how widely they are used, the percentage of plastics they process, or the fact that certain materials or products can best be handled in a particular manner.

In the molding section, injection molding is the most widely used plastics process, making the widest variety of products, and having the largest number of machines in the industry. Extrusion processes the largest volume (in tons) of plastics per year. Blow molding, which is considered an extension of the extrusion process, is one of the most rapidly growing processes. The principle of extrusion and blow molding may both be demonstrated with one machine, however. A logical learning progression from plunger injection to screw injection, extrusion and blow molding can be established. These processes are highly interrelated. They are presented in that order. Compression and transfer molding still process a high percentage of the thermosetting plastics. Compression molding equipment can also be used for lamination.

In forming and lamination, plastics are often formed from, into, or onto sheetlike products. Thermoforming is one of the leading thermoplastics processes, is especially important to the packaging industry, and is simple and inexpensive to use in the school laboratory. Roll forming is more difficult to practice in the classroom, but the theory should be understood. Lamination of thermosetting plastics may often be accomplished with minimal equipment, especially for fiberglass-reinforced plastics. Thermoplastic lamination requires a compression molding or thermoplastic lamination press. If either is available, this process requires little other than the press and thermoplastic laminating sheets.

The casting processes are simple in nature. Often equipment from other processes may be adapted to casting. Expansion molding requires minimal equipment. Thermofusion may be accomplished with either a simple oven or specialized laboratory equipment. Casting processes often represent the only practical method of producing certain products.

Many finishing and decorating operations may be done with woodworking, metalworking, or graphic arts equipment.

Industrial production is encouraged in this book because plastics products are rapidly mass-produced by many processes. Most industrial plastics processing is accomplished by high speed, automated, mass-production equipment. Production organization and techniques should be included in plastics education.

A typical six-week junior-high introductory plastics unit might include an introduction to plastics materials, injection molding, extrusion/blow molding, compression molding, thermoforming, and thermofusion processes. A short mass-production unit might logically follow. Longer units of courses in plastics should include more processes and a more in-depth study of all the processes. Decorating and finishing operations should be included where necessary.

This book is designed to provide the students with the maximum number of hands-on experiences with all the plastics processes for which laboratory materials and equipment are available. Each process is described with appropriate examples of industrial equipment, materials, and products. Laboratory equipment and materials suitable for school use are also discussed. Each process chapter includes one or more operation sequence charts which contain the action, reason for the action, and troubleshooting to aid the student in efficient operation of equipment or processes. These chapters all follow a model outline to aid in the location of information.

Field tests indicate that students can learn to use the various processes, make corrections, and consistently produce good products without constant teacher supervision when using this text. The troubleshooting guides included are similar to those used in the plastics industry. Guides for material selection, storage suggestions, safety precautions, and approximate processing temperatures are also included. Each process chapter also features typical moldmaking procedures.

Finally, metric conversions have been provided throughout the book because the United States is rapidly moving towards metrication. Exact metric conversions have been computed for critical dimensions and rounded off conversions given in noncritical dimensions or where standards will probably indicate these rounded off dimensions. Because many of the metric standards have not yet been set, some of these metric dimensions may well change in the future.

Gerald L. Steele
Muncie, Indiana

Acknowledgments

I am deeply indebted to many people and plastics firms who helped in the preparation of this book. Listed below are those who were especially helpful:

For field testing portions of the book — Gerald Martin and Jay Nolley and the Monroe Community School Corporation, Cowan, Indiana; Tim Jackson and the New Castle Community School Corporation, New Castle, Indiana; Jack Callahan and the Western Wayne Schools, Lincoln High School, Cambridge City, Indiana.

Russ Meritt and William Johnson, Guide Lamp Division, General Motors Corporation, Anderson, Indiana, for generous assistance and access to production facilities for photographic purposes.

Dr. Donavon D. Lumpkin, Director of the Reading Laboratory, Ball State University, Muncie, Indiana, for analysis of reading level.

Dr. Fredrick K. Ault, Associate Professor of Chemistry, Ball State University, for generous assistance in preparation of the chapter on chemistry of plastics.

Charles Creech of Modern Cote, New Castle, Indiana, for help in preparing the roll forming chapter.

George M. Prall, Chemplex Company, Rolling Meadows, Illinois, for valuable assistance in blown film extrusion technology.

A. M. Mascari and B. H. Beckett, Western Electric Company, Indianapolis, Indiana, for evaluating sections of the manuscript for technical accuracy.

Special recognition is also given to Kenneth Snader, Michael McIntyre, Michael Hopkins, John Monroe and many class members for posing under hot studio lights.

The Society of the Plastics Industry, Inc. for materials and editorial assistance.

The Society of Plastics Engineers, Inc. for materials and assistance.

U. S. Industrial Chemicals Company for generous assistance by supplying many diagrams, pictures, and editorial help.

Ball State University for generously allowing me freedom to work and a special leave of absence to complete the manuscript and photography.

Maurice Keroack, The Plastics Education Foundation, Albany, New York, for valuable assistance and constructive criticism on the entire manuscript.

J. Harry DuBois, Morris Plains, NJ

Alfred M. Johnson, Egan Machinery Co., Somerville, NJ

Robert P. Burrington, Uniloy Group, Hoover Ball and Bearing Co., Saline, MI

Greg Klebanoff, Plasti-Mac., Inc., Longwood, FL

F. Tokarz, Dake Corp., Grand Haven, MI

John L. Hull, Hull Corp., Hatboro, PA

Richard D. Brett, McNeil-Femco-McNeil Corp., Cuyahoga Falls, OH

Alan S. Haisser, Sterling Extruder Corp., South Plainfield, NJ

John H. Hudson, National Automatic Tool Co., Inc., Richmond, IN

James L. Warren, Newbury Industries, Inc., Newbury, OH

Jane Blankenship, Ransburg Corp., Indianapolis, IN

Bill McConnell, McConnell Corp., Fort Worth, TX

Robert E. Kostur, Comet Industries, Elk Grove, IL

The cooperation of many individuals, companies, and organizations in supplying illustrations is deeply appreciated. These sources of illustrations are listed on the pages following the Appendices.

A sincere "thank you" also to my family for their understanding during the preparation of this publication.

Table of Contents

═══════ SECTION I INTRODUCTION ═══════

Chapter 1   **Introduction to Plastics** 1

Part 1 Introduction — **1,** Plastics Materials — **1,** Growth of Plastics Industry — **4,** Student Activities — **7, Part 2 Description and Classification of Plastics** — **7,** The Makeup of Plastics — **7,** Two Types of Plastics — **8,** Student Activities — **9, Part 3 Properties of Plastics** — **9,** Hardness and Softness — **9,** Impact Strength — **11,** Electrical Insulation — **13,** Weight — **13,** Heat Insulation — **14,** Temperature Resistance — **15,** Friction — **16,** Elasticity — **16,** Transparency — **18,** Cushioning — **19,** Color — **19,** Thickness — **19,** Plastics Are Many Things — **19,** Student Activities — **20, Part 4 Different Forms of Plastics** — **22,** Film and Sheeting — **22,** Rods, Tubes, and Profile Shapes — **23,** Fibers and Filaments — **24,** Liquids and Adhesives — **24,** Powders, Pellets, and Granules — **24,** Hollow Beads — **25,** Student Activities — **26, Part 5 Common Molding Methods** — **26,** Molding — **26,** Thermoforming — **26,** Lamination — **26,** Casting — **26,** Expansion — **26,** Thermofusion — **26,** Roll Forming — **27,** Finishing Operations — **27,** Outline of Plastics Processes — **27, Part 6 Plastics, Environmental Management, and Energy Use** — **28,** Plastics and Environmental Management — **28,** Reusing Plastics — **29,** Consumption of Natural Resources — **31.**

Chapter 2   **Testing and Identification of Plastics** 32

Individual Nondestructive Tests — **32,** Instructions for Making Nondestructive Tests — **32,** Demonstrations of Destructive Identification Tests — **34,** Instructions for Making Destructive Tests — **39,** Testing of Plastics — **44.**

Chapter 3   **Chemistry of Plastics** 50

Chemical Bonding — **50,** Polymerization — **54,** Achieving the Different Properties of Plastics — **56,** Types of Plastic Chains — **57,** The Future in Plastics — **58,** Student Activities — **58.**

SECTION II MOLDING

Chapter 4 Injection Molding 59

Process Description — **59,** Plunger Injection Molding — **59,** Screw Injection Molding — **60,** Importance of Injection Molding — **61,** Industrial Equipment — **64,** Materials Used in Industry — **65,** Thermoplastic Injection Molding — **65,** Thermoset Injection Molding — **65,** Types of Molding Cycles — **67,** Injection Molds — **67,** Laboratory Injection Molding Equipment — **69,** Hand Molds — **70,** Mold Design — **70,** Materials for Laboratory Injection Molding — **73,** Time, Temperature and Pressure Relationship — **73,** Operation Sequence — **74,** Checklist of Materials and Equipment Needed — **74,** Operating Temperatures — **74,** Tooling up for Production — **75,** Materials and Products for Injection Molding — **83,** Student Activities — **85,** Questions Relating to Processes and Materials — **86.**

Chapter 5 Extrusion 87

Process Description — **87,** Extruder Dies, Cooling and Take-Off Equipment — **88,** Rod and Profile Extrusion — **89,** Sheet and Film Extrusion — **89,** Tubing and Wire Coating — **94,** Strand or Ribbon Extrusion — **95,** Importance of Extrusion — **95,** Industrial Extrusion Equipment — **95,** Materials Used in the Industry — **96,** Laboratory Extrusion Equipment — **96,** Safety Requirements — **97,** Operating Requirements — **97,** Operation Sequence — **97,** Checklist of Materials and Equipment Needed — **101,** Materials for Laboratory Extrusion — **102,** Operating Temperatures — **103,** Tooling Up for Production — **103,** Recycling — **105,** Student Activities — **106,** Questions Relating to Process and Materials — **107.**

Chapter 6 Blow Molding 108

Process Description — **108,** Importance of Blow Molding — **112,** Industrial Equipment — **113,** Materials Used in Industry — **118,** Laboratory Blow Molding Equipment — **119,** Operating Requirements — **119,** Operation Sequence — **119,** Checklist of Materials and Equipment Needed — **119,** Design of Blow Molded Products — **124,** Tooling Up for Production — **124,** Indirect Blow Molding in the Laboratory — **129,** Materials and Products for Blow Molding — **129,** Student Activities — **130,** Questions Relating to Process and Materials — **131.**

Chapter 7 Compression and Transfer Molding 132

Process Description — **132,** Compression Molding — **133,** Transfer Molding — **135,** Importance of Compression and Transfer Molding — **137,** Materials Used in Industry — **138,** Industrial Equipment — **140,** Compression Molds — **140,** Laboratory Compression and Transfer Molding Equipment — **141,** Laboratory Compression Molds — **142,** Operation Requirements — **142,** Operation Sequence — **142,** Checklist of Materials and Equipment Needed — **142,** Operating Pressures and Temperatures — **143,** Tooling Up for Production — **148,** Material Specifications — **150,** Student Activities — **152,** Questions Relating to Processes and Materials — **153.**

SECTION III forming and lamination

Chapter 8   Thermoforming 155

Process Description — **155,** Vacuum Forming — **155,** Pressure Forming — **156,** Free Blowing — **157,** Mechanical Stretch Forming — **157,** Process Combinations — **159,** Thermoforming Techniques for Packaging — **159,** Monoforming — a Combination Process — **161,** Importance of Thermoforming — **161,** Materials Used in Industry — **164,** Laboratory Thermoforming Equipment — **165,** Operating Requirements — **166,** Operation Sequence — **167,** Checklist of Materials and Equipment Needed — **167,** General Operating Hints — **171,** Tooling Up for Production — **172,** Materials and Products for Laboratory Thermoforming — **175,** Student Activities — **175,** Questions Relating to Thermoforming Processes and Materials — **176.**

Chapter 9   Roll Forming 178

Calendering — Process Description — **178,** Importance of Calendering — **180,** Materials Used for Calendering — **182,** Industrial Equipment — **182,** Calender Coating — Process Description — **182,** Importance of Calender Coating — **182,** Materials Used for Calender Coating — **183,** Industrial Equipment — **184,** Knife Coating — Process Description — **184,** Importance of Knife Coating — **184,** Industrial Equipment — **185,** Materials Used for Knife Coating — **185,** Vinyl Fabric Casting — **185,** Operation Sequence for Vinyl Fabrics — **186,** Extrusion Coating — Process Description — **187,** Importance of Extrusion Coating — **188,** Materials Used for Extrusion Coating — **189,** Industrial Equipment — **189,** Student Activities — **189,** Questions Relating to Roll Forming — **189.**

Chapter 10   Lamination 190

Reinforced Plastics — Process Description — **191,** Fiberglass Reinforced Plastic Lamination — **192,** Importance of Reinforced Plastics Process — **193,** Industrial Equipment — **194,** Materials Used in Industry — **196,** Laboratory Fiberglass Laminating Equipment — **198,** Fiberglass Laminating in the Laboratory — **199,** Operation Sequence for Fiberglass Lamination — **201,** Checklist of Materials and Equipment Needed — **201,** Thermoplastic Lamination — Process Description — **207,** Importance and Advantages — **208,** Laboratory Equipment Needed — **208,** Materials for Laboratory Thermoplastic Lamination — **208,** Operation Sequence for Thermoplastic Lamination — **209,** Checklist of Equipment Needed — **209,** Compreg Lamination — Process Description — **214,** Checklist of Equipment and Materials Needed — **214,** Operation Sequence for Compreg Lamination — **214,** Student Activities — **215,** Questions Relating to Materials and Processes — **216.**

SECTION IV CASTING

Chapter 11 **Casting Plastic Materials** **217**

Importance of Casting — **217**, Industrial Equipment — **218**, Process Description — **219**, Laboratory Casting of Polyester Resins — **220**, Clear Polyester Casting Resins — **220**, Operation Sequence for Polyester Casting Resins — **221**, Checklist of Materials and Equipment Needed — **224**, Water-Extended Polyester Casting Resins — **224**, Operation Sequence for Water-Extended Polyester Casting — **225**, Checklist of Materials and Equipment Needed — **229**, Cast Synthetic Wood — **229**, Operation Sequence for Cast Synthetic Wood — **229**, Tooling Up for Production — **230**, Student Activities — **232**, Questions Relating to Processes and Materials — **232**.

Chapter 12 **Expansion Processes** **233**

Characteristics — **234**, Importance of Foams — **235**, Industrial Equipment — **235**, Resin Foam Casting — **239**, Checklist of Materials and Equipment Needed — **240**, Operation Sequence for Resin Foam Casting — **241**, Expandable Polystyrene Bead Foam Molding — **246**, Laboratory Bead Foam Molding — **247**, Operation Sequence for Bead Foam Molding — **248**, Checklist of Materials and Equipment Needed — **249**, Rejuvenating Expandable Polystyrene Foam Beads — **249**, Student Activities — **254**, Questions Relating to Processes and Materials — **254**.

Chapter 13 **Thermofusion** **255**

Process Description — **255**, Thermofusion Materials — **256**, Dip Casting and Coating — **256**, Importance of the Process — **256**, Materials Used in Industry — **257**, Industrial Equipment — **257**, Laboratory Equipment — **257**, Operation Sequence for Dip Casting and Coating — **261**, Checklist for Materials and Equipment Needed — **261**, General Operating Suggestions — **261**, Full Mold Plastisol Casting — **261**, Checklist for Materials and Equipment Needed — **262**, Operation Sequence for Full Mold Bait Casting — **262**, Slush Casting — **264**, Operation Sequence for Slush Casting — **264**, Checklist of Materials and Equipment — **266**, Manual Rotocasting of Plastisols — **266**, Checklist of Materials and Equipment Needed — **267**, Operation Sequence for Manual Rotocasting — **267**, Static Powder Molding (Engle® Process) — **268**, Importance of the Process — **268**, Materials Used in Industry — **268**, Industrial Equipment — **269**, Laboratory Equipment — **269**, Checklist of Materials and Equipment — **269**, Operation Sequence for Static Powder Molding — **269**, Static Molding of Thermoplastic Pellets — **269**, Single-Axis Rotational Molding (Heiser® Process) — **272**, Importance of the Process — **272**, Materials Used in Industry — **273**, Industrial Equipment — **273**, Laboratory Equipment — **273**, Checklist for Materials and Equipment Needed — **273**, Operation Sequence for Single-Axis Rotational Molding — **273**, Two-Axis Rotational Molding — **275**, Importance of the Process — **275**, Materials Used in Industry — **276**, Industrial Equipment — **276**, Laboratory Equipment — **277**, Checklist of Materials and Equipment Needed — **278**, Operation Sequence for Two-Axis Rotational Molding — **278**, Student Activities — **282**,

SECTION V Finishing Operations

Chapter 14 Fabrication and Bonding — 283

Machining Plastics — **283**, Cutting Plastics — **284**, Drilling — **289**, Filing — **290**, Scraping — **290**, Sanding — **290**, Buffing and Ashing — **290**, Solvent Polishes — **291**, Cementing Plastics — **292**, Mechanical Fastening — **295**, Heat Sealing Plastics — **303**, Welding Plastics — **307**, Student Activities — **314**, Questions Relating to Materials and Processes — **316**.

Chapter 15 Coating and Decorating — 317

Hot Foil Stamping — **317**, Laboratory Equipment and Materials — **320**, Operation Requirements — **321**, Checklist of Materials and Equipment — **321**, Operation Sequence for Hot Foil Stamping — **321**, Silk Screen Decorating — **322**, Laboratory Silk Screen Decorating — **322**, Painting Plastics — **323**, Plating and Vacuum Metalizing — **326**, Fluidized Bed Coating — **328**, Importance — **329**, Materials Used in Industry — **329**, Industrial Equipment — **329**, Laboratory Equipment — **330**, Operation Sequence for Fluidized Bed Coating — **330**, Checklist for Materials and Equipment Needed — **330**, Electrostatic Powder Coating — **334**, Importance and Advantages — **334**, Materials Used in Industry — **334**, Industrial Equipment — **334**, Laboratory Equipment — **334**, In-the-Mold Decorating — **336**, Two-Color Molding — **336**, Engraving Plastics — **336**, Student Activities — **338**, Questions Relating to Coating and Decorating — **338**.

SECTION VI Production

Chapter 16 Industrial Production — 340

Mass Production — **340**, Interchangeability — **340**, The Assembly Line — **341**, Types of Manufacturing Systems — **343**, Production and Plastics — **345**, The Plastics Industry — **345**, The Five Divisions of Industrial Organization — **347**, Organizing for Industrial Production — **349**, Forming the Organization — **351**, Application for Incorporation and Bylaws — **353**, Financing a Corporation — **358**, Organizing the Operating Divisions — **361**, Research and Development — **362**, Manufacturing — **362**, Marketing — **372**, Finance and Control — **376**, Closing the Business — **377**, Student Activities — **379**, Questions Relating to Mass Production — **379**.

SECTION VII Appendices

Appendix A References — 381

Appendix B Resources — 399

Bibliography — 405

Sources of Illustrations — 408

PART I

INTRODUCTION

For thousands of years, humans have made things for daily use from materials that can be **molded** or shaped. Moldable materials are those which can be shaped into useful things by hand or in a mold. Heat and pressure are often added to help mold these materials. Sometimes heat is also needed to cure (harden) these moldable materials. These molded things keep their shape until they are broken or molded again. Two kinds of moldable materials are used:
1. Natural (made by nature) and
2. Synthetic (made by humans).

Natural materials are those which come directly from nature. Clay, glass, natural rubber, and shellac are all natural materials that can be molded. They have been in use for many years. Prehistoric people learned how to make pots from clay and fire them to make them hard. Later, they learned how to make glass from silica sand and other materials.

Synthetic materials are those **new** materials that are made from natural materials already available. Old materials are changed into new and different materials (synthetic materials). These synthetic materials are then made into products that people can use. Plastics is one of several groups of synthetic materials as shown in the chart of material types on page 2. Examples of other synthetic materials include foods, drugs, grease, and artificial flavoring.

Stated another way, **plastic materials** are **synthetic materials** that can be molded or formed into useful products. This book deals with the processes used in molding synthetic materials and the characteristics of these synthetics.

Plastics Materials

More than 100 years ago, the first commercial synthetic plastic material was made. Its invention was in answer to the need to replace ivory which comes from elephant

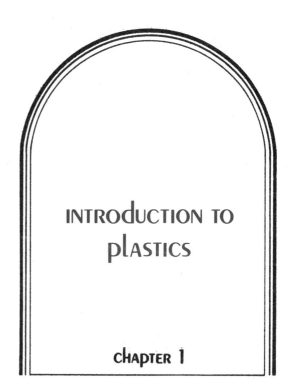

INTRODUCTION TO PLASTICS

CHAPTER 1

tusks. Ivory was in great demand for billiard balls, piano keys, combs, and many other products. The demand was so heavy that game preservation laws were passed in Africa to end the killing of elephants. This caused a shortage of ivory.

During this time, billiards was a popular sport. A $10,000 prize was offered for the invention of an artificial ivory suitable for making billiard balls. In 1868, **John Wesley Hyatt** and his brother Isaah of Albany, New York, patented cellulose nitrate, which they called **Celluloid.**® John W. Hyatt is called "the father of the plastics industry" by many people. Hyatt combined cellulose fibers and nitric acid (nitrocellulose or gun cotton) with solid camphor. Other people had worked with cellulose nitrate and solid camphor before he did. It was Hyatt, however, who developed the right combination to make the first plastic.

Hyatt's Celluloid was used for the first billiard balls made of synthetic material. It was also used for shirt collars and cuffs, movie film, and buggy and auto windows. Because Celluloid® burned very easily, it

Types of Materials

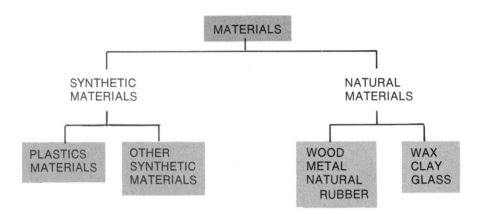

John W. Hyatt is often called the "father of the plastics industry."

caused many movie house fires. The Celluloid® film often broke in the movie projector and caught fire from the great heat of the arc lamp

About 41 years after Hyatt's success with cellulose nitrate, another plastic material was developed. In a search for a synthetic electrical insulator, Dr. Leo H. Baekeland of Yonkers, New York, developed **phenolic** plastic resin in 1909. Baekeland's plastic, called **Bakelite®** phenolic (phenol-formaldehyde), was soon molded into bobbin ends, electrical insulators, distributor caps, timers, and other automobile ignition parts. It was also used for camera parts as early as 1914 by Eastman Kodak Company. John Hyatt, who founded the Hyatt-Burroughs Billiard Ball Company, even changed his billiard balls to Bakelite® phenolic resin in later years.

A number of plastics have been invented since these first discoveries. A selected group of the most important materials, their uses, and the date of discovery is listed in Table 1-1.

Introduction to Plastics 3

A. Baby rattles

B. Cuticle knife with celluloid handle

C. Celluloid horse bridle ornament

D. Dust goggles with celluloid eye windows

E. Smokers' accessories

F. Letter openers

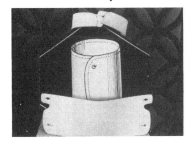
G. Celluloid shirt collars

H. Spanish comb hair ornament

I. Rain curtains of cellulose nitrate-coated cloth with celluloid windows in early autos

Fig. 1-1. Early Celluloid Products

TABLE 1-1
Historical Development of Selected Plastics
(See Appendix A for complete list.)

Date	Material	Main Uses
1868	Cellulose Nitrate (Celluloid®)	Little use today.
1909	Phenolic (Bakelite®)	Distributor caps, insulators.
1927	Polyvinyl Chloride (Vinyl or PVC)	Raincoats, auto upholstery fabric, shoes, notebook covers.
1936	Acrylic (Plexiglas®)	Airplane and snowmobile windshields, taillight lenses.
1938	Polystyrene (Styrene)	Squirt guns, toys, plastic car and airplane models.
1942	Polyethylene	Squeeze bottles, toys, bread wrappers, food containers, carpet fibers.
1942	Polyester	Fiberglass boat hulls, car bodies, coin embedments.
1943	Fluorocarbon (Teflon®)	Non-stick coatings, wire insulation for high-temperature uses.
1943	Silicone	Bathtub seal, rubber molds, space boots and gloves.
1947	Epoxy	High-strength glues, industrial molds and tooling.
1957	Polypropylene	Water ski tow ropes, carpet fibers.

Fig. 1-2. These pots and pans have Bakelite® handles.

Growth of Plastics Industry

The plastics industry has grown more than 20 times (2000%) since 1945. In recent years it has grown at a rate of 15% per year. This has been a higher growth rate than any other industry, although it leveled off at about 2% during the oil crisis in 1973 and 1974. The growth rate is not expected to climb rapidly again until the crude oil problem is solved.[1]

A whole series of plastics has been developed since 1945. Many will be discussed later in this book. A list of plastics is included in Appendix A.

Today products made from plastics are found everywhere. People wear shirts, pants, sweaters, dresses, and other clothes made from plastic fibers. They ride in cars that

[1]Later statistics can be obtained from The Society of the Plastics Industry and **Modern Plastics** magazine.

Introduction to Plastics 5

have all plastic interiors, electrical parts, and bodies. Most new cars now use about 150 or more pounds of plastics. Cars of the future will use much more plastic to make them lighter and more efficient. People use plastic telephones to talk with their friends. They buy detergents, milk, car waxes, and other liquids in plastic bottles and containers. Many of the liquid waxes and cosmetics contain plastic-based materials.

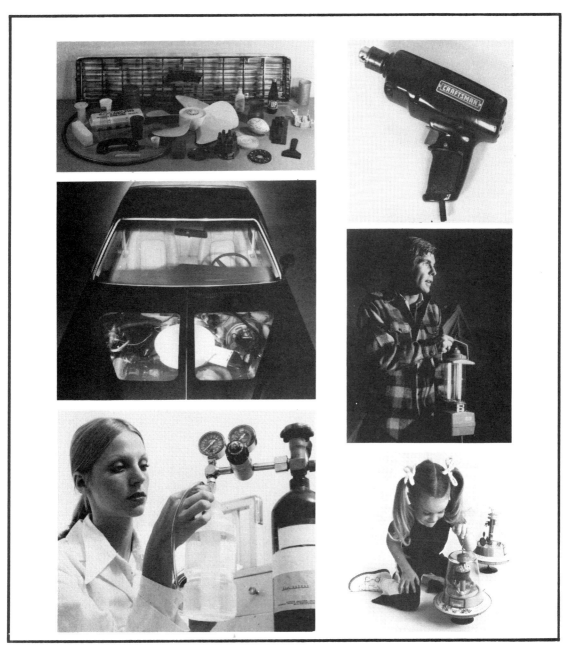

Fig. 1-3. Modern Plastic Products

6   Section I   INTRODUCTION

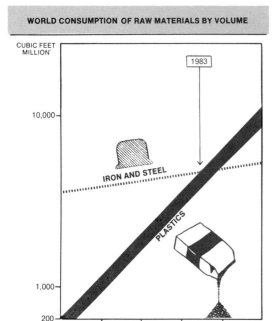

Fig. 1-4. Age of Plastics

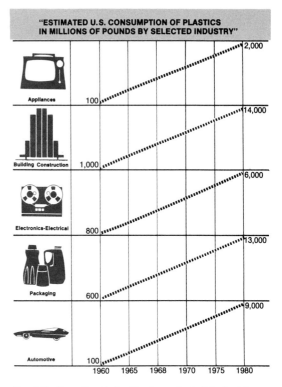

Fig. 1-5. Growth of plastics by selected industries.

Plastics is a billion-dollar industry. It makes several billion pounds of plastics per year. The 1974 U.S. production was over 29 billion pounds or 13.35 million metric tons.[2] This is more than 135 pounds (62 kg) of plastics for each person in the United States. Of these 135 pounds (62 kg) used yearly, 43 pounds (19.5 kg) are for packaging items bought, 30 pounds (13.6 kg) for buildings, 28 pounds (12.7 kg) in items used at home (appliances, furniture, housewares, and toys), 18 pounds (8.2 kg) for electrical products, and 16 pounds (7.25 kg) used on the farm and for transportation.

Plastics have already passed iron and steel in cubic volume. They are much lighter in weight than iron and steel. Plastics weigh about 55 to 185 pounds per cubic foot (881 to 2692 kg/m³). The same amount of iron or steel weighs about 485 pounds (220 kg). Thus, it takes about 3 to 9 pounds (1.36 to 4.08 kg) of plastic to equal the volume of 1 pound (0.45 kg) of iron or steel. In other words, it takes a lot more pounds of plastic than iron and steel to form equal cubic volume.

By 1983, plastics production is expected to exceed the production total of all metals. This will be at least four times more than the 1972 production of plastics. When this happens we will enter the age of plastics. United States Government statisticians report that by the year 2000 plastics use will be three and one-half (3½) times the volume of all metals.[3] The production of plastics materials in the United States is

[2]Data from **Modern Plastics** magazine, January, 1975. For up-to-date information on plastics production and consumption in the U.S.A., see the most recent January issue of **Modern Plastics**.

[3]According to Government statistics quoted by J. Harry DuBois, President of Molecular Dialectrics, Clifton, New Jersey.

predicted by Stanford Research Institute to grow at the annual rate of 8.6 percent to the year 2000. It is expected to grow from 29 billion pounds (13.35 metric tons) produced in 1974 to about 227 billion pounds (102.96 metric tons) in the year 2000. The total sales volume of the three parts of the industry — plastics materials, plastics processors, and plastics machinery manufacturers — is expected to grow from about $23 billion in 1970 to about $634 billion in 2000.[4]

STUDENT ACTIVITIES

1. Make a list of plastic products and how they are used in everyday life. Try to make your list different from your classmates'.
2. Collect samples of plastic materials or products and bring your collection to class. Try to find unusual uses for plastics.
3. Discuss the samples in class. Note the use of each plastic.
4. Make a list of some new uses of plastics. Try to find articles in newspapers and magazines on new plastics products just coming onto the market.
5. List all the materials in an automobile (or some other large product). Of all the materials listed, how many or what percent were plastics?

PART 2

DESCRIPTION AND CLASSIFICATION OF PLASTICS

The Makeup of Plastics

Many new synthetic materials have been developed to meet people's changing needs. One of these synthetic materials is plastics. Plastics meet many needs which used to be met by natural materials. They also meet new needs such as in space travel, communications, undersea exploration, and many others.

In order to make synthetic materials, chemists have taken **atoms** and **molecules** from natural materials such as coal, oil, natural gas, air, water, plant fibers, and salt. The chemists have rearranged these atoms and molecules to form plastics.

Atoms are the smallest particles of a chemical element (a substance broken down in its simplest form). The atoms keep the properties of that element. Two or more atoms linked together form a molecule. A **molecule** is the smallest part of a chemical compound (combination of elements) that keeps the properties of that compound. The more atoms that join together to form a molecule the larger the molecule gets. Several small molecules may also be joined together to form a large molecule. The small molecules used to make plastics are called **monomer** (single unit) molecules. The large plastics molecules formed from many monomer (small, single unit) molecules are called **polymer** (many unit) molecules. The monomer molecules link together like a chain to form the polymer molecules. Polymer molecules are made up of hundreds to thousands of monomer molecules. A polymer molecule chain is usually less than 1/10,000th of an inch (0.00025 mm) long, however. Plastics are made up of many polymer chains (polymer molecules). Chapter 3 gives more details on the chemistry of plastics.

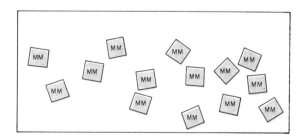

Fig. 1-6. Block diagram of several monomer molecules. Monomer molecules are small molecules that are not connected to each other.

---

[4]Taken from a report issued by the Society of the Plastics Industry, Inc. dated June, 1973.

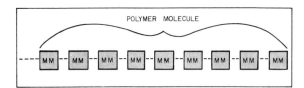

Fig. 1-7. Polymer Molecule
A polymer molecule is composed of hundreds to thousands of monomer molecules joined in a chain.

Plastics, then, are synthetic materials made up of long chains of small molecules. During manufacture, plastics get soft and moldable under heat and pressure. The plastic then hardens (solidifies) into the molded shape.

The name "plastics" comes from the fact that these synthetic materials can be formed or molded into shapes. Thus, plastics has come to mean a moldable or formable synthetic material. Natural materials such as glass, clay (ceramics), and wax do not belong to this category. They are moldable but not synthetic.

Two Types of Plastics

All plastics can be put into one of two groups:
1. Thermoplastic (meaning heat softening) and
2. Thermosetting (meaning heat curing).

Thermoplastics are synthetic materials which become soft when they are heated and hard when they are cooled. They act like candle wax or ice when they are heated or cooled. When wax and ice are heated, they make a **physical** change (change in state) from a solid to a liquid. When they are cooled, they make a **physical** change from a liquid to a solid. The physical change takes place **twice** in thermoplastics during a molding cycle. The solid plastic particles are heated until they are soft and molded into a shape (a physical change). The soft molded plastic is cooled into a solid state to hold the molded shape (the second physical change).

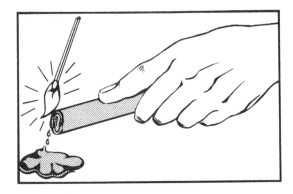

Fig. 1-8. Thermoplastics act like candle wax when heated or cooled.

Thermoplastic materials may be softened and hardened many times. Each time they are reheated, however, some of the additives such as colorants and lubricants may be lost. This loss limits the number of times thermoplastics may be reused. The plastics industry recycles (reuses) scrap thermoplastics in their plants to conserve natural resources and prevent a rapid waste build-up. Recycling of thermoplastics from the home and community also contributes to a cleaner world.

If too much heat is applied to plastics, they may be degraded (destroyed). Plastics materials are degraded when they are burned. They are broken up into basic elements and cannot be reused.

Some familiar thermoplastics are acrylic, cellulose nitrate, polyvinyl chloride (vinyl), polystyrene, polyethylene, fluorocarbon, and polypropylene. A complete list of thermoplastics is shown in Appendix A.

Thermosetting plastics are cured (set) into permanent shape during molding. Heat or heat and pressure cause a physical change followed by a **chemical** change in solid-form thermosetting plastics. In liquid thermosets, only a chemical change is caused by heat. A chemical change is a permanent change. This change is similar to curing cement or baking a cake. Thermosetting plastics cannot be softened or recycled by reheating. They may, however, be

Fig. 1-9. Thermosetting plastics act like concrete when set.

degraded by extremely high temperatures. They may also be cut or broken.

Some familiar thermosetting plastics are phenolic, polyester, silicone, and epoxy. See Appendix A for a complete list.

STUDENT ACTIVITIES

1. On the basis of what you have learned, try to classify as thermoplastic or thermosetting plastic the samples you collected in Part 1. Tag each sample and save it for later use.
2. Make a list of scrap plastic materials from the home that can be recycled. Try to think of ways in which these materials can be reused.
3. Look for sources of scrap plastic materials in the community or around school. Some of these materials may be used in your classroom for processing.

PART 3
PROPERTIES OF PLASTICS

Plastics have many outstanding properties, or qualities. These properties change greatly from family to family of plastics. They also change widely within each family of plastics. A family of plastics is a **group of plastics which is chemically the same.** Each family of plastics has a name such as polyethylene, polypropylene, acrylic, or phenolic.

Chemists take atoms and molecules from air, water, coal, oil, natural gas, plant fibers, and salt and make them into plastics. They can break atoms away from molecules and add other molecules and atoms with heat, pressure, and catalysts. **Catalysts** are chemicals which speed up a chemical reaction.

Different plastics are created by the chemist to have certain properties. Sometimes two or three families of plastics are **chemically** mixed together. When two families are chemically mixed together the plastic is called a **copolymer.** An example is the mixture of polyethylene and polyvinyl chloride. The resulting mixture is called **ethylene vinyl copolymer.** The best properties of both families usually show up in a copolymer. When three plastic families are chemically mixed together, the plastic is called a **terpolymer.** ABS (acrylonitrile-butadiene-styrene) is an example of a terpolymer. Such mixtures often give plastics better or new sets of properties.

Properties such as hardness, softness, impact strength, electrical insulation, heat insulation, and weight are important in plastics. Other properties are temperature resistance, chemical resistance, elasticity, friction, transparency, and the ability to be permanently colored.

Hardness and Softness

Plastics can be made hard and stiff or soft and flexible, or they can be made to have qualities in between these. Usually, lightweight plastics are soft and flexible. Often, heavy plastics are hard and stiff. Hard, stiff plastics may break easily. Some very hard plastics have been made which do not break easily, however.

Phenolic plastics (Bakelite® is one trade name) are hard and stiff. They are quite strong but can be broken. Phenolic is used for hard, smooth, and shiny auto distributor caps, washing machine agitators, and pot handles. Phenolic plastic is almost always dark in color — usually black, brown, or red.

Another hard, stiff plastic is **polyester**. Polyester plastic is used with fiberglass to make boat hulls, car bodies, suitcases, and fishing rods. The glass fibers give strength, while the polyester plastic makes them rigid. Polyester resins may also be used for clear castings. An example is a coin embedment casting for book ends. Polyester plastic is hard, shiny, and naturally clear. However, it may be colored any color.

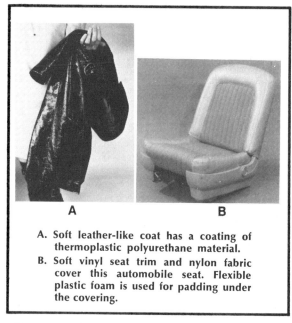

A. Soft leather-like coat has a coating of thermoplastic polyurethane material.
B. Soft vinyl seat trim and nylon fabric cover this automobile seat. Flexible plastic foam is used for padding under the covering.

Fig. 1-12. Soft Plastic Products

A. Hardness is like a diamond.
B. Softness is like a pillow.

Fig. 1-10. Hardness and Softness

Fig. 1-11. Polyester/Fiberglass Boat

One of the softest plastics is low density **polyethylene**. It is also one of the most flexible. Polyethylene, the squeeze bottle plastic, is lightweight and feels soft and waxy. It is used for flexible detergent bottles, sandwich bags, bread wrappers, dry cleaners' bags, and toys. Another soft, flexible plastic is **polypropylene**. It is often used for rope and indoor/outdoor carpet fibers such as Ozite®. Polypropylene is also used for one-piece plastic hinges.

Impact Strength

Most plastic products should be able to withstand being dropped without breaking. This is called **impact strength**. A combination of impact strength and flexibility is often wanted in a product.

Most thermoplastics have acceptable impact strength. One plastic which provides very good impact strength and flexibility is low density **polyethylene** (the squeeze bottle plastic). A detergent bottle should not break if it is dropped. Most of them are now made from polyethylene.

Polyvinyl chloride (vinyl) has very good impact strength also. It is the imitation leather that is used for notebook covers, furniture and automobile upholstery, and food packaging. Vinyl is also used for shower curtains and raincoats.

Ultra-high molecule weight (UHMW) **polyethylene** is a top choice for impact strength. This plastic has very large, compact polymer molecules. It is sometimes used for runners on snow skis. Another plastic of very high impact strength is **polycarbonate. Ionomer plastic** also has high-impact strength.

Impact strength is also important in thermosetting plastics. A **combination of epoxy plastic and fiberglass** probably gives the best impact strength, on a strength to

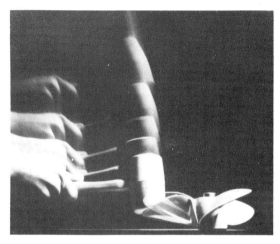

Fig. 1-14. Showing the impact strength of a Lexan® polycarbonate outboard motor propeller.

Fig. 1-15. A professional golf ball has an almost indestructible cover of Du Pont's Surlyn® ionomer resin.

Fig. 1-13. Impact Strength

Fig. 1-18. Teflon® electrical insulation is used in this F-4 fighter plane.

Fig. 1-16. Glass-filled thermoplastic injection molded automobile fan shroud being ejected from the mold.

Fig. 1-17. Electrical Insulation

gardless of strength to weight ratio. Fiberglass-filled polyester resins are used for boats, car bodies, and snowmobile front covers. **Glass-filled phenolics** are used for power tool handles and rocket engine parts. Thermosetting resins are often mixed with glass fibers to give them better impact strength.

Many common thermoplastics are also mixed with glass fibers to give them better impact strength, too. These thermoplastics are called **glass-filled thermoplastics.** Glass-filled polypropylene is used for automobile inner fenders and fan covers.

weight ratio, of all thermosetting plastics. Epoxy/glass materials are used for space rocket parts such as fuel tanks, covers, and heat shields. **Fiberglass-filled polyester resins** have the highest impact strength re-

Introduction to Plastics

Fig. 1-19. Double-insulated saber saw using insulating value of a thermoplastic body.

Fig. 1-20. Phenolic parts are used as electrical insulators.

Electrical Insulation

Plastics, in their natural state, are good electrical insulators. That is, they will not carry an electrical current. The electrical insulation value varies from plastic to plastic. This value depends upon the molecular structure of the plastic. Plastics with balanced molecular forces (nonpolar molecules) within themselves are better electrical insulators. Some plastics may be made **conductive** (capable of transmitting electricity) when metallic fillers are added to the resin.

An important thermoplastic electrical insulator is **low-density polyethylene.** Its balanced molecular forces are not affected by high-frequency electrical currents. Its insulation value and its good flexibility make polyethylene an excellent insulator for electrical wires. These properties made television and radar cables possible. **Fluorocarbon** (Teflon®) insulation is also excellent for electrical wires and cables. It can withstand very high temperatures. Fluorocarbon is more expensive than polyethylene, however. It is used for wire insulation in large electrical transformers and other products where long life at high temperatures is necessary. **Polyvinyl chloride** is widely used for low-frequency electrical insulation. It has a higher melting point than polyethylene.

A rigid thermosetting plastic like **phenolic** is important for its electrical insulation value,

Fig. 1-21. Lightweight

Fig. 1-22. Lightweight plastics float on water.

too. It is used for rigid parts such as automobile distributor caps and rotors. This hard and shiny plastic is also used for electric coffee pot and toaster bases.

Weight

Plastics are generally classified as lightweight solid materials. They are lighter than most metals. Some plastics, such as polypropylene (the featherweight plastic) and polyethylene (the squeeze bottle plastic), are light enough to float on water. Poly-

14 Section I INTRODUCTION

Fig. 1-23. Polypropylene water ski tow rope will float.

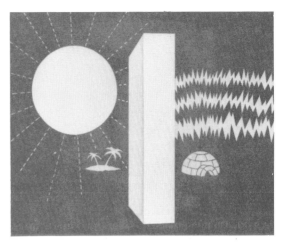

Fig. 1-25. Heat Insulation

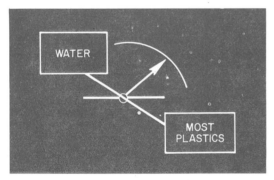

Fig. 1-24. Most plastics are heavier than water.

styrene and polyvinyl chloride are slightly heavier than water and will sink. Thermosetting plastics, such as phenolics, polyesters, and epoxies, are usually heavier than water. They will sink in it.

Plastics have weights as high as 3¼ times that of water, for equal volumes. However, most plastics are less than 2¼ times the weight of water. Compare this with the weights of some common metals. Aluminum is about 2¾, iron is 7¾, and lead is 11¼ times the weight of water.

Heat Insulation

Plastics are good heat insulators. In fact, this good heat insulating quality slows down most plastics processes. The plastic materials are slow to heat. They are just as slow to cool once they are formed. When plastics are foamed or mixed with fillers like asbestos, they are much better heat insulators.

Foamed plastics are the best heat insulators. Because of this, they take 10 times longer to cool in a mold than solid plastics. Foamed polystyrene is one of the best heat insulators. It is often used for picnic coolers and ice buckets (expandable polystyrene bead foam). Throw-away hot drink cups and large building insulation panels are made from polystyrene bead foam. Polystyrene is also foamed into a rigid spongelike structure. One trade name for this is **Styrofoam®**. It is formed into boards for building insulation and home decorations. Styrofoam® is also used for insulating refrigerators and freezers. Rigid polyurethane foam is considered the best heat insulating plastic.

Phenolics mixed with asbestos are good heat insulators where strong, rigid products are needed. Pot and pan handles and electric coffee pot bases are good examples.

Introduction to Plastics 15

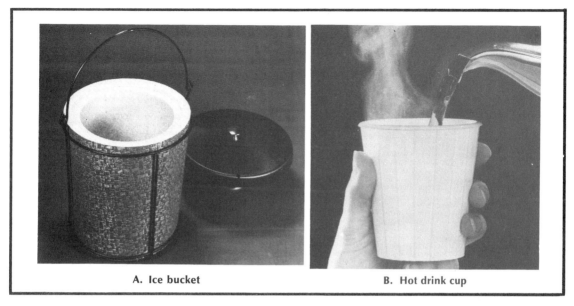

A. Ice bucket   B. Hot drink cup

Fig. 1-26. Insulated Products

Fig. 1-27. This phenolic part of a pressing iron is heat-insulated.

Fig. 1-28. High-temperature resistance of silicone rubber.

Temperature Resistance

Materials are needed which will take extreme temperatures, both hot and cold. They are needed for special gaskets in extreme temperature uses, for molds in which to pour hot metals, and for space flight. They are also used for common home appliances such as electric frying pans or clothes irons.

Silicone (the bathtub seal plastic) is used where a flexible material is needed. It can be used for gaskets, glues, sealers, and rubberlike molds. Silicone, a thermosetting plastic, will stand up to 900°F (482°C) for a few minutes. Silicone boots and gloves were worn by astronauts on the surface of the moon.

Fig. 1-29. Silicone Bathtub Seal

Fig. 1-31. Space suits worn on the moon have Teflon® parts.

Fig. 1-30. An astronaut on the moon wears a boot with a sole constructed of silicone.

Phenolic plastics have good high-temperature resistance when mixed with asbestos. Asbestos-filled phenolic is often used for electric frying pan and iron handles.

Fluorocarbon plastics will stand both very low and very high temperatures, ranging from —400°F to +500°F (—240°C to +260°C). They are thermoplastics which are often used for insulation on wires in such places as space rockets where temperature ranges from very hot to very cold. Fluorocarbon threads have also been made into cloth for space suits used on moon flights.

Friction

Several plastics are low in friction (slippery), Fig. 1-33. Fluorocarbon plastics (the nonstick plastics) are among the best. Fluorocarbon (Teflon®) plastics are so slippery that most things will not stick to them. Molecules of two materials have to intermingle (mix together) to stick together. Teflon's tight molecular bonding keeps the molecules from intermingling so other things will not stick to it. Teflon coatings are applied to frying pans, electric irons, saw blades, and other products.

Polyethylene has the same nonstick quality, but it melts at a much lower temperature than fluorocarbon (400°F or 204°C less). It cannot be used as a nonstick coating for things used in high-temperature places. It can, however, be used as a mold for **casting** polyester plastic resins.

Elasticity

Some plastics can be stretched like a rubber band. They are called **elastomer** (stretchy) **plastics**. Elastomers can be stretched to at least twice their length at

Introduction to Plastics 17

Fig. 1-32. In space rockets, Teflon® is used in electrical insulation, heat shields, and fuel tanks.

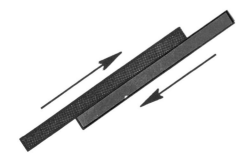

Fig. 1-33. Friction

Fig. 1-34. Pots and pans may be Teflon®-Lined.

Fig. 1-35. Elasticity is the capacity to return to original shape after being stretched.

Fig. 1-36. Hoses used by space-walking astronauts are made of flexible and temperature-resistant silicone rubber.

Fig. 1-37. Elasticity of silicone rubber is useful in making molds for plastic and metal products.

Fig. 1-39. Clear acrylic plastic lens for a car.

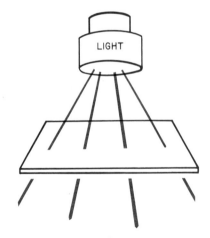

Fig. 1-38. Transparency (Clearness)

72°F (22°C) and will return to their original shape. Silicone (the bathtub seal plastic) is an excellent example. It is used as a flexible glue for bathtubs and showers, airplanes, and space rockets. Silicone molds are used for casting metal and plastic parts.

Transparency

Transparency, or clearness, varies with plastics. Some are as **clear** as glass. Some cannot be seen through and are called **opaque** plastics. Plastics that can be partly seen through are called **translucent.**

Natural polystyrene is crystal clear, but is often colored as desired. Clear polystyrene is used for food containers. It is inexpensive, odorless, and tasteless and the food can be seen in it. Polystyrene is often used for such toys as squirt guns. It should not be used for outdoor products since it weathers quickly and shatters like glass.

Acrylic plastic (Plexiglas® or Lucite®) is used for many clear, outdoor products. It weathers well and is stronger than polystyrene. Airplane windshields, lighted transparent signs, and automobile taillight lenses are good examples of acrylic plastic products. Low-cost camera lenses and contact lenses are molded from acrylic plastic.

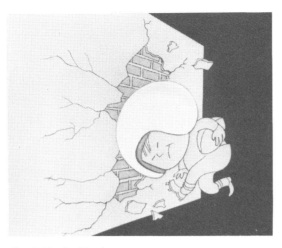

Fig. 1-40. Cushioning

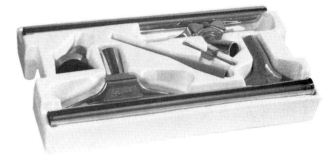

Fig. 1-41. Products are often packed in bead foam plastics that have cushioning characteristics.

Cushioning

Cushioning is the ability of a material to absorb a shock or impact. It is usually used to protect something from damage. Cushioning materials are used to protect TV sets, typewriters, radios, and all kinds of other products from damage during shipment. Cushioning materials are used on the dashboard, steering wheel, and other places in cars to protect people. They are also used in chairs, seats, and mattresses. Most cushioning materials are now made from plastic foams. They may be either rigid or flexible. Foams such as **expandable polystyrene bead foam, polyurethane** (either flexible or rigid), and **polyethylene foam** are popular examples.

Color

Plastics should not be identified by color except in the natural state since different plastic families may be colored alike. Most plastics are colored before they are made into products. Color additives (colorants) are put into the plastic either before or during the molding process. The main advantage in coloring plastics is that the color goes all the way through the plastic. It is usually not surface colored (painted or stained) as is wood and metal.

Thickness

Most plastic families can be processed in a wide variety of thicknesses and the plastic products made into almost any thickness as well. Thickness should not be used as a means of identifying plastic families.

Plastics Are Many Things

Plastics have many properties. Their properties vary from plastic family to plastic family. Plastics are hard, soft, or in between; strong or weak; stiff or flexible; and elastic or nonelastic. Some plastics are clear. Some cannot be seen through while others let some light through. Most plastics are light in weight, good electrical insulators, and good heat insulators. Some plastics are very slippery. The chemical makeup determines the properties of a plastic. The chemist matches the needs of a product or use for it with the properties of a material. If an existing material does not meet these needs, a new material is built. The chemist arranges atoms and molecules to suit certain needs. These new materials are the synthetics.

Manufacturers of plastic products can usually make their choice of material from several different types of plastics. The plastic chosen often depends on how great the product need is (consumer demand) and

TABLE 1-2
Properties of Selected Plastics
(As compared to other plastics. All materials in natural unaltered state.)

	Property / Family	Hardness	Impact Strength	Electrical Insulation	Weight	Heat Insulation	Temperature Resistance	Friction	Elasticity	Transparency
THERMOPLASTICS	ACRYLIC (Lens)	H	M	M	M	M	M	M	M	H
	FLUOROCARBON (Nonstick)	H	H	H	VH	VH	VH	VL	M	L
	POLYETHYLENE, HI DENSITY (Milk bottle)	H	L	H	L	H	L	L	L	M
	POLYETHYLENE, LO DENSITY (Squeeze bottle)	L	VH	VH	L	M	VL	VL	H	M
	POLYPROPYLENE (Featherweight)	M	M	H	L	M	M	L	M	L
	POLYSTYRENE GEN. PURPOSE (Squirt Gun)	H	L	M	L	L*	VL	H	L	VH
	POLYVINYL CHLORIDE (Imitation leather)	M	H	M	M	L	VL	M	M	M
THERMOSETTING	EPOXY (Glue)	H	H	H	L to M	M	H	M	L	H
	PHENOLIC (Distributor cap)	VH	H	VH	M	H	H	H	VL	None
	POLYESTER (Boat)	H	L	H	M	M	M	L to M	VL	H
	SILICONE (Bathtub seal)	L to M	H	H	M	VH	VH	L	H	L

VL = Very Low
L = Low
M = Medium
H = High
VH = Very High

how many other products there are like it of similar price (market competition). When lots of similar products are on the market and there is much competition, cheaper materials are often used so that the product price can be kept low. When a special quality product is offered and competition is low, a better grade of material is often used. Selecting plastics for product manufacture on the basis of demand and competition, however, is not the best way to determine which plastic should be used. Plastics (or any material) should be chosen for use by their **properties**. Thus, if the product is a child's toy, an unbreakable plastic is most desirable. Also, any product should have a reasonable period of usefulness. Often, product usefulness is shortened by a poor material choice. Better materials will usually make products last longer. The properties of several popular plastics can be quickly compared in Table 1-2.

STUDENT ACTIVITIES

1. Discuss the outstanding properties of the sample plastics collected for class.
2. Discuss why each product needs the properties of the plastic from which it is made.
3. Try to find ways in which these products can be improved by changing the plastic from which they are made. What properties should be improved and why?
4. If the plastic product is a single-service product (to be used only once), why was the plastic it was made from chosen?

Introduction to Plastics 21

	P	L	A	S	TIC
1	MATERIALS	MOLDED	FORMED	PRESSURE	MOLDING
2	PRODUCTS	SYNTHETIC	NATURAL	ATOM	MOLECULE
3	MONOMER	POLYMER	PLASTIC	THERMOPLASTIC	THERMOSETTING
4	PROPERTIES	PLASTIC FAMILY	CATALYST	HOMOPOLYMER	COPOLYMER
5	TERPOLYMER	CHEMIST	PRODUCED	CELLULOID	HYATT
6	BAEKELAND	BAKELITE	PHENOLIC	POLYVINYL CHLORIDE	ACRYLIC
7	POLYETHYLENE	CURE	HARDNESS	IMPACT	INSULATION
8	RESISTANCE	ELASTICITY	TRANSPARENCY	OPAQUE	TRANSLUCENT

	P	L	A	S	TIC
1					
2					
3			FREE		
4					
5					

Fig. 1-42. Plastics Game

5. Play the game **Plastic**. Study Fig. 1-42. You will quickly see that this game is played much like Bingo. To play the game, prepare your own game sheet. **Do not write in this book.** First, study the five columns of words near the top of Fig. 1-42. Then, on a sheet of paper write the word "plastic" in separated letters as shown so that they are headings for five columns. Select five words from those listed under the letter **P** and write them in a column under the **P** on your paper. Select five words from those listed under **L** and write them in a column under **L** on your paper. Continue the same procedure for each column. Next, rule off a form like the one shown, **numbering** the five rows of spaces **vertically** and **lettering** the five columns of spaces **horizontally**. Write "free" in the center space as shown. Your teacher will call out a word from those listed in Fig. 1-42 and will designate a particular space that it may be written in on your form, such as "P—1" or "L—2." Your teacher will then ask a student what the word means. If the student's answer is correct, all those who have listed that word on their paper can write it in the designated space. If the answer is wrong, no one can use the word until someone correctly defines it. This continues until a student has five words in a row — horizontally, vertically, or diagonally. Remember, the word can only be written in the place on your form that the teacher designates.
6. Design new plastic game cards.

PART 4

DIFFERENT FORMS OF PLASTICS

Plastics are produced in many different forms. They are made into film and sheeting; rods, tubes and profiles; fibers and filaments; liquids and adhesives; molding powders and pellets; and hollow beads. A brief description of each kind follows. Descriptions of the processes used to form these plastics into usable products are given later.

Film and Sheeting

Flexible plastic film is .010" (0.254 mm) or less in thickness. A postcard is about .010" (0.254 mm) thick. Some film, such as dry cleaners' bags, is as thin as .0005" (0.0127 mm). Plastic film is also used for shower curtains, raincoats, farm and building film, paint tarps, and sandwich bags. A new plastic paper has recently been introduced that is also a plastic film. It may soon compete with wood pulp paper.

Plastic film is usually formed in wide pieces and rolled into very long lengths. It is then cut into shorter lengths and widths to be sold. Plastic builders' film (Visqueen®) is sold in 100 foot rolls (30.48 m), .004" (0.101 mm) to .006" (0.152 mm) thick and up to 22 foot (6.7 m) widths.

Plastic sheet is more than .010" (0.254 mm) in thickness. It is a flat section of plastic that is longer than it is wide. Plastic sheets are often made in 40" to 48" (1016 mm to 1219 mm) widths and 72" to 96"

Fig. 1-43. Plastic film is placed over cement at a construction project.

Introduction to Plastics

(1829 mm to 2438 mm) lengths. Plastic sheets as wide as 120" (3048 mm) are now available on special order. Standard sheet sizes vary with different suppliers, but 40" x 72" (1016 mm x 1829 mm) and 48" x 96" (1219 mm x 2438 mm) are common. Sheets are often cut into smaller stock sizes for sale.

Plastic sheeting is used for airplane and snowmobile windshields, windows in public buildings, storm door windows, and lighted sign panels. A large amount of plastic sheeting is formed into refrigerator liners and door panels, toys, suitcase sides, and sporting goods.

Rods, Tubes and Profile Shapes

Plastic rods and tubes may be made by several different methods. They may be cast, machined, or extruded (pushed through a specially shaped opening). Profile shapes are similar to plastic rods and tubes. They are all long, continuous plastic shapes. Tubes are used for things like soda straws and garden hoses. Rods are used as parts for toys and other products. Hot Wheels® track is an example of a plastic profile.

A. All-terrain vehicle body made from ABS sheet plastic.

B. Sailboat with ABS sheet plastic hull.

Fig. 1-44. Sheet Plastic Products

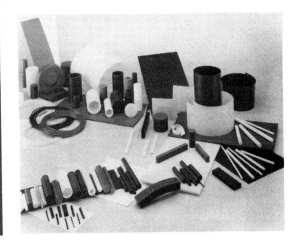

Fig. 1-45. Plastic Rods, Tubes, and Profiles

Fig. 1-46. This display shows the uses of plastic fibers and filaments.

Fig. 1-47. Silicone auto polish is an example of liquid plastics.

Fibers and Filaments

A plastic filament is a long strand of plastic, usually very small in diameter — about the size of a hair. It is often made by pushing and pulling soft plastic through a very small hole. The plastic is pulled through the hole rapidly and stretched as it comes out. This stretching makes the plastic fiber smaller than the hole. Other plastic filaments are made by cutting thin plastic film into many strands.

Plastic fibers are usually short strands (½″, or 12.7 mm or somewhat longer). They are made by chopping up the longer plastic filaments.

Plastic fibers are used for broom and brush bristles and in making plastic rugs. Filaments are used for making plastic rope, cloth, and carpets. Ozite® indoor-outdoor carpets are a good example of this use. Plastic fibers and filaments are also used in many clothing materials, such as dacron (polyester), rayon (cellulose) and nylon.

Liquids and Adhesives

Many plastics are liquids — white glue (polyvinyl acetate), epoxy, and polyester plastic resin are good examples. Vinyl plastisol (polyvinyl chloride-acetate), another form of liquid plastics is used to make such things as hollow plastic balls and coin purses. Many of the glues and finishes now used for woods and metals are liquid plastics. Some silicones are produced in liquid form and are used as waxes for cars and furniture and as plastic mold releases.

Powders, Pellets, and Granules

Most plastics to be molded are powders, pellets, or granules. Those plastics used for

Introduction to Plastics 25

Fig. 1-48. Plastic Powders

Fig. 1-49. Plastic Pellets

injection molding, extrusion, and blow molding are usually sold in pellet form. Plastics for rotational molding, fluidized bed coating, and electrostatic spray coating are in powder form. Plastic powders are finely ground dry plastics. Solid-form thermosetting plastics (phenolics and others) are made in granular form. Many granular plastics look like sand.

The standard size container for most solid plastic powders, pellets and granules is the 50 pound (22.68 kg) bag. Large containers are 250 (113.4 kg) and 1000 pound (453.6 kg) tote bins and 130,000 pound (58968 kg) rail cars. Large users save money by buying plastics in the 130,000 pound (58968 kg) rail cars.

Hollow Beads

Expandable polystyrene plastic is available in small spheres with porous centers. They look like a sponge inside. Polystyrene beads are expanded by heating. After expansion they are placed in a mold and expanded more until they form a solid part. The part is removed from the mold after cooling.

Polystyrene beads can be expanded to nearly 50 times their original size. They make light, strong, shock absorbing, heat insulating materials. See Figs. 1-50, 1-51.

Fig. 1-50. Unexpanded Polystyrene Beads

Fig. 1-51. Expanded Polystyrene Beads

> **STUDENT ACTIVITY**
> 1. Collect samples of various forms in which plastic materials appear. Look for them in local plastics industries, hobby shops, and your school industrial arts laboratory.
> 2. Heat a few pieces of each sample in an electric frying pan which has been coated with fluorocarbon mold release. Look for differences in the melting temperatures and melting rates.

PART 5
COMMON MOLDING METHODS

Plastics are molded or formed in a number of different ways. All together, these different ways of molding or forming are called **processes.** Some of the processes are related to each other. A very brief description of each of the process groupings follows. More detailed information is given in each of the process chapters. These chapters are grouped together by the relationship of the processes to each other. Each group is a section of the book.

Molding

Molding processes are those in which the plastic material is compressed in or forced into or through a mold or die. Plastic materials are formed or finished **under pressure.** Molding includes injection molding, extrusion molding, blow molding, compression molding, transfer molding, and cold molding.

Thermoforming

Thermoforming is a group of processes in which hot plastic sheet or film stock is formed over or into a mold by vacuum, air pressure, or mechanical force. Vacuum forming is divided into two kinds — drape forming and cavity forming.

Lamination

Lamination is a group of processes in which layers of plastic materials or plastics combined with other materials are formed into products. It includes high- and low-pressure lamination. Low-pressure lamination products are made from fiberglass and thermoplastic sheets. Low-pressure lamination is any pressure up to 1000 pounds per square inch (6900 kPa). High-pressure lamination can be divided into two groups: (1) high-pressure fiberglass lamination and (2) high-pressure thermosetting lamination. High-pressure thermosetting laminates are products like table, desk, and counter tops. Formica® and Micarta® materials are high-pressure laminates. High-pressure lamination is above 1000 pounds per square inch (6900 kPa).

Casting

Casting processes are those in which the liquid or granulated plastic material enters the mold by gravity. In other words, it is poured into the mold. In casting, no force other than gravity and atmospheric pressure is applied to the material in the mold. Casting processes use both thermoplastic and thermosetting plastic materials.

Expansion

Expansion is a group of processes in which the plastic material is poured (cast) or forced (molded) into or onto a mold and then expanded. The expanded plastics are made up of cells. Both open and closed cell structures are formed. Plastics may be molded by injection, compression, and extrusion; sprayed on; poured; and formed from expandable beads. Both thermoplastics and thermosetting plastics may be expanded.

Thermofusion

Thermofusion is a group of processes in which plastic materials are formed by fusing (melting) them together on or in a mold or product. Thermofusion means **heat fusion.** No force, other than gravity and atmospheric pressure, is applied to the plastic during processing. Plastic powders, pellets, and vinyl plastisols are used in this group of processes. Thermofusion includes several **static**

Introduction to Plastics

(at rest) and **dynamic** (in motion) molding, casting and coating methods. Processes such as vinyl dip casting and coating, stationary powder molding, slush casting, and rotational molding are included.

Roll Forming

Roll forming is a group of processes in which the plastic is formed with the aid of rollers. The plastic is either formed onto or into a sheet. The sheets may be backed with a substrate (fabric, paper, or fiberglass material) or unbacked. Roll forming includes calendering, calendar coating, knife (spread) coating, and extrusion coating.

Finishing Operations

A number of secondary processes and finishing operations are used to put plastics and other parts together, seal packages, or decorate plastics. These processes include fabrication, bonding, and decorating. In addition, some coating operations are considered finishing operations.

Outline of Plastics Processes

An outline of the plastics processes as described here and the types of materials each uses are shown in Table 1-3.

TABLE 1-3
Plastics Processes and Types of Materials Used with Them

Process	Materials		Process	Materials	
MOLDING			EXPANSION		
Injection	TS	TP	Foam Molding	TS	TP
Extrusion	TS	TP	Injection	TS	TP
Blow		TP	Compression	TS	TP
Compression	TS	TP	Extrusion		TP
Transfer	TS		Resin Foams	TS	TP
Cold	TS		Bead Foam		TP
THERMOFORMING			THERMOFUSION		
Vacuum		TP	Static	TS	TP
Drape		TP	Dynamic	TS	TP
Cavity		TP	ROLL FORMING		
Pressure		TP	Calendering		TP
Free Blowing		TP	Coating		TP
Mechanical Stretch		TP	Calender		TP
LAMINATION			Knife		TP
Low Pressure			Extrusion		TP
Fiberglass	TS		FINISHING OPERATIONS		
Thermoplastic Sheet		TP	Fabrication	TS	TP
High Pressure			Bonding	TS	TP
Fiberglass	TS		Decorating	TS	TP
High Pressure Laminates	TS				
CASTING					
Thermoplastic Resins		TP			
Thermosetting Resins	TS				

TS = Thermosetting
TP = Thermoplastics

PART 6

PLASTICS, ENVIRONMENTAL MANAGEMENT, AND ENERGY USE

Plastics and Environmental Management

Three and one-half billion tons (3.17 billion metric tons) of all types of solid waste are disposed of (gotten rid of) each year in the U.S. Of this, only about 360 million tons (326.5 million metric tons) are collected. This solid waste comes from all sources, municipal (city) and industrial, according to the U.S. Bureau of Solid Waste Management. A large portion of this is agricultural and mining waste. All of this disposed of waste averaged about 6 pounds (2.72 kg) of trash per day per person in 1970. By 1980, it will rise to about 8 pounds (3.62 kg) per day per person. This is an annual increase of 4%.

Plastics make up about 2% of the solid waste collected nationally. Plastic waste may run as high as 5% in some large urban areas. This amounted to about 3 million tons (2.7 million metric tons) of plastic waste in 1970. By contrast, paper makes up about 50% of the collected solid waste load. Sources of plastic wastes are divided as follows: packaging ⅔; industrial waste ⅙; and housewares, toys, furniture, and transportation ⅙.

The **convenience and design aspects of plastics** many times overshadow the less obvious benefits of plastics. The use of plastics as a replacement for glass, metal, paper, and other materials often results in such benefits as these:

1. Lighter weight packaging.
2. Reduced shipping costs.
3. Reduced breakage.
4. Improved product distribution.
5. Reduced solid waste load.
6. Increased safety.
7. Improved health.
8. Improved consumer convenience.
9. Aid to truth in packaging.
10. Reduced contamination and infection through medical disposables.
11. Easier disposal by sanitary landfill or incineration.

Many of these tend to reduce costs of consumer products.

There are a number of methods of solid waste disposal. They are (1) open dumping, (2) sanitary landfill, (3) composting, (4) incineration, (5) biodegradation, (6) feeding to pigs, and (7) dumping at sea. The following may be used for plastics.

Open Dumping

Open dumping is the most widely used solid waste disposal method, accounting for three-fourths of all municipal waste. It is an eyesore, health hazard, fire hazard, and waste of valuable land and should be stopped as soon as possible.

Sanitary Landfill

Sanitary landfill is a cleaner, more economical means of waste disposal. In this method, solid wastes are covered with dirt regularly which often helps to improve poor land. Sanitary landfill is gradually replacing open dumping. It accounts for about 10% of present solid waste disposal. Sites are scarce, however.

Plastics make suitable sanitary landfill as they do not decompose rapidly. They do not release odors, gases, or liquids to pollute the land, air, or water. Plastics take up space and help to keep the landfill from sinking.

Composting

Composting is a method of pulverizing (breaking down into small particles) and chemically and biologically treating solid wastes. The composted material is often used to improve the soil. Plastic materials may be used for this alone or with other wastes. Very little plastic or any other waste is disposed of in this way presently.

Incineration

Incineration, or burning, is a practical way of disposing of plastic wastes. However, most U.S. incinerators are out of date. Better incinerators are used in Japan and Europe. Incinerators now used in the United States are capable of handling the present levels of plastics in solid waste (2%). They are capable of the higher levels expected as far in the future as 1980-1990, an estimated 6%. By that time, undoubtedly, more modern incinerators will have replaced those in use today. In the future it is expected that incinerators will recover the energy released during combustion to aid in solving our energy problems.

Biodegradation

Biodegradation is a breaking down of materials. It is possible that plastics can be developed for commercial use which will break down under ultraviolet light, water, or soil enzymes. This process may, however, tend to discourage wide scale recycling of plastics and may present other problems that must be resolved before this process can be used.

Reusing Plastics

Returnable Containers

One way in which to reuse plastic containers is to return them to manufacturers as glass containers are. They could either be sanitized and refilled or recycled and made into new containers. However, returning of glass containers is not popular and returning plastic containers may not be either.

Recycling

Recycling is generally a process of regrinding plastics and returning them to the production system. Most recycling of plastics is done in-plant where quality control can be maintained.

In-plant recycling generally is the grinding up of plastic scraps — cutoffs, sprues, and runners. This ground up material is then fed into an extruder, a machine that shapes plastic by forcing it out. The plastic is extruded into ⅛″ (3.17 mm) rods and cut up into ⅛″ (3.17 mm) long pellets. One of the largest scale operations of this kind is in the Indianapolis Works of the Western Electric Company. About 4.5 million pounds (2041 metric tons) of ABS (acrylonitrile-butadiene-styrene) are recycled there annually.

Off-color or off-specification stock may also be reprocessed by direct in-plant recycling. It may be mixed with other stock or recolored in the process to give a uniform end product.

Post-consumer recycling is another way plastic waste could be reused. In this process, used plastics would be returned to a central plant for reprocessing. However, there are many problems that will have to be solved before this system could be used. Once these problems are solved, this may become an economical and useful method of reclaiming plastics.

One of the problems is that industry usually specifies that products be made from 100% virgin (new) material to assure top performance. This discourages the development of procedures that would make the reuse of plastic materials possible. A second problem is sorting plastics (1) from other solid waste and (2) by family (generic type). Quality plastic products cannot be made from mixed types of plastics.

A number of solutions to the sorting problem are being experimented with. They are (1) flotation, (2) coded tags, and (3) low-grade plastics from mixed scrap.

Flotation is a technique in which the plastics are sorted by their different specific gravities. Each plastic will float or sink in a different specific gravity solution. The plastic is first cleaned and ground into pieces of uniform size less than ½″ (12.7 mm) in dimension. The lighter plastic floats off the first solution (water, $D=1.0$) and the balance sinks. That which sinks is put in a higher density solution (salt water, $D=1.2$) where part floats and the rest sinks. This process is continued with solutions of different density until the plastics are separated as shown in Fig. 1-52.

Coded tags have been used experimentally to separate plastic scrap. The tags on the plastic have an electronically sensi-

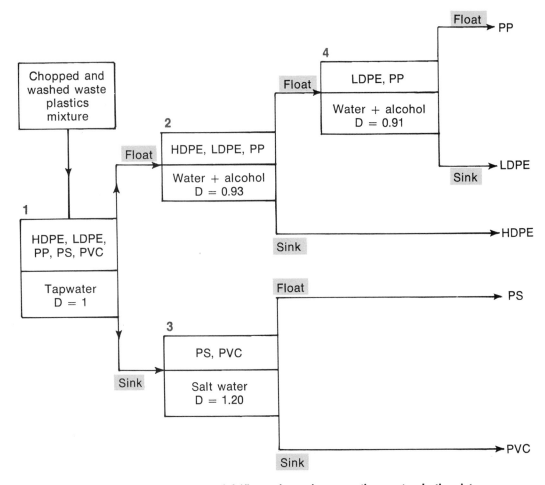

Fig. 1-52. Theoretical four-stage sink/float scheme for separating waste plastic mixture.

tive coating. An electronic sorter detects the plastic type and sorts it automatically.

Low-grade plastics products have been made from a mixture of plastic types. Some projects have produced agricultural pipe, rods, and containers from mixed recycled plastics. In some experiments the scrap is mixed with binders or virgin plastics to get a better grade of product. The main disadvantage is in quality control. Mixed plastics often vary in composition from batch to batch.

Other uses of unsorted plastic have been as an aggregate for concrete and other materials, mixing with sawdust to make charcoal, and as an agricultural mulch.

Energy Recovery

Incineration, or burning, has usually been considered a solid waste disposal method. Recently, it has been used for energy recovery. Newer incinerators in the United States have come equipped with heat exchangers. The heat is used to drive electricity generators or heat buildings. Older incinerators could also be converted for this use.

Plastics materials have a higher energy content (12,000-19,000 BTU per pound or 26000-41800 BTU per kilogram) than mixed municipal waste (5,000 BTU per pound or 11000 BTU per kilogram). Plastics in municipal waste help to improve the combustion efficiency of solid waste incinerators.

A recent study conducted at New York University showed that incinerator furnaces of conventional types will perform satisfactorily with normal refuse containing up to 6% or more plastics. It is estimated that the 6% level will not be reached until 1980 or later. In 1970 about 2% of the solid waste was plastics.

Plastics contain **latent** energy. This means that they contain energy which can later be used for other purposes. In other words, the energy contained in the raw materials from which plastics are made is kept in the plastics. It can be released later by burning or by other means.

Pyrolysis

Pyrolysis is a process by which a material is heated in the absence of oxygen. It may also be done under pressure. The material may be municipal waste that contains plastics. The process, sometimes called **destructive distillation,** drives off volatile components. Often, part of the dry organics in the solid waste can be converted into oil. A char consisting of mainly carbon and quite often a fairly large ash content is left. Gases such as hydrogen, methane, and carbon dioxide are released. Liquids such as acetic acid, acetone, and methanol are also given off.

The char residue, or remains, might be used for polishes and inks. The gases and liquids may be used as fuels or to produce new synthetic materials. Some of the fuel given off can be used to heat the furnace to continue the process.

Consumption of Natural Resources

Most plastics are made wholly or partly from petroleum or natural gas products or byproducts. The petroleum industry is the main source of supply for the plastics industry. The two industries are highly interrelated. Other sources of materials are phenol from coal tar; chlorine from salt; acetic acid from coke, limestone, and water; formaldehyde from coal, air, and water; and benzene from coal.

Shortages of many materials are becoming a worldwide problem. Petroleum is and will be in short supply for years to come. In 1975 the plastics industry used only about 1½% of the petroleum supply in the United States, according to the Society of the Plastics Industry, Inc. (SPI). If petroleum products are used wisely, there should be an adequate supply of plastic products for many years to come. As other natural sources of energy are developed to replace petroleum as a fuel — such as nuclear, geothermal, solar, hydroelectric, and tidal energy — petroleum should be thought of as a nonfuel resource. This resource can be used to make products like plastics, synthetic fibers, pharmaceuticals (drugs), agricultural chemicals, and synthetic rubber. Eventually, new plastics will probably be developed that use other raw materials. Until these new materials become available, people will need to conserve, recycle, and use plastics materials wisely.

For additional information on this subject, the SPI may be contacted at the address below:

The Society of the Plastics Industry, Inc.
355 Lexington Ave.
New York, NY 10017

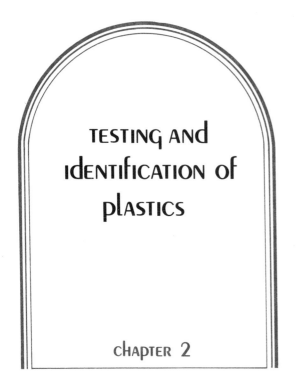

TESTING AND IDENTIFICATION of PLASTICS

CHAPTER 2

A number of tests, from simple to more complex, may be used to identify different plastics. Some tests do not need any equipment while others do. Many times one can guess the type of plastic that a product is made of by knowing how the product is used.

The tests described in this chapter are divided into two groups:
1. **Nondestructive tests** and
2. **Destructive tests.**

The nondestructive tests are designed for all students to do. The destructive tests should be done **by the teacher or under the teacher's direct supervision only.**

Not all the tests need to be done for each plastic. If it is perfectly clear after a few tests what family the plastic belongs to, there is no need for further tests. Students need not test each sample they bring in. One or two well planned tests by each student will provide the needed experience.

Remember that some plastics may be hard to identify. Some of them have chemicals added to them which will change the results of the tests. While the tests provided here are only a guide, they will help identify many plastics.

Individual Nondestructive Tests

A wide selection of plastics should be available to the class for testing. Each student should pick out one or two samples to test at his or her work station. The plastics identification flowchart, Fig. 2-1 on page 33, can be used as a guide for the sequence of tests.

Before nondestructive and destructive tests are begun, the following work sheets should be prepared for recording test results. **Students are cautioned against writing on the sample work sheets in this book.**

1. To record test findings, prepare forms like the Identification Work Sheets, Figs. 2-2 and 2-3 on pages 34 and 36. Use them to record features of the plastic sample being tested.
2. To record the results of the solubility test, prepare a work sheet like the one shown in Fig. 2-9 on page 42.

Use the identification charts, Tables 2-1, (page 35), 2-2 (page 37), and 2-3 (page 38) for help in finding the right answers to the plastics tests.

Instructions for Making Nondestructive Tests

When doing the nondestructive individual tests, make the first but not final guess using the following procedure:
1. Find out the use of the plastic.
2. Look at the plastic.
3. Feel the plastic.
4. Bend the plastic.
5. Smell the plastic.

Take one test at a time for a given sample. Using the prepared work sheet similar to Fig. 2-2, write down the identifying features suggested by the test. Then write down a guess as to what the plastic is. Move to the next test. Write down the identifying features found and make another guess as to what plastic it is. Move through all the tests in this way. If your guess remains the same, you may have the right answer. Answers can be checked by making destructive tests, but these must be made by the teacher or only under the direct supervision of the teacher.

Testing and Identification of Plastics

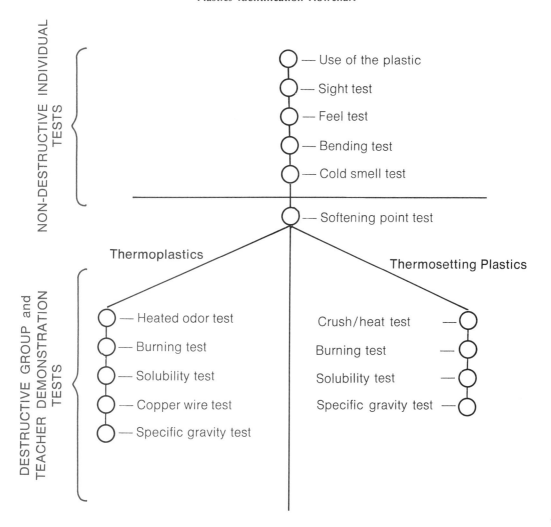

Fig. 2-1
Plastics Identification Flowchart

Destructive tests should only be used when there is a doubt about the answer.

Use of the Plastic

If possible, find out what the plastic sample was used for. Knowing this will often help you identify the plastic. You will find that several tables located in Appendix A provide information on plastics, — their trade names and common uses as well as characteristics and other data. Also, try to find out whether the sample is a thermoplastic or a thermosetting plastic. Knowing this will help narrow your choices for identification.

Sight Test

Each plastic has certain identifying features. Some may be sorted out by their appearance. Look at the plastic sample and compare its appearance with the descriptions of common plastics in Table 2-1.

**Fig. 2-2
Nondestructive Identification Test Work Sheet**

Test	Identifying Features	Answer: Plastic You Think It Is	What It Actually Is
Use			
Sight			
Feel			
Bending			
Smell (cold)			

SAMPLE NUMBER _____

NOTE: The teacher may prepare a work sheet similar to Fig. 2-2 and duplicate it for classroom use.

Feel Test

Feel the plastic. Compare how it feels with the descriptions of common plastics in Table 2-1. Scratch the surface of the plastic. Check the color of the scratch. It may help in identifying the plastic. Hit the plastic to check it for a metallic ring. If it has the metallic ring, it is probably polystyrene.

Bend Test

Bend the plastic at room temperature. Compare your findings with the bending descriptions in Table 2-1.

Smell Test

Scratch the surface of the plastic. Use a knife or a piece of sandpaper. Smell the freshly scratched or sanded plastic. Compare the smell of the plastic with descriptions listed in Table 2-1.

Not all the plastics are listed in Table 2-1. A more complete identification chart is located in Appendix A. Use it for those plastics you have trouble identifying.

Demonstrations of Destructive Identification Tests

Safety Note

The following tests should be done by the **teacher or under the teacher's direct supervision only. Do not try to do them alone.** These tests should be used for plastics that have not been positively identified by the nondestructive tests or, if necessary, for checking the results of those tests. BE CAREFUL.

Compare the results of each **destructive** test with those given in the thermoplastics

TABLE 2-1
Nondestructive Identification Tests
Thermoplastics and Thermosetting Plastics

THERMOPLASTICS

Test \ Material	ACRYLIC The "Lens" Material	FLUOROCARBON The "Nonstick" Material (Teflon®)	POLYETHYLENE (Natural) The "Squeeze Bottle" Material	POLYPROPYLENE (Natural) The "Featherweight" Material	POLYSTYRENE (Gen. Purpose) The "Squirt Gun" Material	POLYVINYL CHLORIDE The "Imitation Leather"
Sight	Clear as glass	Dull, slippery surface	Milky white waxy look (natural color) dull	Milky white (natural) shiny, hard wax	Crystalline, glass-like	Leather-like, shiny or dull
Feel	Tough — hard to scratch	Hard and waxy	Soft and waxy	Very hard and slightly waxy, shiny	Hard, smooth Metallic ring	Dry slipperiness
Bending	Stiff — breaks under strong loading	Usually stiff	Low density — very flexible; High density — stiff but pliable	Flexible in thin sections — one-piece hinges	Stiff — breaks like glass — shatters	Film — flexible Solid — stiff
Smell (cold)	Sweet	None	Candlewax	None	None	New car

THERMOSETTING

Test \ Material	EPOXY Gen. Purpose	PHENOLIC Gen. Purpose	POLYESTER Gen. Purpose	SILICONE RTV Rubber
Sight	Naturally clear; slight brown tint takes any color	Orange (natural). Usually colored dark black, brown, red or dark green. Never pastel.	Naturally clear. Slight pink from catalyst system. May be colored. White mark when scratched.	Milky white or colored, shiny surface
Feel	Hard, solid	Hard, solid	Hard, solid	Rubbery, slick
Bending	Almost none	Almost none	Almost none	Very flexible
Smell (cold)	—	Carbolic acid or phenol	—	—

Fig. 2-3
Destructive Identification Test Work Sheet

Test	Questions			Results			Conclusion: What Plastic Is It?
Softening Point	1. Did it melt? 2. At what temperature?						
Heated Odor	1. What does it smell like?						
Burning Test	1. Does it light easily? 2. What color was the flame? 3. Does it smoke? 4. Does it make soot? 5. Does it drip as it burns? 6. What is the smell? 7. Does it keep burning?						
Solubility Test	Sample Number?	Solvent Used?		Dissolve? Yes / No			
Copper Wire Test	1. Does the flame turn **green**?			Yes	No		
Specific Gravity Test	SG = $\frac{(1.0)\ C}{C\text{-}D\text{-}F}$ or $\frac{(.810)\ C}{C\text{-}D\text{-}F}$ or $\frac{(.789)\ C}{C\text{-}D\text{-}F}$ What specific gravity is it?						

NOTE: The teacher may prepare a work sheet similar to Fig. 2-3 and duplicate it for classroom use.

and thermosetting plastics identification charts, Tables 2-2 and 2-3, or from another **known** plastic material. Check to see whether the results from the nondestructive tests change.

Use only those tests needed to get an answer to plastics identification. There is no need to use a test that will not help find the answer. As each test is finished, answer the list of questions on the Destructive Identification Test Work Sheet (similar to Fig. 2-3, prepared before tests were begun). **Do not write in the book.** The following tests should be used in the order listed.

TABLE 2-2
Destructive Identification Tests, Thermoplastics

Test \ Material	ACRYLIC The "Lens" Material (Plexiglas®)	FLUOROCARBON The "Nonstick" Material (Teflon®)	POLYETHYLENE (Natural) "Squeeze Bottle" Plastic	POLYPROPYLENE (Natural) "Feather Weight" Plastic	POLYSTYRENE "Squirt Gun" Plastic	POLYVINYL CHLORIDE "Imitation Leather"
Softening Point	374°F (190°C)	Softens slightly at 621°F (327°C)	LD - 221°F (105°C) (softens in boiling water) HD 248°F (120°C)	334°F (168°C)	374°F (190°C)	302°F (150°C)
Heated Odor	Fruit-like	None	Paraffin	Sweet	Illuminating gas	Hydrochloric acid
Burning	Burns: blue flame — yellow top, spurts, black smoke fruit-like or floral odor.	Won't burn: will deform in thin sections, curls, chars slightly, weak burned hair odor.	Burns rapidly: blue flame — yellow top; melts, drips. Burn area swells and gets transparent candle wax odor.	Burns rapidly: blue flame — yellow top, burn area transparent, swells, drips, odor - diesel fumes.	Burns: orange/yellow flame, black soot, sweet odor (marigold)	Burns: Self-extinguishing yellow flame, green mantle; spurts green and yellow, softens white smoke. Chlorine odor
Solubility	Acetone Benzene $CHCl_3$ Ethyl acetate	Swells but does not dissolve in acetone.	Some hot aromatics: benzene toluene	Some hot aromatics: benzene toluene	Acetone Benzene $CHCl_3$ Esters Ether	Acetone, carbon tetrachloride $CHCl_3$, cyclohexanone
Copper Wire Test	No reaction	No reaction	No reaction	No reaction	No reaction	Green flame
Specific Gravity	1.18 - 1.19 Sinks H_2O	2.10 - 2.30 Sinks H_2O	LD - .910-.925 MD - .926-.940 HD - .941-.965 Floats H_2O	.85-.90 Floats H_2O	1.05-1.09 Sinks H_2O	1.16-1.72 Sinks H_2O

TABLE 2-3
Destructive Identification Tests, Thermosetting Plastics

Test	EPOXY Gen. Purpose The "high strength glue."	PHENOLIC Gen. Purpose "Pot & pan handle" resin.	POLYESTER Gen. Purpose The "fiberglass" resin.	SILICONE RTV Rubber The "bathtub seal" resin.
Softening Point	Does not soften.	Does not soften.	Does not soften.	Does not soften.
Crush/Heat Test	No film on aluminum. Not degradable in 20-30% sodium hydroxide.	No film on aluminum. Degrades in 20-30% sodium hydroxide.	White film on aluminum. Degrades in 20-30% sodium hydroxide.	Not applicable.
Burning Test	Burns readily: Yellow flame, some soot. Amine odor.	Difficult to burn: Self extinguishing; Cracks. Phenol or carbolic acid odor.	Burns: Yellow flame, Black soot. Sour cinnamon odor.	Burns: Self extinguishing; Leaves white ash.
Solubility	Slight effect by acetone or methyl ethyl ketone.	Insoluble	Slight attack by carbon tetrachloride.	Insoluble
Specific Gravity	1.11 - 2.40 Sinks in H_2O	1.55 - 1.90 Sinks in H_2O	1.30 - 1.50 Sinks in H_2O	1.05 - 1.23 Sinks in H_2O

1. Softening point test.
2. Heated odor test (for thermoplastics, if the sample melts in the softening point test).
3. Crush/heat test (for thermosetting plastics, if the sample does not melt in the softening point test).
4. Burning test.
5. Solubility test.
6. Copper wire test.
7. Specific gravity test.

Instructions for Making Destructive Tests

The following tests are to be done **under direct teacher supervision only.**

Softening Point Test

The softening point test will show (1) whether the plastic is a thermoplastic or a thermosetting plastic and (2) the temperature at which the plastic softens (if it does). If the plastic does not soften, it may be a thermosetting plastic. Several procedures may be used, depending on the equipment available.

(a) Fisher-Johns Melting Point Apparatus

This is a precision melting point instrument. If one is available, a complete set of instructions is included with it. See Fig. 2-4.

(b) Electric Frying Pan or Griddle

A used electric frying pan or griddle in good working order may be used as a substitute for the Fisher-Johns instrument for measuring the softening point. A small hand pyrometer or an accurate thermometer should be used to check the temperatures of the frying pan. Most electric frying pans have a 25°F (14°C) to 50°F (28°C) range between on and off in the temperature control. Some have a variation in temperature across the surface of the pan.

Clean the surface of the pan with fine steel wool. This will help remove all grease from the pan. Coat the inside of the pan with two or three coats of **fluorocarbon** (Teflon) type mold release. This will keep the samples from sticking to the pan. Put small slivers in the pan or use plastic pellets of the sample no larger than 1/8″ × 1/8″ × 1/8″ (3.175 mm³). Heat the pan to about 200°F (93°C). When the pilot light goes out, use a small stick to check the samples for softness. Raise the temperature about 10°F (5.5°C) if none have softened. Check the samples again. Be sure to check the samples just as the pilot light goes out. This is the highest temperature for that setting of the pan. The pan will cool 25°F to 50°F (14°C to 28°C) before it will begin to heat again. If none of the samples soften at that temperature, turn the pan up about 10°F (5.5°C) again. Check the samples when the pilot light goes out. Do this several times. Write down the

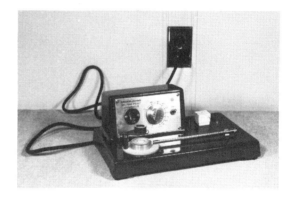

Fig. 2-4. Fisher-Johns Melting Point Apparatus

Fig. 2-5. Softening Point Test — Frying Pan Method

softening point of each sample in the pan. Temperatures of the pan may be most accurately recorded if they are checked with a hand pyrometer.

Several samples may be tested at the same time in the pan. Be sure to **label** each sample in the pan for proper identification.

If the plastic does not soften in this test, it may be a thermosetting material. There is only one limitation to the frying pan in this test. The temperatures of most frying pans do not exceed about 425°F (218°C). Some thermoplastics melt at above that point.

Record the softening points in the proper place on the Destructive Identification Test Work Sheet. Compare the results with Tables 2-2 and 2-3.

Heated Odor Test

Use the heated odor test for thermoplastics — those samples that melted in the softening point test.

Safety Note

Do this test over a metal or asbestos-covered bench top. Keep a fire extinguisher handy. Do test only under the DIRECT supervision of the teacher. Do not hold the test tube directly under your nose. Draw the heated vapors to your nose with your hand.

Place a small sliver or ⅛" cube (3.175 mm³) (plastic pellet) in a Pyrex-type test tube. Hold the test tube with a spring clamp. Heat the bottom of the test tube slowly over an alcohol lamp or Bunsen burner. Wave the bottom of the test tube through the flame slowly. When the plastic gets hot, smell the vapors by drawing the vapors to your nose with a cupped hand. On the prepared Destructive Identification Test Work Sheet, write down what the plastic smells like. Compare the results with Tables 2-2 and 2-3.

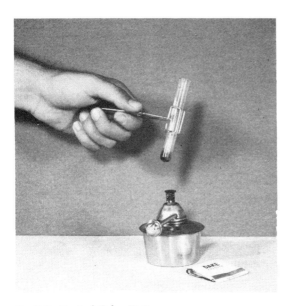

Fig. 2-6. Heated Odor Test

Crush/Heat Test

The crush/heat test is for thermosetting plastics only. It can be used to separate phenolic, epoxy, and polyester plastics from each other. If the plastic did not melt in the softening point test, use this test.

Crush the thermosetting plastic into fine pieces. Put them on a frying pan or hot plate. Bring the hot plate to 500°F (260°C). Hold a piece of aluminum foil ½" (12.7 mm) above the sample of plastic. Hold one end of the aluminum foil with two pieces of metal clamped to it to cool the end. The

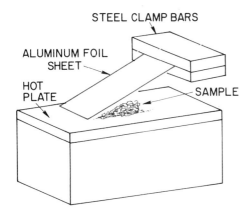

Fig. 2-7. Crush/Heat Test

other end of the foil should be left free. If a thin white cloudy film forms on the aluminum, the sample is a polyester resin. If a film does not form, the sample is either an epoxy or a phenolic plastic.

Phenolic and epoxy plastics may be separated by soaking them in a 20% solution of **sodium hydroxide.** If the sample degrades in the sodium hydroxide solution, it is probably phenolic. If it does not degrade in the sodium hydroxide solution and does not produce a film on the aluminum foil, it may be assumed to be an epoxy plastic.

Safety Note

Sodium hydroxide is a dangerous chemical and should be handled only by your teacher.

Remember that this test is only for phenolic, epoxy, and polyester plastics.

Burning Test

Safety Note

Do this test over a metal or asbestos-covered bench. Avoid plastic drippings; they burn. Have a fire extinguisher handy. Make test only under direct teacher supervision. Work in a well-ventilated room.

Take a thin, long strip of plastic about 1/32" thick x 1/2" wide x 6" long (0.792 mm x 12.7 mm x 152.4 mm) for this test. **Caution: Hold the plastic with a pair of pliers to keep from getting burned.** Hold the plastic strip over the open flame. A Bunsen burner, alcohol burner, or propane torch may be used to light the plastic. A match or cigarette lighter may be used for thinner strips of plastic. Hold the plastic at the edge of the flame. Hold it there until it lights, but not more than 10 seconds. Watch how easily the plastic lights. Record what you notice about this on the paper you have prepared as the Destructive Identification Test Work Sheet. Do not write in this book.

Check the following features about the plastic in this test: (1) how easily the plastic lights, (2) the color of the flame, (3) whether smoke or soot comes from the flame, (4) the smell, if any, from burning, and (5) whether the plastic keeps on burning or stops when it is taken from the flame.

Safety Note

Be careful about smelling the burning plastic. Wave your hand over the plastic **after** it has been taken from the flame. This will draw **some** of the smell to your nose.

Record what you notice about the burning plastic on the prepared Destructive Identification Test Work Sheet. Compare this with Tables 2-2 and 2-3.

Solubility Test

Safety Note

Many solvents are dangerous in large quantities. Use only very small amounts of each solvent in the bottom of a test tube for each test. Put in just enough solvent to cover the sample. Handle all solvents carefully.

Always use the solvent which is easily available, least expensive, and safest. Materials and help for this test can be obtained from the science department.

Place a small sliver or pellet of plastic in a small test tube, and put in just enough solvent to cover the sample. Put a cork

Fig. 2-8. Burning Test

Fig. 2-9
Solubility Test Work Sheet

Sample Number	Solvent Used	Dissolve? Yes	Dissolve? No	Conclusion: What Plastic Is It?

NOTE: The teacher may prepare a work sheet similar to Fig. 2-9 and duplicate it for classroom use.

Fig. 2-10. Solubility is determined after plastics samples are allowed to remain in solvents for 12 to 24 hours.

stopper in the top of the test tube. Label the test tube the sample is in. Write down a sample identification number and the name of the solvent. Use the prepared Solubility Test Work Sheet similar to Fig. 2-9. Record the results on the Destructive Identification Test Work Sheet.

Put only one sample in a test tube and use a solvent which **should** dissolve the sample in it. In order to decide between two plastic families for an unknown plastic sample, use two test tubes. Put a sample of the unknown plastic in each test tube. Use a different solvent for each of the two test tubes. Place the test tubes in a test tube rack. Let them stand for about 12 to 24 hours. Slight agitation may be necessary. **Do not shake the test tubes violently.**

Do not expect all the samples to dissolve completely. Some solvents will just swell or otherwise slightly change the sample. Compare the results with the information given in Figs. 2-5 and 2-6.

Copper Wire Test

─────── **Safety Note** ───────
Keep the wire, sample, and the flame away from you.

The copper wire test should be used **only** if other tests seem to show the sample may have chlorine in it. It may be polyvinyl chloride or some other chlorinated plastic.

Heat a **copper** wire in an open flame until it is red and the flame clears. Touch the plastic sample with the wire. Put the wire back in the flame. If the flame turns **green**, the plastic has chlorine in it. Thus, it may be polyvinyl chloride. Compare the results of this test with Table 2-2. Record the results on the work sheet.

Specific Gravity

Specific gravity is the relationship of the weight of any material to the weight of water. A few plastics weigh less than water. Most weigh more than water. This test will show (1) whether a plastic weighs more or

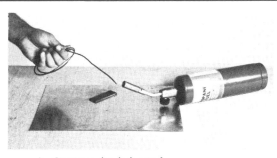

A. Copper wire is heated.

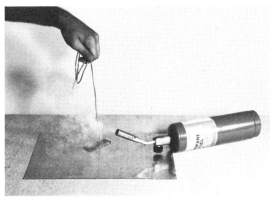

B. The plastic sample is touched with the wire.

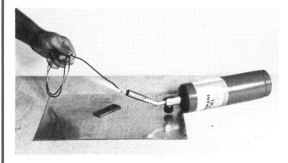

C. Wire is returned to the flame to determine the presence of chlorine in the plastic.

Fig. 2-11. Copper Wire Test

than water, it will sink in it. A plastic with a specific gravity of 0.91 weighs 0.91 times or about 9/10 the weight of water. It is slightly lighter, so it will float on water.

Two common methods for measuring specific gravity may be used. Use the one for which equipment is available.

Cut a sample piece of plastic about 1/8" thick x 1/2" x 1/2" (3.175 mm x 12.7 mm x 12.7 mm). Make sure the edges are very clean and smooth. Air bubbles will stick to a rough surface and change the results of the test.

1. **Direct Reading Specific Gravity Scale.** Several direct reading specific gravity scales are available commercially. No calculations are necessary in their use as readings are taken directly from a scale. Directions for their use are included with the scale. The same general precautions and procedure as given below are used. See Fig. 2-12A.

2. **Laboratory Balance Scale, Triple Beam.** A triple-beam laboratory balance scale as shown in Fig. 2-12B is used. It may be borrowed from the science department. A small beaker or glass jar is also needed. Follow the procedure below.

a. Attach a small metal wire to the pan of the balance scale. Weigh the wire in the air, letting the wire swing freely. Write this weight down as weight "A".

b. Tie the plastic to the wire and weigh BOTH the plastic and the wire together in the air. Write this weight down as weight "B".

c. "B" minus "A" equals "C" ($B-A=C$), the weight of the plastic in the air.

d. Fill the beaker or jar with distilled water (H_2O). Add 1% liquid dishwashing detergent or 1% photoflo solution (a chemical photographic wetting agent) to the water. The detergent or photoflo solution will help get rid of air bubbles on the surface of the plastic. Put the beaker of water under the plastic hanging on the wire. Raise the beaker of water up until the plastic is completely covered with water.

e. Weigh the plastic while it is under water. Record this as weight "D".

less than water and (2) how much more or less it weighs. Thus, if a plastic has a specific gravity of 1.75, it weighs 1.75 (1 3/4) times as much as water. If it does weigh more

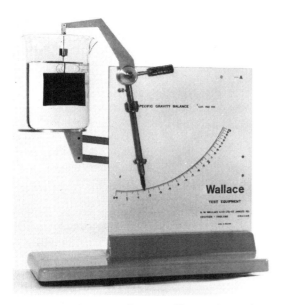

Fig. 2-12A. Direct Reading Specific Gravity Scale

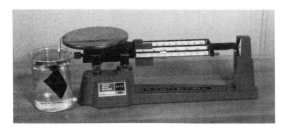

Fig. 2-12B. The triple-beam balance scale is used in the specific gravity test.

f. Remove the plastic from the water and take it off the wire. Weigh the wire again **in the water** to the same depth. Record this as weight "E", the **weight of the wire in the water**.

g. "A" minus "E" equals "F" (A−E=F), the loss of the weight of the wire in the water.

h. Record the figures as shown in the example below.

"A" = _____ "B" = _____

"A" − "B" = "C" "C" = _____

"D" = _____ "E" = _____

"A" − "E" = "F" "F" = _____

For plastics that sink in water, use the following formula for figuring the specific gravity of the plastic sample.

$$\text{Specific gravity} = \frac{(1.0)\,C}{C-D-F}$$

If the sample floats in water, use alcohol as the liquid to weigh the sample in. Three different types of alcohol may be used. If the alcohol is pure, the figures given below may be used in the formula. If it is not pure, it must be checked with a **hydrometer** to find the exact specific gravity of the alcohol. That specific gravity must then be used in the formula.

$$\text{Specific gravity (methonal alcohol)} = \frac{(.810)^1 C}{C-D-F}$$

Testing of Plastics

There are a number of controlled tests designed specifically to determine the exact characteristics of plastics materials. All of these tests are fully described in **The American Society for Testing and Materials (ASTM) Annual Books of Standards.** Three or more volumes of the complete set are currently devoted to plastics.[2] Several of these tests are briefly described here to acquaint you with their general procedures and significance.

Tensile Testing (ASTM D-638)

Tensile test specimens can be injection molded, compression molded, machined, or die cut from sheet plastic. Typically, the test

---

[1]Note: This figure is the specific gravity of methonal (wood) alcohol. Use .789 for **pure** isopropyl (rubbing) alcohol and **pure** ethyl alcohol. Use the specific gravity of diluted alcohol.

[2]These volumes may be obtained from The American Society for Testing and Materials, 1916 Race St., Philadelphia, PA 19103.
A condensed version of these tests can be found in "Standard Tests for Plastics," Bulletin GIC, Celanese Plastics Company, Newark, NJ. The description of the tests given here is from that source.

specimen is 1/8" thick and formed to the shape shown below.

The test specimen is tightly clamped into the jaws of a universal testing machine. The jaws are moved apart at a uniform rate of 0.2, 0.5, 2 or 20 inches per minute (5.08 mm, 12.7 mm, 50.8 mm or 508 mm/minute), pulling the sample from both ends. The stress (force) developed is automatically plotted against the strain (elongation) on graph paper. This is called a **stress-strain curve**.

The data on the stress-strain curve is the most important single indication of strength in a material. It shows the force that is necessary to pull the specimen apart. It also shows how much the material will stretch before it breaks. This data can be valuable for engineers in the selection of the right material for a part. The tensile strength of the material must be higher than the part requires in order to prevent part failures. The amount of elongation before failure is important because it is an indication of how much shock (or rapid impact) a material can absorb.

Impact Testing (ASTM D-256)

Two methods of impact testing are used. One is called the **Izod impact test** and the other is the **Charpy test**. The Izod impact test usually uses a 1/8" x 1/2" x 2" (3.175 mm x 12.7 mm x 50.8 mm) notched specimen clamped on one end. The Charpy test uses a 1/8" to 1/2" thick x 1/2" wide x 4" (3.175

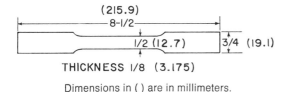

Dimensions in ( ) are in millimeters.

Fig. 2-13. Typical Tensile Test Specimen

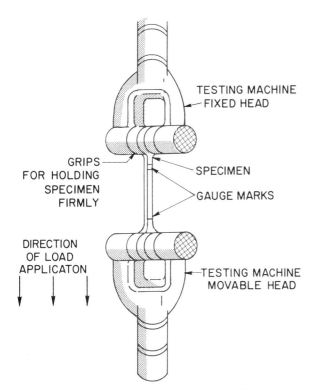

Fig. 2-14. Mounting a specimen for tensile testing.

Fig. 2-15. Tensile test is made on bonded thermoplastic samples in a hot/cold cabinet.

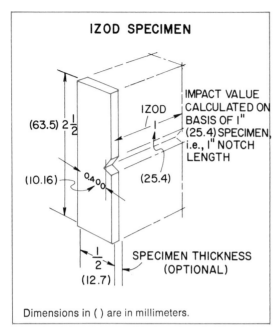

Fig. 2-16. Typical Izod Impact Specimen

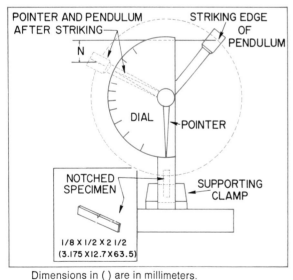

Fig. 2-17. Impact Testing Machine

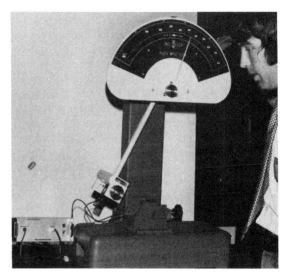

Fig. 2-18. Izod impact test on a plastic specimen.

Many impact testing machines are designed to make both tests. Of the two, the Izod impact test is the most used for plastics.

The Izod impact specimen is tightly clamped by one end in the base of the testing machine. A pendulum on the machine is released and the force absorbed in breaking the sample is recorded by the distance the pendulum swings in the follow-through.

The Izod impact test indicates the energy required to break the notched specimens. It is usually calculated in foot pounds per inch (ft.-lb./in. or kilogram force per meter or joule) of notch, although the specimen may be thinner than 1 inch (see Fig. 2-16).

The Izod value is useful in comparing various types and grades of a plastic. As some plastics are notch sensitive, this test should not be considered a reliable indicator of overall toughness or impact strength. It will, however, indicate the need for avoiding sharp corners in notch sensitive materials such as nylon and acetal.

The Charpy impact test is conducted in a manner similar to the Izod test except the test bar is laid between two supports as shown in Fig. 2-19. The pendulum impacts the test specimen between the two supports.

mm to 12.7 mm x 12.7 mm x 101.6 mm) long, notched specimen supported on each end. In some impact tests, both Izod and Charpy, unnotched specimens are used.

Testing and Identification of Plastics   47

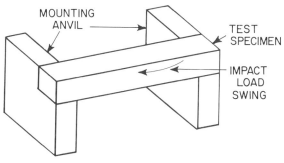

Fig. 2-19. Charpy Impact Method

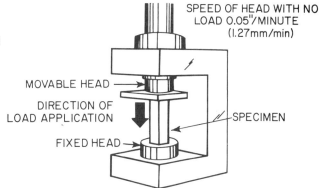

Fig. 2-20. Compressive Testing

The energy absorbed on impact is recorded in the same way. The Charpy test reveals bending impact strength while the Izod tests impact under shear.

Compressive Properties of Rigid Plastics (ASTM D-695)

Compressive test specimens are usually ½" x ½" x 1" (12.7 mm x 12.7 mm x 25.4 mm) prisms or ½" x 1" (12.7 mm x 25.4 mm) diameter cylinders. The test specimens may be machined from sheet, rod or tube; or they may be compression or injection molded.

The test specimen is placed in a compression tool between the universal testing machine heads. The heads exert a constant rate of compression on the specimen. An indicator registers the loading. The compressive strength is calculated as the pounds per square inch (psi) or newtons per square meter (pascals) needed to rupture or deform the specimen at a given percentage of its height. It can be expressed as psi (kPa) either in rupture or at a given percentage of deformation.

The compressive strength is usually limited to a design value, since most plastics products (except foams) seldom fail from compressive loading alone. Compressive strengths can be used to compare different materials or grades of materials for an application. Some products, like the ones shown in Figs. 2-21 and 2-22, depend on compressive strength. These plastic step blocks are shown being tested for compressive strength.

Fig. 2-21. Compression-testing polycarbonate plastic step blocks.

Fig. 2-22. Universal test machine being used for a compressive test.

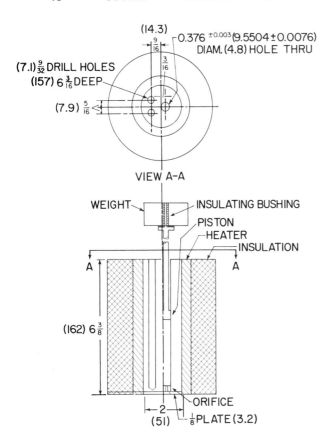

Dimensions in ( ) are in millimeters.

Fig. 2-23. Melt Indexer (Extrusion Plastometer)

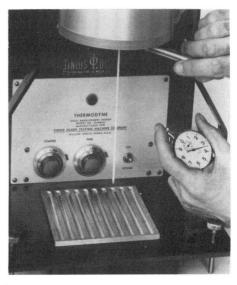

Fig. 2-24. Melt indexer in use.

Fig. 2-25. Melt Indexer

Flow Rate or Melt Index (ASTM D-1238)

Plastic particles such as powder, granules, and strips of film may be tested for flow at a given temperature in this test. The plastic is placed in a cylinder which has been preheated to from 125°C to 275°C, depending on the plastic being tested. A common temperature is 190°C, which is used for polyethylene. A load of 2160 grams (43.25 psi or 298.4 kPa) is normally used on the piston to force the melted plastic through a 0.0825" (2.095 mm) hole. The amount of plastic, in grams, forced through the hole in ten minutes is the flow rate, or **melt index.**

Melt index values are primarily used by raw material manufacturers for controlling material uniformity. They are also often used to determine flow in processing. Plastics with low flow rates (low melt index) are often used for extrusion and blow molding. Higher flow rates are often used for injection molding. Exact melt index or flow rate is dependent on the process and product being made.

Flexural Properties (ASTM D-790)

Flexural test specimens may be cut from sheets, plates or molded shapes or molded to the desired finished dimensions. Specimens are usually ⅛" x ½" x 5" (3.175 mm x 12.7 mm x 152.4 mm), but sheets as thin as 1/16" (1.59 mm) may be used. The span and width depend upon the thickness of the specimen.

The specimen to be tested is placed on two supports. A load is applied in the center at a specified rate. The loading at failure (psi or kPa) is the flexural strength. It is the amount of load that can be applied before the specimen will break. For materials that do not break, the flexural stress at 5% strain is calculated. This is the loading in psi or kPa necessary to stretch the outer surface 5%.

Flexural strength is important to designers and engineers for the proper specification of materials that may break under stress.

A number of other tests is used by various firms that use plastics. Each is described in detail in the **American Society for Testing and Materials Handbooks.**

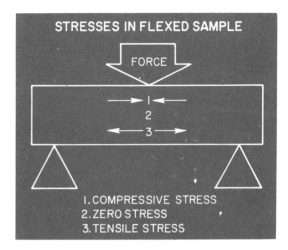

Fig. 2-27. Stresses in a flexed sample.

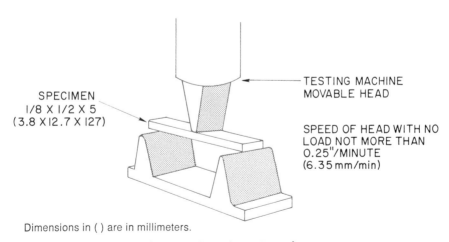

Dimensions in ( ) are in millimeters.

Fig. 2-26. Flexural Test Procedure

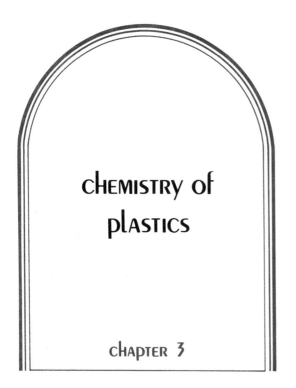

chemistry of plastics

CHAPTER 3

Plastic materials are made up of long chains of carbon backbone (center) molecules. These large **polymer** (many unit) molecules are made from small **monomer** (single unit) molecules. Each small monomer molecule is made from two or more atoms. The center, or backbone, atoms of the monomer molecule used for plastics are usually carbon. Other atoms may be mixed in with the carbon atoms in the backbone of some plastics. The monomer molecule is often called the "building block of plastics." Figure 3-1 shows the monomer molecule in three forms: a block diagram, a ball and stick model, and a formula diagram.

The monomer molecule shown in Fig. 3-1 is an ethylene monomer. It has two carbon atoms in its backbone and four hydrogen rib atoms. Figure 3-2 shows the parts of an ethylene monomer molecule. The backbone, made up of carbon atoms, is held together by a **double chemical bond**. The rib atoms are attached to the backbone atoms by a **single chemical bond**.

Chemical Bonding

Chemical bonding (covalent bonding) is a process in which atoms link together by sharing **electrons** in the outer region of the atom. In a **single chemical bond**, one electron from **each** joining atom is shared. A

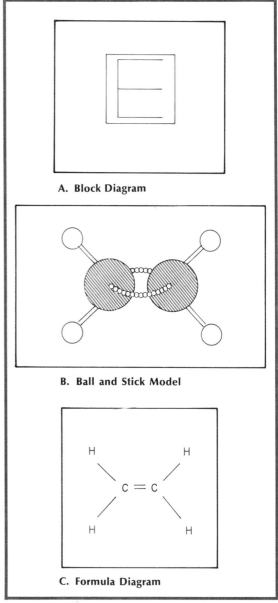

A. Block Diagram

B. Ball and Stick Model

C. Formula Diagram

Fig. 3-1. Ethylene Monomer Molecule

double chemical bond is formed when two electrons from **each** joining atom are shared. Thus, a single chemical bond is formed when two joining atoms share two electrons, and a double chemical bond is formed when two joining atoms share four electrons.

The **hydrogen atom**, shown in Fig. 3-3, has one electron in its outer region or **K** shell. The valence sign +1 indicates the number of electrons in the outer region. However, this outer region, or shell, has the capacity to hold **two** electrons. Thus, the hydrogen atom may accept or share one electron from another atom. An atom is said to be **active** if the outer region or shell is not filled.

A helium atom also has only one shell, the K shell, but it has **two** electrons which fill the shell and make helium inactive, or **inert**. Atoms that are inert will not bond chemically to other atoms.

The ability of an atom to fill an incomplete outer shell lets it combine with other atoms, either like or unlike. They will, however, combine only with other atoms that have incomplete outer shells. This combination of two or more atoms, by sharing electrons in their respective outer shells, is called **chemical bonding** (covalent bonding). Thus, when two or more atoms chemically bond together they form a **molecule**.

When two **active** atoms come close together, they may link up by **sharing** electrons in their respective outer shells. Note that the **carbon** atom in Fig. 3-4 can share or take

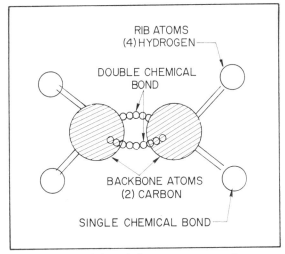

Fig. 3-2. Parts of the ethylene monomer molecule.

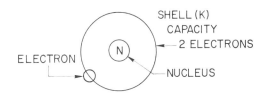

Fig. 3-3. Diagram of a hydrogen atom, showing its parts.

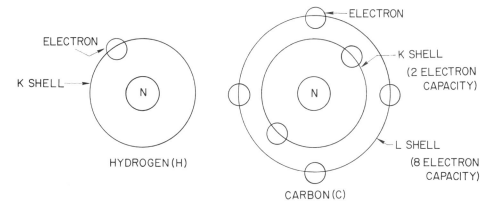

Fig. 3-4. Two Unlike Active Atoms, Carbon and Hydrogen

52 Section I INTRODUCTION

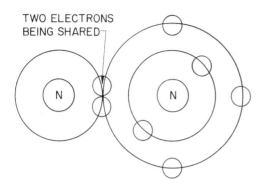

Fig. 3-5. Carbon-Hydrogen Bond (CH)

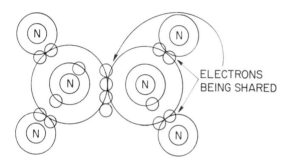

Fig. 3-6. Ethylene Monomer Molecule ($CH_2 = CH_2$)

Fig. 3-7. Polyethylene reactor system is shown in the foreground. In the background are silos where finished plastic resins are stored.

on four electrons. It has four electrons in its outer shell (**L** shell), but that shell can hold eight electrons. Thus $8 - 4 = 4$, which is the number of electrons it can share. When these two atoms bond together, they look like those in the diagram, Fig. 3-5. They probably will not be found as a single CH unit, however, as the carbon atom will quickly combine with other atoms to form a larger molecule.

By sharing the electrons, the atoms bond together to form a stable molecule. Figure 3-6 represents the chemical bonding in the **ethylene monomer** molecule.

The chemical (covalent) bond is very strong. A large amount of heat energy must be absorbed to break a chemical bond. A similar bond joins the monomer molecules together to form a **polymer chain.**

Other rib atoms that are used to form monomer molecules to make plastics are (1) fluorine, (2) chlorine, (3) nitrogen, (4) carbon, and (5) oxygen. Each of these atoms is chemically active, so they can easily join by chemical bonding. The rib atoms in a molecule may be all the same kind or a mixture of several kinds in a single monomer molecule.

Ethylene monomer molecules are made from natural gas or crude oil by **distillation,** a process by which gas or vapor is driven from solids or liquids. Monomer molecules, such as ethylene monomer, can be joined together end-to-end to form large molecules. The monomer molecules, which are either a liquid or a gas depending on their makeup, are pumped into a large tank. This tank is called an **autoclave,** or **reactor,** depending on its design, and is often like a big pressure cooker. The autoclave, or reactor, is sealed after it is loaded with monomer molecules and a suitable catalyst. A **catalyst** is a chemical which speeds up a chemical reaction. By contrast, an **inhibitor** slows down a chemical reaction which is why it is often

Chemistry of Plastics 53

Fig. 3-8. Polymerization

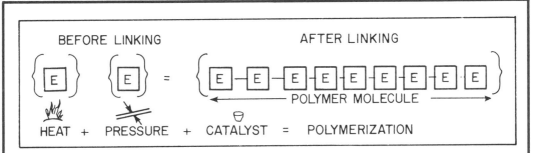
A. Block Diagram

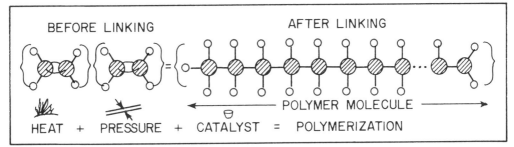
B. Ball and Stick Model

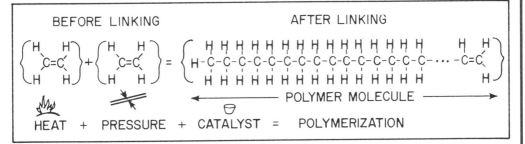
C. Formula Diagram

D. Paper Clip Model

used to extend the storage life of plastics resins. Figure 10-22 in Chapter 10 shows two views of a typical autoclave.

Heat is added to the sealed autoclave which usually raises the pressure inside. Pressures from less than atmospheric pressure, a vacuum, to as high as 45,000 psi (310 500 kPa) are used, depending on the plastic being made. The **heat, pressure** or lack of pressure, and **catalyst** in the autoclave cause the monomer molecules to join together. They form long chains called **polymer molecules.**

Polymerization

In order to form the polymer chains (polymer molecules), one of the **double chemical (covalent) bonds** in the monomer molecule must be broken. This chemical bond between the backbone atoms of the monomer molecule is broken by heat, pressure, and a catalyst. A single chemical bond is left between the backbone atoms in the monomer molecule. The chemical bond that was broken is then free to look for other atoms or molecules to link up with. In most cases, the atoms will link up with another backbone atom just like themselves. This will occur until the linking reaction is stopped by a stray atom linking onto the end of the chain. When they link up this way, they form long chainlike molecules. This chemical **chain-linking** is called **polymerization** (the act of forming a large or polymer molecule). Polymerization is similar to the linking up of a large group of box cars to form a railroad

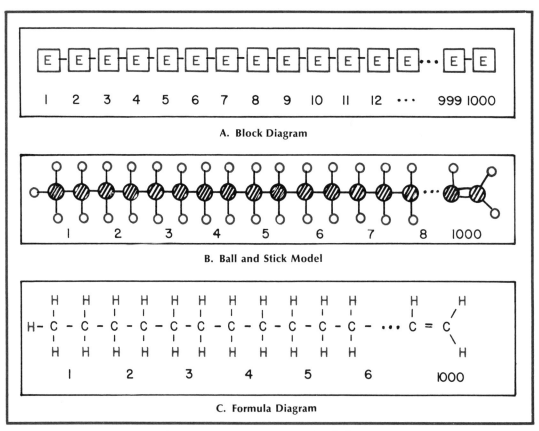

Fig. 3-9. Polymer Molecule (Polyethylene)

Chemistry of Plastics 55

train. It is also like linking up a box full of paper clips into a long chain. Figure 3-8, page 53, shows how polymerization works.

The chain linking process takes place under the heat and pressure until hundreds to thousands of monomer molecules have joined together. A loose rib atom will usually link onto each end of the chain, stopping the process. Giant polymer molecules are seldom longer than 1/10,000 of an inch (0.00254 mm). One can understand, then, that each atom and molecule is extremely small. Atoms and molecules cannot be seen through an ordinary microscope. Thousands of them

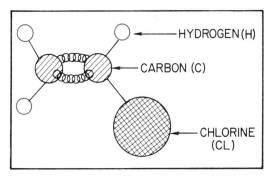

Fig. 3-10. Ball and stick model of vinyl chloride.

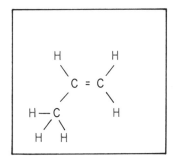

Fig. 3-13. Formula diagram of propylene.

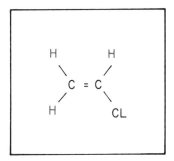

Fig. 3-11. Formula diagram of vinyl chloride.

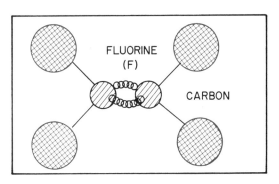

Fig. 3-14. Ball and stick model of fluorocarbon.

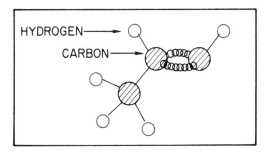

Fig. 3-12. Ball and stick model of propylene.

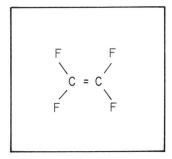

Fig. 3-15. Formula diagram of fluorocarbon.

are needed to make up the point of a pin. Diagrams of a polymer molecule (polyethylene) are shown in Fig. 3-9.

There are a number of other monomer molecules used for building polymer molecules. Several common ones are (1) vinyl chloride, (2) propylene, and (3) fluorocarbon. When they are polymerized (linked), they will become plastics. Note the different rib atoms in them, Figs. 3-10 through 3-15. Others are shown in the polymer chemistry section of Appendix A.

Achieving the Different Properties of Plastics

In order to make different properties in plastics, two or three different kinds of monomer molecules can be chemically joined together. When two kinds of monomer molecules are mixed together to make a polymer molecule, the resulting plastic material is called a **copolymer** (meaning two-monomer polymer). An example of a copolymer plastic is ethylene-vinyl copolymer, diagramed in Fig. 3-16.

Ethylene-vinyl copolymer is made up of **alternating** ethylene monomer and vinyl chloride monomer molecules in the polymer chain. Compare this chain with either polyethylene or poly-vinyl chloride chains.

A combination of three kinds of monomer molecules in the polymer chain is called a **terpolymer** (meaning three-monomer polymer). Many such combinations are in use today. Many more will be made in the future.

Fig. 3-16. Ethylene-Vinyl Copolymer

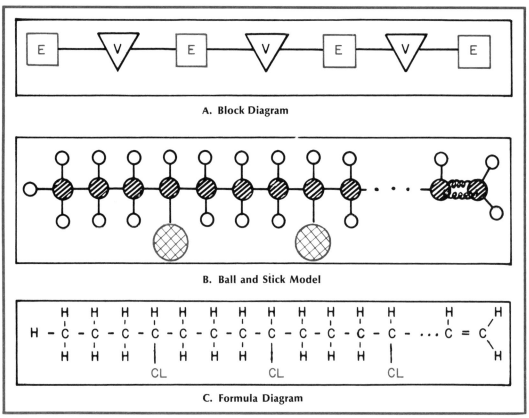

A. Block Diagram

B. Ball and Stick Model

C. Formula Diagram

The key to these combinations is chemistry. Chemists have to find out how to chemically combine these monomer molecules to form the polymers of tomorrow. Figure 3-17 shows a block diagram of ABS (acrylonitrile-butadiene-styrene) terpolymer. ABS is one of the most common terpolymers.

Types of Plastic Chains

The monomer molecules within the polymer chains are held together end-to-end by chemical bonds. These chains lay next to each other in layers. They are often piled up like a stack of wood. The chains that lay next to each other are held together to make a solid mass. Different kinds of forces hold the adjacent chains together in thermosetting plastics and thermoplastics.

Thermosetting Plastic Chains

Thermosetting plastic chains are held to their next door neighbor chains by **chemical** bonds. This is the same kind of a bond that holds all plastic chains together along their backbone. The chemical bonds **between** chains in thermosetting plastics are called **cross-links.** Heat will not loosen this chemical bonding without destroying (degrading) the plastic material. Cross-linking makes the plastic very solid. Figure 3-18 shows a paper-clip model of the cross-linking of a thermosetting plastic.

Thermoplastic Chains

Thermoplastic chains laying next to each other **do not** have the chemical bonds between the chains. Instead, the chains next to each other are **held together by an electrical attraction.** When the chains are laid side by side this electrical attraction causes the chains to bond together. The electrical attraction, or **electrostatic force,** is not as strong as the covalent bond that links the atoms within the molecules. Since the electrostatic force between the chains is weak, thermoplastics are often more flexible than thermosetting plastics. Figure 3-19 shows a paper clip model of two **thermoplastic** chains next to each other. Imagine that each clip has a very small magnet on it. It will then pull the next chain to itself. This attraction is

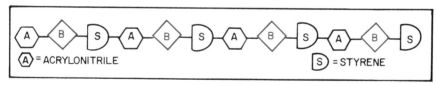

Fig. 3-17. Block diagram of ABS terpolymer.

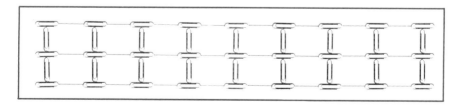

Fig. 3-18. Paper clip model of the cross-linking of thermosetting plastic chains.

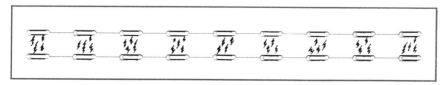

Fig. 3-19. Paper clip model of thermoplastic chains held together with very small magnets.

Fig. 3-20. Loading Hopper Cars
Plastic pellets are loaded into special hopper cars for shipment to manufacturers of plastics products. These hopper cars hold 180,000 pounds (81648 kg) each. Plastic pellets are also shipped in 1000-pound (453.6 kg) tote bins and 50-pound (22.68 kg) bags. Liquid plastic resins may be shipped in tank cars, barrels, or cans.

not a solid linkage since the linkage can be moved. This is similar to the way thermoplastic chains are held next to each other.

Thermoplastics may be softened by heating which weakens the electrostatic forces that hold the neighboring chains together. When enough heat is added to the thermoplastic materials, they soften into a flowable state, a physical change. In the softened state, neighboring thermoplastic chains can slip, and the plastic may be formed. This may be done by a number of different methods which will be discussed later in this book. Cooling the thermoplastics makes them a solid. This is also a **physical change**. The thermoplastic then holds itself in its new shape until it is heated again. A physical change occurs each time the thermoplastic is softened or solidified. Because the electrostatic force between chains can be weakened by heat, thermoplastics can easily be reprocessed (recycled). Thermosetting plastics cannot be recycled because they are permanently bonded. Recycling is discussed further in Chapter 5.

The Future in Plastics

Chemists will develop many new polymers in the future. They will juggle atoms and molecules to form new materials with new and different properties. This promises an exciting future. New materials will create new processing problems. These new processing problems will lead to new processing methods. New products will also result from the new materials. The new materials and new processes may solve some of the old problems, too. All this will provide opportunities for new jobs. You may become interested in the many career opportunities offered by the plastics industry. They will be in management, engineering, research, planning, toolmaking, processing, and sales. Some day **you** may help to invent or process the new materials of the future.

All of the plastic materials must be processed to make them into the products people use in their daily lives. The next section of this book will deal with the processing methods and their relationship to plastics materials.

STUDENT ACTIVITIES

1. Find ball and stick model kits in the science department and make ball and stick models of different monomer molecules from the diagrams provided here and in Appendix A.
2. Build ball and stick models of **polymer** molecules from the monomer molecules.
3. Build **copolymer** ball and stick models.
4. Build **terpolymer** ball and stick models.
5. Build ball and stick models of polymers and monomers from styrofoam balls and toothpicks. Paint the different atoms different colors, using a water base paint. Be sure to use different size balls to represent the different size atoms.

Process Description

Injection molding is a widely used process in which hot, soft plastic is injected (pushed or forced) into a closed mold. The material is cured or hardened in the mold. Both thermoplastic and thermosetting plastics can be injection molded. Thermoplastic injection molding is the most widely used. The thermoplastic material is hardened by cooling it in the mold.

Thermosetting injection molding is a fairly new process and needs special injection unit parts. The thermosetting plastic is cured by heating it in the mold. There are two types of injection molding machines: (1) plunger injection and (2) screw injection. Both kinds of injection molders are divided into two sections: (1) the injection section and (2) the mold operating section.

Plunger Injection Molding

Plunger injection molding (sometimes called **ram injection**) is the older of the two types of injection molding. It has been used

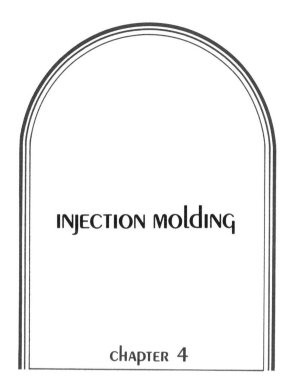

INJECTION MOLDING

chapter 4

since about 1922 in the United States. Plunger injection molders mold all types of thermoplastics. They may also mold certain thermo-

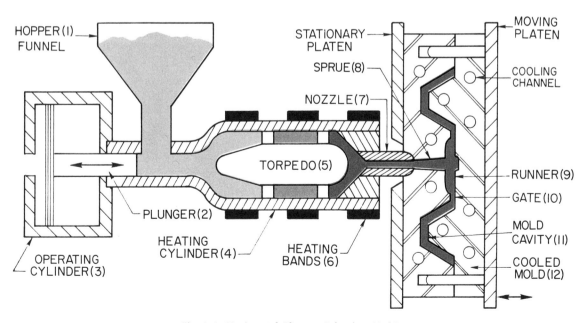

Fig. 4-1. Horizontal Plunger Injection Molder

sets under some conditions. Laboratory-size injection molders for schools are often plunger machines. However, laboratory plunger injection molders should not be used to mold thermosets in schools. Plunger machines are simpler and lower in cost than screw injection molders. They are also simpler to operate.

To follow the operation of the plunger injection molder, study Fig. 4-1. Thermoplastic molding material is placed in the **hopper funnel** (1) and is dropped by gravity into the **feed throat**. It is pushed forward by the **plunger** (2). The plunger is moved back and forth by the **operating cylinder** (3). As the plastic is moved forward, it is pushed against the **heating cylinder** walls (4) by the **torpedo** (5). The cylinder is usually heated with electric **heating bands** (6). The torpedo acts both as a heat storage and a spreader. The larger heating cylinder allows a greater surface for heat transfer into the plastic. This is needed because plastic is a good heat insulator. The thin stream of plastic passing between the torpedo and the heating cylinder walls also aids in quick heat transfer. The hot plastic leaves the cylinder through the **nozzle** (7). The nozzle fits tightly against the **sprue** (8). The sprue carries the plastic into the mold. The hot plastic then flows through the **runners** (9), **gate** (10), and into the mold cavity (11). Refer to Figs. 4-1, 4-20, and 4-21.

The plastic cools and hardens in the mold under pressure. The mold is then opened and the part is pushed out (ejected). The product of one injection molding cycle is called a **shot**. One shot may have one or more parts. It includes the sprue, runner(s), and gate(s).

Screw Injection Molding

Screw injection molders are similar to plunger injection molders. The main difference between the two kinds of machines is in the heating cylinder. See Fig. 4-2. The plastic is fed by gravity into the **cylinder** (2) from the **hopper funnel** (1). A turning, auger-like **screw** (3) moves the plastic forward in the heating cylinder. The screw works like a meat grinder screw when it moves the plastic forward. Frictional heat is caused in the plas-

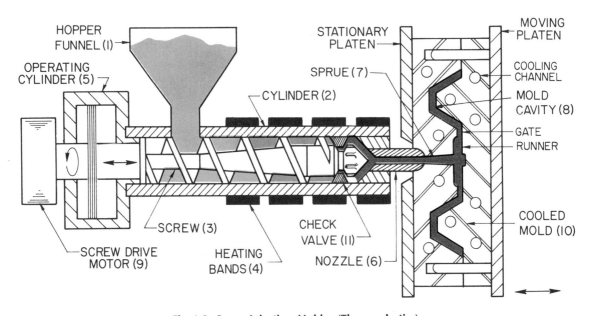

Fig. 4-2. Screw Injection Molder (Thermoplastics)

tic as it is moved forward by the screw. Heat is also added to the outside of the cylinder by electric **heating bands** (4). The heat from both the screw and heating bands melts the plastic as it moves forward and is mixed by the screw. The plastic moving forward pushes the screw backward. The screw stops turning when enough plastic is built up ahead of it. The screw is then pushed forward by the **operating cylinder** (5) to inject the plastic. A **check valve** (11) is usually put on the front end of the screw. The check valve keeps the plastic from sliding backward on the screw during injection. The plastic enters the **mold cavity** (8) through the **nozzle** (6) and **sprue** (7). As the plastic cools in the mold, the screw is turned by the **drive motor** (9) and heats more plastic as the screw moves backward again. The next cycle (time and procedure from start to finish) is then ready to begin. The mold opens after the screw stops turning. The part is then ejected (removed) from the mold.

The screw injection molder melts the plastic with less chance of degradation (harm) to the plastic. Less heat is added to the material in the screw injection molder. Often, the total cycle time is shorter than for plunger injection molding. The screw injection molder also mixes the plastics and colors better than the plunger injection molder.

The screw injection molder is the newer kind of machine. It has almost replaced the plunger injection molder. Its heating cylinder works much like an extruder in which the heated plastic materials are pushed forward by a turning screw. (Chapter 5 has information on extrusion.) The extruder screw turns all the time that the plastic is being processed. The injection molder screw, however, starts and stops turning during each cycle.

Importance of Injection Molding

Injection molding is considered the most important plastics process in terms of the numbers of and different kinds of products made. It is probably the most widely used plastics process. There are more injection

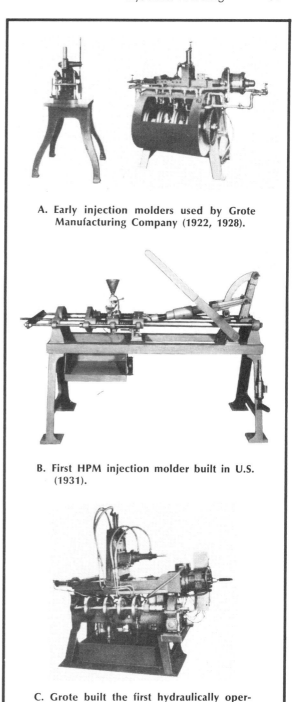

A. Early injection molders used by Grote Manufacturing Company (1922, 1928).

B. First HPM injection molder built in U.S. (1931).

C. Grote built the first hydraulically operated automatic injection molding press (1933).

Fig. 4-3. Injection Molders 1922-1934

molding machines in use today than any other type of plastics processing machine. Nearly 50,000 injection molders are now in use in the United States. About 5000 plants do injection molding. More than 27% of the plastics used in the U.S. are injection molded. By 1980, it is expected that 32% of the plastics will be injection molded. Extrusion is the only plastics process that uses more plastics than injection molding. Measured in pounds of plastics materials processed per year, injection molding would then be considered the second most important plastics process.

Injection molding can be traced back to 1872 and John Wesley Hyatt's stuffing machine. Machines now in use date back to developments made since about 1922. At that time, the Grotelite Company of Bellevue, Kentucky, (now Grote Manufacturing Company of Madison, Indiana) imported 12 injection molders from Germany. In 1928, Grote built the first automatic injection molder. The Foster Grant Company imported several injection molders from Germany about 1930. HPM Corporation built an injection molder in the U.S.A. in 1931.

Injection molding is used to produce long production run parts at a very fast rate. Parts can be made that are as small as the telephone parts shown in Fig. 4-4 or as large as the Nosco Maxi-Pallet shown in Fig. 4-6. A wide range of thermoplastics may be used.

Fig. 4-4. Precision Molded Telephone Dial Parts
Some of these parts require fits of ±.001" to function properly. They were molded on machines like the one shown in Fig. 4-5.

Fig. 4-5. Small production injection molder can make precision parts such as those shown in Fig. 4-4.

Fig. 4-6. Nosco Beverage Can Pallet Production
Pallets are shown stacked in the foreground.

Injection Molding

Part size	4½" x 44" x 56"
	(114.3 x 1117.6 x 1422.4 mm)
Part weight	28 pounds (12.7 kg)
Part cost (1970)	about $9.50
Expected part life	Greater than wooden pallets
Machine size used to mold it	
Clamp force	2500 ton (2267 Mt)
Shot size	640 ounce (18144 g)
Machine cost (1970)	about $500,000
Mold	
Weight	40 tons (36 Mt)
Cost (1970)	about $250,000
Number of runners (hot)	60
Dimensions in ( ) are in millimeters.	

Fig. 4-7. Nosco Beverage Can Pallet Mold

Fig. 4-8. This large injection molded part is a filter plate that weighs 100 pounds (45.36 kg).

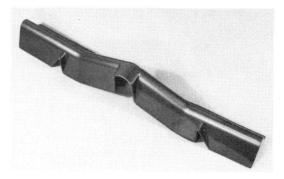

Fig. 4-9. Injection molded thermoplastic rubber part for the automotive industry. The part is painted with a flexible urethane paint.

In addition, many thermosetting plastics are now screw injection molded.

The Nosco pallet is said to be one of the world's largest injection molded parts. It is made on one of the world's largest injection molding machines. Parts such as plastic drinking tumblers, telephone parts, automobile grills, home appliances, and automobile battery cases are also injection molded.

Advantages

Injection molding has many advantages. It is a fast process. It can be used to form many different kinds of parts as well as very complex shapes. Products can be made which do not vary much in size from part to part. One thousandth of an inch (.001") (0.0254 mm) or less variation from exact size is possible. A great range of part sizes is possible as is shown by the can pallet and telephone parts, Figs. 4-4 and 4-6. Costs for plastic materials are generally lower than for thermoforming, casting, and reinforcing (fiberglass).

64   Section II   MOLDING

Fig. 4-10. Large single-stage screw injection molder with 2700 ton (2449 Mt) clamp and 300 ounce (8505 g) styrene injection capacity. Note worker's size as related to machine.

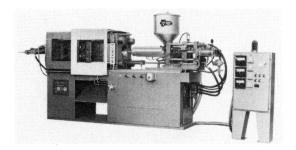

Fig. 4-11. Small industrial injection molder with 150 ton (136.1 Mt) clamp, 4 to 8 ounce (113 g to 226 g) shot and 15″ x 24″ x 15″ (381 mm x 609 mm x 381 mm) mold size.

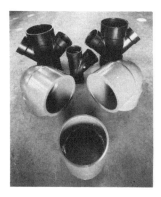

Fig. 4-12. Injection Molded Parts

Disadvantages

Injection molding also has disadvantages. The major one is the size and cost of the machines for the part size produced. It takes strong, expensive machines to make even the smallest parts. Also, mold costs for production injection molding are high. Molds and machines must be built to take high molding pressures. Injection molding is less expensive than most processes for long production runs. For short production runs, such processes as thermoforming, casting, and reinforcing (fiberglass) are less expensive since much less costly molds and machines are used.

Industrial Equipment

Industrial injection molding equipment ranges from about 1 ounce (28.359 g) to 800 ounces (22.64 kg) in shot size or larger. **Shot size** is that amount of general purpose polystyrene plastic that can be injected in one stroke (shot) of the injection plunger or screw.

The mold clamping capacity of an injection molder is important, too. It is given in tons of force which hold the mold halves together during injection. Industrial injection molders range from 25 tons (23 Mt) to more than 3000 tons (2750 Mt) clamp force. The larger the part size and the greater the nozzle pressure of the machine, the greater the clamp force that is needed to keep the molds closed during injection. Pressures from 3000 to 10,000 pounds per square inch (psi) (20 700 to 69 000 kPa) are applied to the inside of the mold by the injected plastic. This pressure is multiplied by the shadow

Injection Molding 65

Fig. 4-13. Injection Molded Table Leg

area of the part that the clamp sees. It equals the number of pounds or kilopascals clamping force required to hold the mold closed.

Materials Used in Industry

Almost all thermoplastics may be injection molded. Common thermoplastics used are acrylic, cellulose acetate, polyethylene, polystyrene, and polypropylene. Some high-temperature and special thermoplastics may need special injection molding machines. These include fluorocarbon (**Teflon®** or **TFE**), nylon, and polyphenylene oxide (PPO).

Many thermosetting plastics are being injection molded with specially designed screw injection molding machines. Examples of injection molded thermoset plastics are phenolic, melamine, urea, rubber, and thermoset polyester.

Thermoplastic Injection Molding

Thermoplastics, such as those listed above, are melted at temperatures of from 225°F (107°C) to over 650°F (343°C) in injection molders. They are injected at nozzle pressures from several hundred pounds per square inch to over 30,000 pounds per square inch (207 000 kPa). Injection pressures in industrial production machines are often higher than laboratory equipment.

Mold temperatures range from less than 100°F (37°C) to over 300°F (148°C) in injection molding. The mold must be cool enough to harden the plastic but warm enough to prevent cooling too fast. Rapid cooling may cause unfilled molds or uneven shrinkage of the plastic. Different thermoplastics need different cooling rates and mold temperatures. Generally, the higher the melting temperature of the plastic, the higher the mold temperature needed. Table 4-1 shows mold temperature ranges for different thermoplastics.

Thermoplastic materials become hard (solidify or freeze) by being cooled. This is a **physical** change. It usually takes place in the mold.

Thermoset Injection Molding

Phenolic plastic as well as other thermosetting plastics is being successfully injection molded in screw injection molding ma-

TABLE 4-1
Mold Temperature Ranges Used for Selected Thermoplastics

Group	Plastic Family	Mold Temperature
I.	Polyethylene, polypropylene, polystyrene.	<100°F (<38°C)
II.	Acrylics, ABS, polyvinyl chloride, cellulosics.	100°F to 200°F (38°C to 93°C)
III.	Nylon, fluorocarbon, polycarbonate, acetal, and other high temperature thermoplastics.	>200°F (>93°C)

chines. Thermoset injection molding is done at a great savings over compression and transfer molding for some parts. The thermoset injection molding screw is designed so that it can be removed from the cylinder if the plastic hardens in the heating cylinder. The screw has a constant root diameter and can be unscrewed from plastic that has accidently cured in the heating cylinder. The cylinder temperature is controlled by a liquid circulating in a jacket around the cylinder. It does not use electrical heating bands. Either a heated or cooled liquid may flow through the jacket to heat or cool the cylinder. **Cylinder temperatures** are kept just high enough to melt the plastic. They must be kept **below** the curing temperatures of the plastic. Thermoset heating cylinder temperatures vary from about 130°F to 240°F (54°C to 116°C), depending on the material being processed. This is lower than for most thermoplastics.

Mold temperatures for thermosetting injection molding are high. The high temperature causes the plastic material to polymerize (cure) in the mold. This polymerization is a **chemical** change. Injection mold temperatures for thermosetting materials of from 250°F to 350°F (121°C to 177°C) are common.

The basic difference between thermoplastic (heat softening) and thermosetting (heat

Fig. 4-14. Mold temperature controlling unit for plastics molding. Water is run through this unit which controls its temperature before it is circulated through the mold.

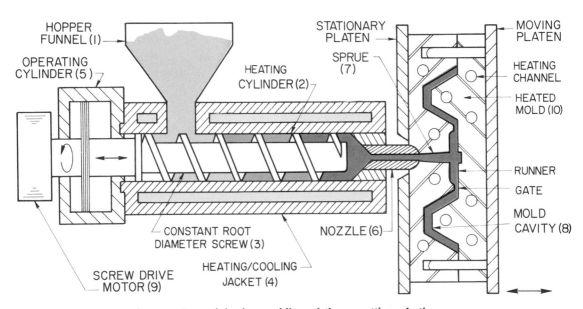

Fig. 4-15. Screw injection molding of thermosetting plastics.

curing) injection molding is the temperature of the mold. For thermoplastics, the mold is cooled to harden (solidify) the material. Thermosetting plastics require the mold to be heated for curing. Other differences are in the method of controlling the heating cylinder temperature and in the screw design.

Safety Note

Thermosetting materials should not be molded in plunger-type laboratory injection molding machines in schools.

Types of Molding Cycles

Injection molding may be cycled automatically, semiautomatically, or manually. In automatic injection molding, each machine runs automatically. One operator may watch several machines and check them for proper operation. In semiautomatic injection molding, one operator is usually assigned to each machine. The operator must open the mold safety door to remove the completed part at the end of each cycle. When the mold safety door is closed, the cycle is started again. In manual or hand molding, the mold is usually removed from the machine to remove the part. Manual or hand molding is usually restricted to small, light weight laboratory injection molds.

Injection Molds

Injection molds are the forms that shape the plastic while it cures or hardens. They are made in two or more parts. One half usually is held in place (stationary half), and the other (the moving half) moves back and forth to open and close the mold. The stationary half is usually bolted to the **platen** closest to the injection nozzle. The platens are the part of the machine clamping mechanism that holds the molds as shown in Figs. 4-1 and 4-2. The moving half of the mold is bolted to and opens and closes with the moving platen of the machine. When the moving half of the mold opens, **ejection pins** move forward, pushing the molded part from the mold. **Stripper plates** sometimes are used to eject the parts. A stripper plate separates the part from the mold in cases where ejection pins may cause damage. Ejection pins or stripper plates are necessary for automatic or semiautomatic injection molding. A diagram of an automatic injection mold is shown in Fig. 4-16.

A system of sprue, runners, and gates guide the soft plastic from the nozzle to the cavities. The **cavity** is the hollow part of the mold that forms the part. The **sprue** is usually a round tapered hole. **Runners** are usually needed only in molds with more than one cavity. The runners carry the hot plastic from the sprue to the one or more cavities. Runners are most often round or half-round in shape, although they may also be square or rectangular. The **gate** is a smaller opening at the end of the sprue or runner. It allows the part to be broken from the sprue or runner easily. Gates also add back pressure to the runner system to give more even filling of the mold cavities. Gates provide a place for the plastic to harden (freeze) **first** in the mold. They help keep the mold from getting too full or the material from decompressing (flowing) back into the sprue or runner. Correctly designed gates will let a smooth flow of plastic into the mold. Molds will often flash (plastic escapes at the parting line) when the injection pressure is too high. Flashing may also occur when the injected plastic is too hot.

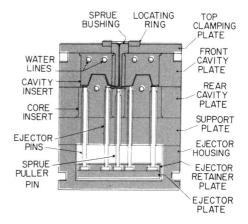

Fig. 4-16. Typical two-plate injection mold showing the various mold component parts.

68   Section II   MOLDING

Fig. 4-17. Two-Piece, Two-Cavity Laboratory Injection Mold

Fig. 4-18. Multiple-cavity family injection mold for clothes pins.

Fig. 4-19. A part from a family injection mold. Note that all the parts for one or more complete units are molded at one time. The runner system carries the plastic to all the cavities.

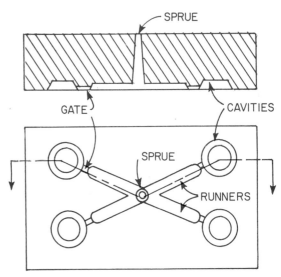

Fig. 4-20. A typical sprue, runner, and gate system for cavity side (rear) injection. Note how the system is balanced. All of the runners are the same length and shape. All the cavities are the same distance from the sprue.

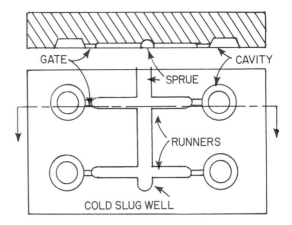

Fig. 4-21. A typical sprue, runner and gate system for parting line injection. This system is not truly balanced as is the system in Fig. 4-20. The top two cavities tend to fill first.

Two types of injection systems are used in injection molding:

1. Cavity side (rear) injection and
2. Parting line injection.

When cavity side injection is used, the plastic is injected through a tapered sprue from the rear of the stationary half of the mold. It then usually follows the parting line through the runner(s) and gate(s) to the one or more cavities. See Fig. 4-20. Cavity side injection is usually used in industrial injection molding. In parting line injection, the plastic is injected from the top of the mold. The parting line sprue is usually cut into both halves of the mold. The sprue follows the parting line to the runner(s), gate(s) and one or more cavities. Parting line injection is quite often used for laboratory injection molding. See Fig. 4-21. Most horizontal injection molders inject the plastic through the cavity side (rear) of the mold. Most small vertical laboratory injection molders inject at the parting line.

Laboratory Injection Molding Equipment

Most laboratory injection molders used in schools are **vertical plunger** machines. They should be used only for **thermoplastics**. Laboratory injection molders normally range in size from about ¼ ounce (6 g) to 1 ounce (28.35 g) shot size. Most of them are air operated. A few are hydraulically operated. The rest are operated by hand levers. Typical examples are shown in Figs. 4-22 and 4-23.

Most laboratory injection molders work on 115 volt electricity. They usually need a 10 ampere or larger electrical circuit, a separate circuit being used for each machine. Air operated machines need from 80 to 100 pounds per square inch (522 to 690 kPa) compressed air. Sometimes molds need water for cooling. This requires a water line and drain. Molds that are air cooled require no drain. Most laboratory molds in schools are not water cooled. They are just allowed to air cool on the bench between cycles.

Fig. 4-22. Simplomatic Injection Molder
This is an air operated vertical plunger machine with a shot capacity of about ¼ ounce (6 g). It can be mounted on a bench or stand and used with an air compressor.

Fig. 4-23. Newbury Wasp Injection Molder
This is an air operated vertical plunger machine with a shot capacity of 1/3 to 1 ounce (9.5 g to 28.35 g).

Fig. 4-24. Injection molding attachment for a Carver compression molding press.

Hand Molds

Molds for school use are usually **hand operated molds** and are not bolted to the machine platens. The molds are taken from the machine **by hand** after each injection and the molded part taken from the mold by hand. These molds usually do not have ejection pins to remove the molded part.

Some laboratory machines can use **semiautomatic molds** which may be bolted to the platens. These molds have **ejection pins** that remove the parts when the mold opens. A plate pushes the ejection pins out as the mold is opened, and the pins eject, or push out, the part from the mold.

Mold Design

Important in the mold design are (1) the draft angle, (2) the place of the parting line, (3) the place for injection, and (4) the types of molds.

Fig. 4-25. Hand Operated Laboratory Injection Molds

Fig. 4-27. The part is removed from the mold and inspected.

Fig. 4-26. The mold is opened to remove the part by hand.

Fig. 4-28. Semiautomatic Injection Mold
This is a six-cavity checker mold. Ejection pins form the core of each cavity. Note the diagram of this mold in Fig. 4-29.

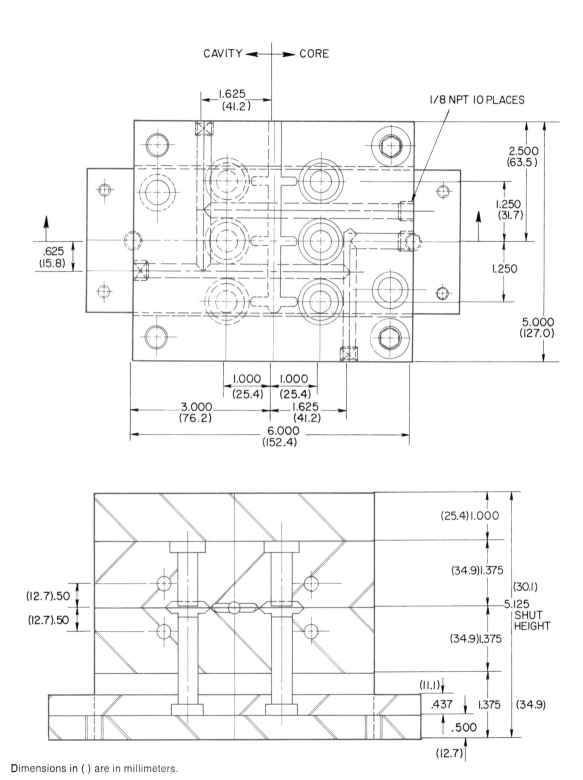

Dimensions in ( ) are in millimeters.

Fig. 4-29. Semiautomatic Injection Mold
This is a diagram of the mold shown in Fig. 4-28.

71

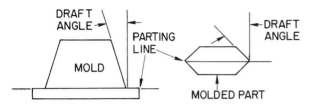

Fig. 4-30. Draft Angles and Parting Lines

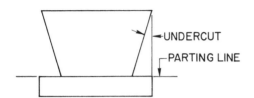

Fig. 4-31. An Undercut Mold

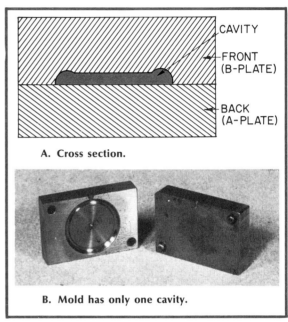

A. Cross section.

B. Mold has only one cavity.

Fig. 4-32. Flat-back mold for an impact washer.

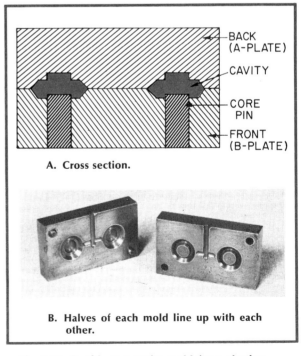

A. Cross section.

B. Halves of each mold line up with each other.

Fig. 4-33. Double two-cavity mold for a checker.

Draft Angle. In order to eject most parts from the mold, the mold must have draft. A draft is the angle or taper designed into the mold so that the part can be easily removed. Undercuts may be used **only** with a very flexible plastic such as **low density polyethylene.** Plastic shrinks when it cools. The draft angle, or taper, of the mold lets the part be taken out. Draft angles vary from ½ degree to several degrees. Usually, a 2 to 5 degree draft angle is best for molds used in schools. See Figs. 4-30 and 4-31.

Parting Line. The parting line is the point at which the two mold halves come together. This line should be located along the edge of a part and should be on one straight line or plane, if possible. Also, the parting line should be placed around the thickest place on the part.

Place for Injection. Usually, on school laboratory equipment the plastic will be injected at the parting line. The gate should be located on the parting line nearest the thickest part of the product.

Mold Types. Two kinds of injection molds are usually used in the school laboratory. One is the **flat-back** mold, Fig. 4-32. It is the simplest to make. The other is the **double-**

Injection Molding

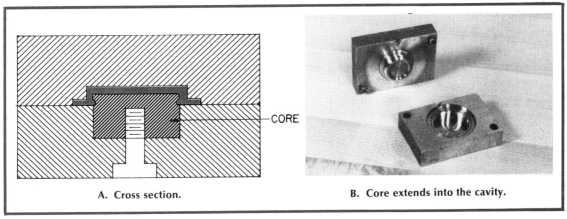

Fig. 4-34. Combination mold for a bottle cap.

cavity mold, Fig. 4-33. It is slightly more difficult to make since it needs accurate layout and machining so that both halves of the cavity line up. A combination of these mold types may also be used. In a combination mold, a part of one cavity (the core) extends into the other cavity half, Fig. 4-34.

Materials for Laboratory Injection Molding

Plastics usually used for laboratory injection molding include polypropylene, polyethylene, acrylic, ABS, and polystyrene. These plastics give a wide range of properties and products. They are all safe to use and easy to mold. The exact plastic used will depend upon the product to be made. Selection of materials for specific products is covered later.

Time, Temperature and Pressure Relationship

There is a very close relationship between **time, temperature** and **pressure** when plastics are molded. This is very important in all plastics processes. When one of three — time, temperature or pressure — is changed, one or both of the others must usually be

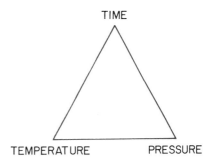

Fig. 4-35. Time, Temperature, and Pressure Relationship

changed also. This is especially true after a certain combination of time, temperature, and pressure has been found that will mold a good part. When one of the three is **increased,** one or both of the others must often be **decreased.** Thus, if the cylinder temperature is increased, less injection pressure is usually needed. Often, the injection forward time (the time that nozzle pressure is kept on the molded part) is also less. This is because the higher cylinder temperature will heat the plastic more rapidly. The total cycle time may then become less. If the material is hotter, however, it may require a longer cooling time.

The plastic may not flow at a given cylinder temperature. An increase in either cylinder temperature or nozzle pressure may correct this and cause the plastic to flow. Often, changing the time between molding each part will change the temperature of the plastic in the cylinder. This is called **residency time**. It is the time that the material spends heating up in the cylinder. If the parts are molded faster, the plastic in the cylinder will not heat up as much. A slightly higher cylinder temperature should then be used. If the parts are more slowly molded (less shots per minute), the plastic may get too hot and flash. A slightly lower cylinder temperature should then be used.

The nozzle pressure may be changed, rather than the temperature, to allow for a change in the cycle time. A slower cycle may require less nozzle pressure because the plastic heats up more between shots. A faster cycle may require more nozzle pressure.

Plastic **particle size** also affects the cycle time. Smaller plastic particles melt faster than large ones. This is because plastic is a natural heat insulator. The thicker the particle, the greater the insulation factor.

Time, temperature, and pressure are equally important considerations. The relationship may be likened to the triangle, Fig. 4-35. Changing one of the three will affect the other two. An increase in one of the three usually will mean a decrease in another. A decrease in one usually means an increase in one of the others.

Operation Sequence

There are five steps in the injection molding process:
1. Melt the plastic thoroughly in the heating cylinder.
2. Inject the right amount of plastic into the mold.
3. Use enough injection pressure to make a good surface finish on the part. The part should be free of weak welds, flow marks, sinks, and other defects.
4. Cool the plastic fast enough to harden it.
5. Open the mold for ejection (removal) of the part. Close the mold for the next molding.[1]

The molding sequence with the **action, reason** for the action and **trouble shooting** is shown in Table 4-2. Pictures of many of the steps are also included. Follow the action sequence carefully. **Look at the machine instruction manual for exact operation information.**

Checklist of Materials and Equipment Needed

Assemble the following materials and equipment for injection molding:
1. Injection molder.
2. Injection mold(s).
3. Injection molding plastic (type depends on mold and product).
4. Mold release — silicone or fluorocarbon.
5. Wrenches for adjusting the machine (if needed).
6. A small wood dowel about 8″ to 12″ (20.3 to 30.4 cm) long.

Operating Temperatures

Chapter 2 listed the softening points for thermoplastics in Table 2-2, Destructive Identification Tests. A complete list of softening points is given in Appendix A. Cylinder temperatures for plunger machines are usually about 100°F (55°C) above the softening point of the plastic used. Temperatures vary slightly from one machine to another. This is because there are differences in the controls and designs of the machines. Table 4-3 will serve as a guide for molding temperatures of several common injection molding plastics. Start on the low end of the range and slowly increase the cylinder temperature. Stop at a point at which the plastic flows easily from the nozzle. Mold several parts. Slowly adjust the cylinder temperature (10°F to 15°F or 5°C to 8°C at a time) until good parts are made. Remember to maintain a uniform

[1]Adapted from **Fundamentals — Injection Molding Technology,** The Dow Chemical Company, Midland, Michigan.

Injection Molding

Fig. 4-36. Equipment and materials are ready for injection molding.

molding cycle. This will keep the residency time of the material uniform. Uniform residency time will help maintain uniform molded parts.

Tooling up for Production

Molds for the production of plastic parts by injection molding may be obtained from a number of sources. Ready-made molds that fit laboratory machines may be bought from many laboratory machine manufacturers. Molds may also be made from ready-made mold blanks which some machine manufacturers sell. Blanks may also be made from bar stock or cast from metal-filled epoxy. The procedure for making molds by these two processes is given here.

Machining Metal Molds

Molds machined from metal bar stock may be made from aluminum, brass, or cold-rolled steel. The steps describing the machining procedure begin on page 81.

TABLE 4-2
Operation Sequence for Injection Molding

Sequence/Action	Reason	Troubleshooting
❶ Adjust the machine to fit the mold. (a) Set the clamp pressure (if needed) for a firm over-the-center snap. (b) Adjust the mold centering screw. See Figs. 4-37 and 4-38.	(a) Mold must be clamped firmly to prevent flashing. (b) Mold must be centered under nozzle to receive material in sprue.	**Flashing:** Check clamp pressure. Reduction of molding pressure and/or temperature will also prevent flashing.
❷ Plug in the electrical cord. Turn on the machine and warm it up to operating temperature. Check the plastic chart for various molding temperatures. (See Table 4-3).	Machine should be warmed up before operation is tried. Partly heated material will pack or plug the cylinder if it is plunged.	**If machine fails to warm up:** Check electrical circuits for power. Be sure switches are on.
❸ (a) Place material in the hopper. Fill the injection cylinder. Allow the material to preheat in the cylinder. (b) **Caution!** Do not overfill the cylinder. No material should stay in the cylinder **funnel** at any time. See Figs. 4-39 and 4-40.	(a) Material must preheat in cylinder before injection can be made. (b) **Overfilled vertical cylinder funnel will scatter material when injecting and will cut the nozzle pressure.**	**If material will not flow through nozzle:** Operating temperature has not been reached. **If material scatters or sticks to cylinder funnel:** Cylinder is being overfilled.
❹ Connect the air line if it is an air-operated press. Purge the cylinder before injection. Material must be at operating temperature (plasticized). See Fig. 4-41.	This checks moldability of material. Old material in cylinder may be degraded or contaminated.	**If cylinder is plugged:** Raise the temperature slightly and purge. Do not overheat machine as the material may degrade.
❺ Put a little mold release on the mold. Apply only as needed. **Do not put on for each shot.** See Fig. 4-42.	This keeps parts from sticking.	**Sticking parts:** Apply mold release. High temperature materials require high temperature mold releases.

(continued on page 78)

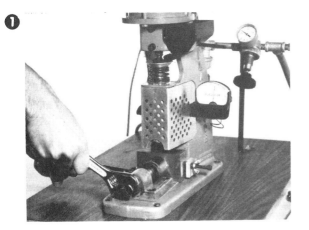

Fig. 4-37. Adjust the machine to fit the mold.

Fig. 4-40. Fill the injection cylinder.

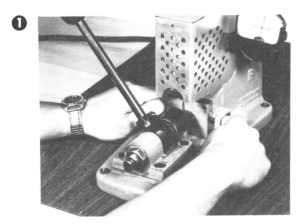

Fig. 4-38. Adjust the mold centering screw.

Fig. 4-41. Purge the injection cylinder.

Fig. 4-39. Place material in the hopper.

Fig. 4-42. Spray on mold release.

TABLE 4-2 (Cont.)

Sequence/Action	Reason	Troubleshooting
6 (a) **Check** sprue and nozzle alignment. (b) Clamp the mold into the machine. See Fig. 4-43.	(a) This prepares for injection. (b) Mold improperly clamped or aligned may not fill.	**Flashing at the nozzle:** May be caused by poor nozzle and sprue alignment.
7 Inject the material into the mold. (a) Operate manual injection lever down. OR (b) Operate 4-way air valve forward or downward (injection direction). See Figs. 4-44A and B.	This fills the mold and allows part to solidify.	**Partially filled mold:** (a) Material too cold. (b) Mold too cold. (c) Low molding pressure. (d) Not enough material in cylinder. (e) Gates too small. (f) Air trapped in mold. (g) Mold needs venting. **Overfilled mold (Flashed):** (a) Material too hot. (b) Mold too hot. (c) High molding pressure. (d) Not enough clamping pressure.
8 Hold pressure on the material for a short time (several seconds). This is known as **dwell time** or **injection forward time**.	This allows additional material to flow into mold to replace any shrinkage. It prevents material from decompressing back through the sprue.	**Shrink marks on the molded part:** (a) Short dwell time. (b) Material too hot. (c) Uneven wall thicknesses on part (poor part design).
9 Retract the plunger. (a) Operate the manual injection lever up. OR (b) Operate the 4-way air valve back or upward (retract direction).	This readies the cylinder for the next shot.	
10 Refill the injection cylinder to the proper height.	This allows new material to preheat for the next cycle.	**Material does not heat fast enough:** (a) Cylinder temperature too low for fast cycling. (b) Material not put in the cylinder soon enough.
11 (a) Open the mold clamping device and remove the part from the mold, OR (b) Open the clamping device and allow the ejector pins to the eject part (semiautomatic). See Fig. 4-45.	This removes the part from the mold.	**Part warps after removal:** (a) Part not cooled in mold long enough. (b) Shrink fixture is needed to control shrinkage.

(continued on page 80)

Injection Molding 79

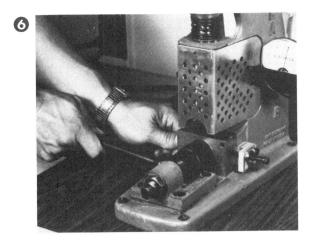

Fig. 4-43. Clamp the mold into the machine.

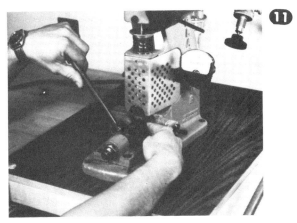

Fig. 4-45. Open the mold clamping device.

A. Injecting.

B. Operating injection lever.

Fig. 4-44. Inject the material into the mold.

Fig. 4-46. Remove and inspect the part.

TABLE 4-2 (Cont.)

Sequence/Action	Reason	Troubleshooting
12. Inspect the part for defects. See Fig. 4-46.	Necessary adjustments in the molding cycle can be made.	**Silver Streaks:** (a) Plastic temperature is too high. (b) Moisture is in material. (c) Injection speed is too fast. (d) Mold temperature is too low. (e) Injection pressure is too high. **Poor Surface Finish:** (a) Mold is too cold. (b) Injection pressure is too low. (c) The mold has a poor surface. (d) Too much mold release has been applied.

Material has been adapted from *Troubleshooting Injection Molding Technology,* Dow Chemical Company, Midland, Michigan. This booklet offers additional information.

TABLE 4-3
Common Injection Molding Temperatures for Thermoplastics

Plastic Material (Family Name)	Molding Cylinder Temperature Range
ABS (Medium to High-Impact)	380° to 525° F (193° to 274° C)
Acrylic	400° to 520° F (204° to 271° C)
Polyethylene, Low Density	225° to 325° F (107° to 163° C)
Polyethylene, High Density	275° to 500° F (135° to 260° C)
Polypropylene	400° to 550° F (204° to 288° C)
Polystyrene, General Purpose	325° to 500° F (163° to 260° C)
Polystyrene, High Impact	350° to 600° F (177° to 315° C)

Procedure for Mold Making

1. Secure metal bar stock large enough to make the mold blanks. Check the machine specifications for blank sizes. Cut bar off to length.
2. Surface and square all six faces until the blank is the correct size for the mold cavity to fit the machine. A metal lathe, milling machine, or surface grinder may be used.
3. Locate and drill holes for the guide pins. Holes should be located at least one half their diameter from the edge. (Their center should be at least their diameter from the edge.) Drill both halves together. Drill the guide pin holes 1/64" (0.396 mm) **undersize**. Ream the tight side holes at least .002" (0.05 mm) **undersize** for a press fit. Ream the holes in the loose side of the mold .001" to .002" (0.025 to 0.05 mm) oversize for a slip fit.
4. Lay out the mold cavities and machine them to size. Cavities may be milled, drilled, machined on the lathe, or electrically discharge machined. Flat surfaces on steel may be ground with a surface grinder. They may also be ground on a flat surface with sandpaper, especially on aluminum and epoxy molds.

Fig. 4-47. Cutting aluminum bar stock.

Fig. 4-49. Drilling guide pin holes.

Fig. 4-48. Squaring stock on the milling machine.

Fig. 4-50. Machining the mold cavity.

Fig. 4-51. Surfacing with sandpaper.

Fig. 4-52. Polishing a mold cavity.

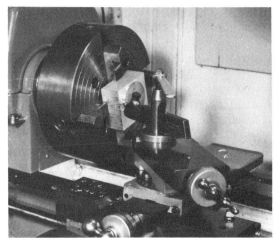

Fig. 4-53. Facing or squaring on the lathe.

Fig. 4-54. Drilling guide pin holes on the lathe.

5. Polish the mold cavities with progressively finer abrasive paper (100, 280, 400, and 600 grit). Crocus cloth will produce an even finer finish. A final polish may be made with polishing compound and a small buffing wheel or brush. The buffer can be mounted in either a drill press or flexible shaft. If scratches remain, remove them with a fine oilstone. Resand and repolish that area.
6. Press in the guide pins. They need stick out only about once the diameter of the pin.

A metal lathe is the only machine that is needed to make very simple metal molds. It can be used for turning, drilling, milling, and grinding. A drill press or milling machine is helpful in drilling guide pin holes and sprues, but this may also be done on the metal lathe.

Casting Epoxy Molds

Molds for injection molding may be cast from metal-filled or mass-casting epoxy. They are easier to make than machined metal molds when fine details and irregular shapes are desired. Such molds should be used for parts for which it would be difficult to machine a mold. A medallion is a typical example, Fig. 4-55.

Epoxy cast molds should be cast into a metal frame or hollow mold blank such as that shown in Fig. 4-56. Details for casting the mold are given in Chapter 6, Blow Molding.

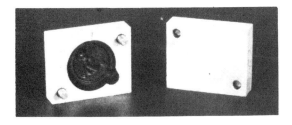

Fig. 4-55. Cast Epoxy Medallion Mold

Materials and Products for Injection Molding

Plastics commonly used for laboratory injection molding are listed in Table 4-3, along with their molding (cylinder) temperatures. These plastics are easy to use in the laboratory. (The properties of some of them are described in Chapter 1.) These plastics are easy to use because they melt at low temperatures as compared to other plastics. They generally flow easily at lower injection pressures, and they are all safe materials to use.

Probably the best all around injection molding plastic for the school use is **polypropylene**. It molds easily, flows well, has good surface gloss, and is inexpensive.

Polystyrene is also inexpensive, but it shrinks more than most plastics and is brittle.

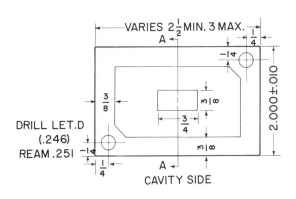

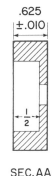

NOTE: DRILL BOTH HALVES TOGETHER – REAM SEPARATELY

MATERIAL: ALUMINUM
BRASS
1020 STEEL

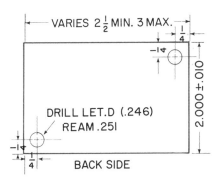

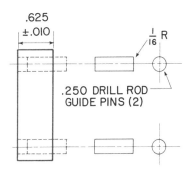

METRIC CONVERSION

U.S. (in.)	S.I. (mm)	U.S. (in.)	S.I. (mm)	U.S. (in.)	S.I. (mm)
.010	0.254	1/16	1.59	2	50.8
.246	6.24	1/4	6.35	2-1/2	63.5
.250	6.35	3/8	9.52	3	76.2
.251	6.37	3/4	19.05		

Fig. 4-56. Complete mold blank for epoxy cast cavities.

This often leads to problems in getting parts out of the mold. High-impact polystyrene may be used successfully in the school.

Polyethylene is a good plastic, especially for flexible products. It molds well at low temperatures. High-density polyethylene may be used for more rigid products. It melts at a slightly higher temperature than low-density polyethylene.

ABS may be used in the classroom, but it needs slightly higher molding temperatures. It is more expensive than polystyrene, polyethylene, and polypropylene. Over-heating ABS will cause a poor surface finish called **streaking**.

Acrylic plastic needs high molding pressures, as it does not flow as easily as most thermoplastics. It does, however, make strong, optically clear parts.

Cellulosics, such as cellulose acetate and cellulose acetate butyrate, mold well but are not used often in the school laboratory. Nylon needs higher molding temperatures. It is not recommended for school use. Polycarbonate plastics do not flow easily and need high molding pressures. Polycarbonate is not recommended for school laboratory molding either.

There are two **plastics that should not be used** in the school laboratory for injection molding. They are **acetal** and **polyvinyl chloride** (PVC). Acetal has a very narrow range between melting and degradation (breakdown). Acetal melts at 347°F (175°C) and will virtually explode at about 400°F (204°C). Polyvinyl chloride gives off HCl (hydrochloric acid) gas when its temperature gets too high. The HCl gas will mix with water vapor in the air and make hydrochloric acid. The acid level in the air will usually not become dangerous, but it will rust the equipment. PVC should be avoided for injection molding in the school laboratory.

Some product ideas are listed here for several plastic materials:

1. **Polystyrene or High-Impact Polystyrene**
   - Golf green markers
   - Golf tees
   - Salt shaker tops
   - Name tags
   - Luggage tags
   - Medallions
   - Jacks (ball and jacks)
   - Forks, spoons, and knives
   - Sewing machine bobbin
   - Coasters
   - Rings
   - Corn-on-the-cob skewers
   - Guitar picks
   - Poker chips

2. **Polypropylene**
   - Swizzle sticks
   - Toothpaste tube rollers
   - Coffee stirrer/sugar spoon
   - Duplex electrical outlet safety covers
   - Hinged pillboxes
   - One-piece hinges
   - Golf tees
   - Golf green markers
   - Corn skewers
   - Pop can tops

3. **Polyethylene**
   - Salt and pepper shaker plugs
   - Salt shaker tops
   - Small toys
   - Pop bottle caps
   - Super balls
   - Key tags

4. **ABS (Acrylonitrile-Butadiene-Styrene)**
   - Checkers
   - Medallions
   - Golf tees
   - Arrow knocks
   - Toys
   - Poker chips
   - Coasters
   - Guitar picks

5. **Acrylic**
   - Small magnifying glass
   - Small lenses

Grades and Specifications

Thermoplastic injection molding materials are of various grades and specifications. Those best suited for the school laboratory are given in Table 4-4 which gives the melt index (rate of flow), particle size, and density (specific gravity). As a general rule, plastic materials for school use should be selected that are of a (1) general purpose, (2) middle range, (3) smaller particle size, (4) higher melt index, and (5) lower density. Plastics of these specifications will generally be easier to work with on small equipment.

The **melt index** is a rating of how fast a thermoplastic flows when it is melted. The higher the melt index number, the more easily the plastic flows when it is melted. Higher melt index plastics do not require high nozzle

TABLE 4-4
General Purpose Injection Molding Grades for Thermoplastics Suitable for Laboratory Use

Material (Family Name)		Melt Index	Density	Particle Size
ABS		Not applicable	0.990 - 1.06	1/8" cube or pellet (3.175 mm^3)
Polyethylene	L.D.	3.0 - 60.0	0.910 - 0.925	1/8" cube or pellet
	M.D.	5.0 - 30.0	0.926 - 0.940	1/8" cube or pellet
	H.D.	5.0 - 30.0	0.941 - 0.965	1/8" cube or pellet
Polypropylene		5.0 - 20.0	0.850 - 0.910	1/8" cube or pellet
Polystyrene	G.P.	2.0 - 11.0	1.04 - 1.05	1/8" cube or pellet
	Hi Imp.	3.0 - 9.0	1.05 - 1.07	1/8" cube or pellet

pressures for molding. For a description of the melt index test, see Chapter 2.

Densities vary according to the need. Usually, the lower densities indicate shorter chain molecules. They also indicate molecules that are less close together. Lower density plastics are often softer and less stiff. Higher density plastics are often stiffer and higher in strength. Most of the **general purpose** injection molding resins are in the middle of the density range for a given plastic family.

Fig. 4-57. Model reciprocating screw injection molder that will mold Play-Doh®.

STUDENT ACTIVITIES

1. Process a variety of materials in each available mold. Use only polyethylene for undercut molds, such as bottle caps.
2. Experiment with plastics materials.

    Note: To record your experimentation, prepare a log sheet similar to the one shown in Appendix A, page 393. Save this record of your experimentation for later reference.

    a. Blend two or more colors of **the same material** to get a variety of color effects.
    b. Blend two different thermoplastic materials together and mold. What differences are noted in the products produced? **Caution:** Do not use undercut molds, such as bottle caps, for this experiment.
    c. Blend two densities of the same material to alter the density of the material. Note how the blend averages to a density between the two parent materials.
3. Experiment with the injection molding process. Change only one variable at a time. Make several shots at a given setting before changing again.
    a. Vary the injection molding temperature. Record and analyze the results.
    b. Vary the injection forward time in injection molding. Record and analyze the results.
    c. Vary the injection molding pressure. Record and analyze the results.
4. Design small products which can be injection molded on laboratory equipment. Consider the following criteria for design:
    a. Draft angle
    b. Placement of parting line

c. Gate placement
d. Flat-back mold or double-cavity mold
5. Design and construct molds for products designed. Consider the following design criteria:
    a. Basic principles of good design
    b. Draft angles
    c. Flash lines or parting plane
    d. Gate location
    e. Mold type — flat back, double cavity or combination
    f. Wall thickness on parts
Molds may be cast or machined.
6. Organize a company and mass produce plastic products.
    a. Organization
        (1) Operations sequence
        (2) Equipment locations
        (3) Supplies
        (4) Safety
    b. Recordkeeping
        (1) Materials used
        (2) Processing variables — temperature, pressure, and time
    c. Assembly of parts
    d. Finishing of parts
7. Build a model reciprocating screw injection molder and molds. Mold Play-Doh®. See Fig. 4-57.
8. Experiment with an educational toy injection molder. Compare the process and products with the laboratory injection molder.
9. Look up material data in the **Modern Plastics Encyclopedia**.

Questions Relating to Materials and Process

1. How are material specifications determined for injection molding?
    a. Melt index
    b. Specific gravity
    c. Molecular weight
2. What melt index-density-material specifications are best suited for the following applications?
    a. Pop bottle caps
    b. Luggage tags
    c. Beverage tumblers
    d. Picnic spoons, forks, and knives
    e. Coffee stirrers
Why was that particular material and grade chosen for each product?
3. Why do certain grades of a given plastic material lend themselves to injection molding more than others?
4. Given a specific material specification, determine whether it is well suited for injection molding. Why or why not would you not choose this material for injection molding?
    a. High-impact polystyrene, density 1.05, flow number 5, particle size 1/8″ (3.175 mm).
    b. Low density polyethylene, density .924, flow number 2.0, particle size powder.
    c. High density polyethylene, density .949, flow number 0.26, particle size 1/8″ (3.175 mm).
    d. ABS, density 1.06, flow number N/A, particle size 1/8″ (3.175 mm).
    e. Polypropylene, density .904, flow number 12.0, particle size 1/8″ (3.175 mm).
5. What is probably the most effective way in which to determine whether a plastic is suited to a specific situation?
6. Relate other categories of materials to the injection molding process, such as rubber, glass ceramics, etc. Why are some feasible (possible) and others not?
7. Record the code numbers from various materials in the laboratory and look up the specifications in the data books available from the resin manufacturers.
8. Use the **Modern Plastics Encyclopedia** as a basic reference for materials, processes, and suppliers to the plastics industry.

Process Description

Extrusion is a very widely used plastics process. More plastic is processed by extrusion than any other way. It is a process in which the plastic materials are heated and pushed forward by a turning screw. The screw pushes the plastic material through an opening or hole called a **die.** The die gives the plastic its shape. Long, continuous shapes that are made by the extrusion process are garden hoses, soda straws, plastic rods, and irregular shapes, called **profiles.** Plastic and rubber wire insulation is applied by extrusion. Thin plastic film for items such as sandwich and dry cleaning bags is made by extrusion. Paper and cardboard are often plastic-coated by the extrusion process. Most blow molders use an extruder to heat and melt the plastic. Many thermoplastics used for injection molding are first extruded into small rods and then chopped into plastic pellets. Thermoplastic materials are now being extruded, as well as some thermosetting plastics, rubber, and food products.

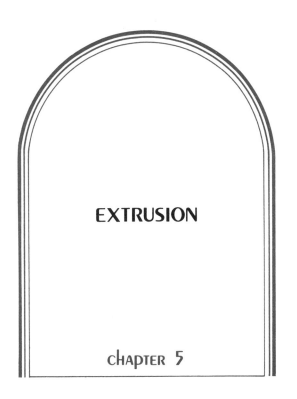

EXTRUSION

chapter 5

Plastic powder or pellets are loaded into the **hopper funnel** (1) of the extruder, as shown in Fig. 5-1, and dropped by gravity into the **barrel** (3). The plastic is moved forward by the turning of the **screw** (2). The screw works like a grain auger or meat grinder screw. It turns all the time that the extruder is working. It does not start and stop like the one in the screw type injection molder (see Chapter 4). As the material moves forward in the **barrel** (3) it is heated. The heat comes from two sources: outside **heating bands** (4) and inside frictional heat. The heat from the heating bands is controlled

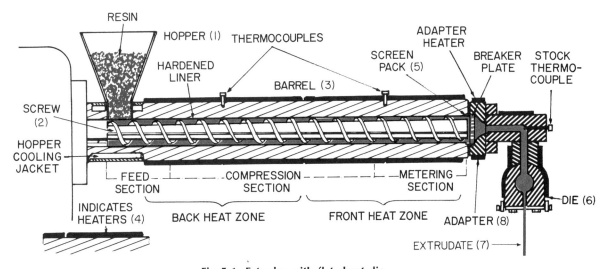

Fig. 5-1. Extruder with flat sheet die.

by temperature controllers (thermostats). The **frictional** heat is caused by the screw pressing and rubbing the plastic particles together in the barrel. Seventy-five percent (75%) or more of the heat added to the plastic in extrusion is from friction.

The screw is designed to compress the plastic as it moves forward. Trapped air and gases escape back through the hopper or a vent in the barrel. The screw usually turns about 20 to 200 revolutions per minute (rpm). The last section of the screw (metering section) meters (measures) the flow of the melted plastic. The metering section makes sure that there is enough back pressure on the material in the barrel to give an even melt. The metering section also helps keep an even flow of plastic going through the die. Pressure inside the barrel of up to 5000 **pounds per square inch (psi) (34 500 kPa) is common.**

The melted plastic passes through a **screen pack** and **breaker plate** (5) which traps impurities in the plastic and straightens out the flow, Fig. 5-1. The screen pack keeps the die from clogging. Without the breaker plate, the plastic tends to spiral as it comes out of the barrel. The breaker plate and screen pack also provide some back pressure.

The plastic enters the **die** (6) and takes on the shape of its opening. The plastic that comes through the die is called **extrudate** (7). The **die adapter section** (8) and die are also heated with heating bands. The heating bands are usually electrically powered.

The extrudate (extruded plastic) coming from the die is taken away and cooled by a **take-off** system. This system is regulated to take the plastic away from the extruder at just the right speed. The speed of both the take-off system and extruder are variable.

The extruder has at least three heating zones:
1. Rear barrel,
2. Front barrel, and
3. Die.

As many as 10 heating zones may be used on some extruders. The temperature of each zone is controlled individually by automatic temperature controllers. The temperature of the material is increased as it moves towards the die. The hopper section of the barrel is cooled with water or air circulation. Cooling the hopper keeps the plastic from sticking to the hopper or the screw as it is fed in. Most large extruder screws have water cooling channels inside them to keep them from getting too hot.

Extruder Dies, Cooling and Take-Off Equipment

Extrusion requires more than just an extruder. Dies are needed to shape the plastic as it comes from the extruder. Cooling and take-off equipment are necessary to cool and properly remove the extrudate.

A number of die types are used for plastics. They may be divided into four groups:
1. Rod and profile,
2. Film and sheeting,
3. Tubing and wire coating, and
4. Strand and ribbon.

Three types of cooling are used for plastic extrudate:
1. Forced air,
2. Water spray and water quench (dip), and
3. Chill rolls.

Extrusion take-off equipment may be divided into two types:
1. Pinch rolls and
2. Caterpillar tread units.

Each extrusion system is made up of a combination of an extruder, a die, cooling,

Fig. 5-2. Small extruder screw 15″ long x ¾″ diameter, or 381 mm x 19 mm (20:1 L/D ratio).

and take-off units. Some typical examples of these systems are described in this chapter and are listed according to die types. Combinations, other than those shown, may be used for specific situations. It would be difficult to list all the different combinations that could be used.

Rod and Profile Extrusion

Rod and profile dies produce rather solid or thick plastic extrudates. A hole in the shape of the rod or profile is cut into the die. Such holes are most often made straight through the dies. This allows the plastic to pass straight through the die without bending. The plastic extrudate coming out of the die is cooled with fans or water and pulled away. Pinch-roll or caterpillar tread takeoff units are most often used to remove rods and profile shapes. See Figs. 5-3A and 5-3B.

Sheet and Film Extrusion

Plastic sheeting (flat plastic over .010" to 2" thick or 0.254 mm to 50.8 mm) is extruded with a flat sheet die. This die is similar to the T-type and coat hanger-type dies shown in

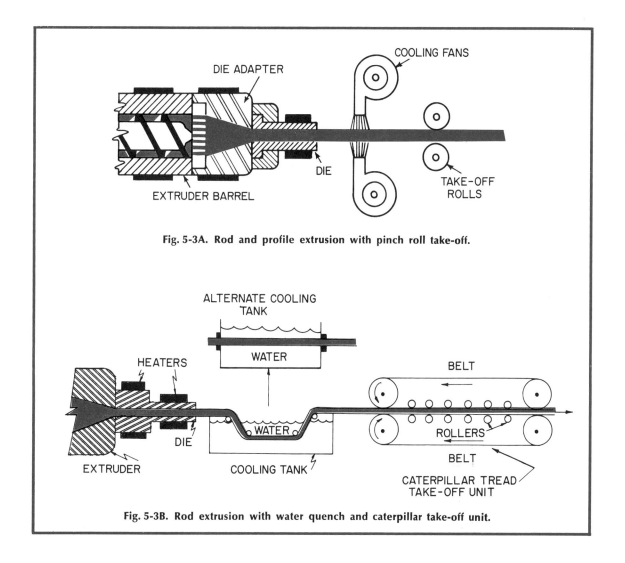

Fig. 5-3A. Rod and profile extrusion with pinch roll take-off.

Fig. 5-3B. Rod extrusion with water quench and caterpillar take-off unit.

90   Section II   MOLDING

Fig. 5-4. Caterpillar tread extrusion take-off unit for tubing and profile shapes. It may also be used to pull extruded rod.

Fig. 5-5. Typical Profile Extrusion Shapes

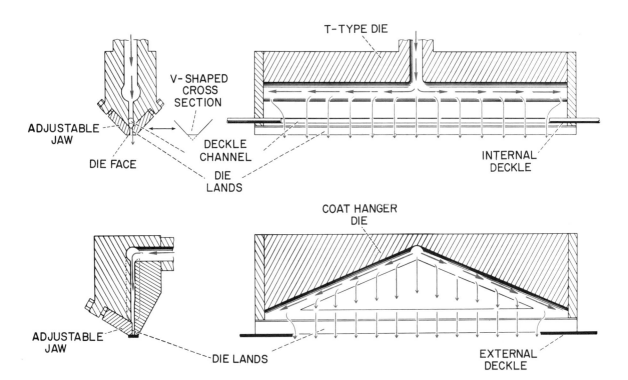

Fig. 5-6. Schematic cross section of T-type and coat-hanger-type extrusion dies. Locations of internal and external deckles are indicated. Deckles are used to adjust the width of the extrudate coming from the die.

Figs. 5-6 and 5-7 and the fan-type die shown in Fig. 5-27. The jaws of the T-type and coat hanger-type flat sheeting dies are often adjustable to control the thickness of the sheet. Plastic sheeting is usually extruded horizontally and taken away with a three-chill roll cooling unit and two rubber pull rollers. The water-cooled chill rolls cool and finish the plastic sheet. Either a smooth or textured (rough) surface may be made on the sheet, depending on the surface finish of the chill rolls. Polished chill rolls give the sheet a polished finish. Textured rolls put a texture on the sheet surface. After the rubber pull rolls pull the plastic sheet from the chill rolls, it is rolled up or cut to length.

Plastic film (flat plastic .010" (0.254 mm) or thinner) is extruded by two methods:
1. Flat film dies and
2. Blown film dies.

The flat film die is designed very much like the flat sheeting die. It is often extruded downward because the thin film is hard to handle. Take-off equipment is similar to that used for flat sheet.

Fig. 5-9. Sheet production line with extruder (left), chill rolls (center), and take-off unit (right).

Fig. 5-7. T-type flat sheeting die 136" (3454.4 mm) wide. This is believed to be the world's largest sheet extrusion die.

A. Extruder produces plastic sheet.

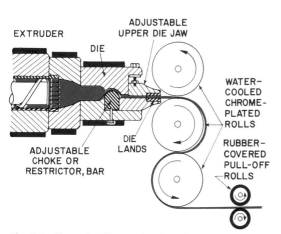

Fig. 5-8. Sheet Cooling and Take-Off Unit (Chill Roll)

B. Closeup of sheet extrusion showing chill roll stack.

Fig. 5-10. Laboratory Extruder

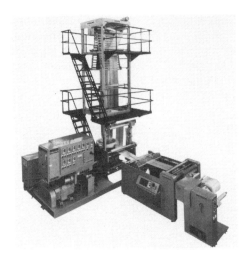

Fig. 5-11. Blown film tubing line in action.

Blown plastic film is usually extruded upward through a **ring die** similar to a tubing die. The blown tube (bubble) walls, however, are usually thinner than extruded tubing. The ring die has a mandrel in the center to produce a hollow tube of plastic. Air is blown into the tube through the **mandrel** (shaft) to expand the tube (bubble) and thin the walls. Once the bubble is formed, the air valve is closed. A stagnant body of air is held inside the bubble by the pinch rolls at the top of the tower. An air ring around the blown tube cools and shapes the bubble. Bubbles from a few inches, or centimeters, to 36 feet, or 10.97 m, in circumference can be made this way. Blown film is used for products such as sandwich bags, dry cleaner's bags, and construction and agricultural sheeting.

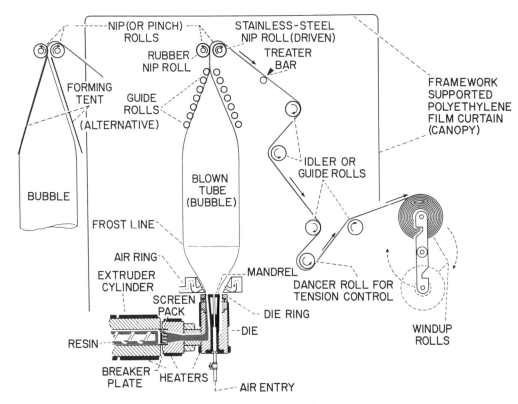

Fig. 5-12. Blown Film Extrusion

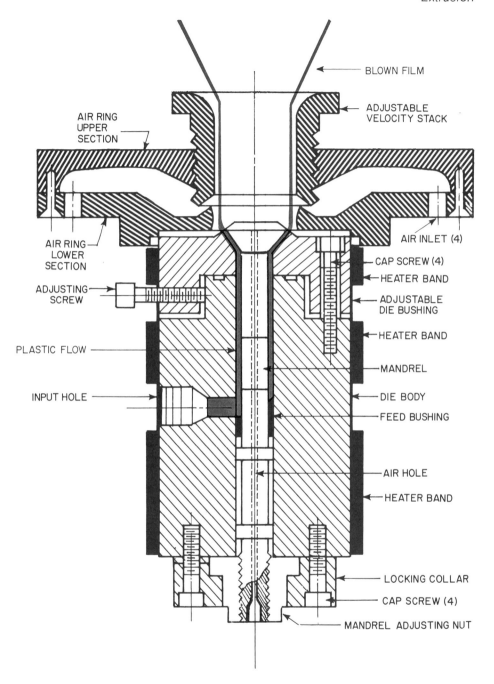

Fig. 5-13. Cross section of a small, adjustable, side-fed blown film die and air ring assembly.

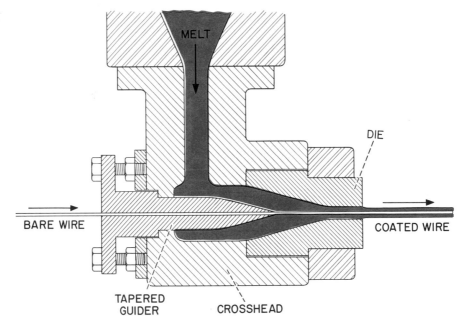

Fig. 5-14. Crosshead Wire Coating Die

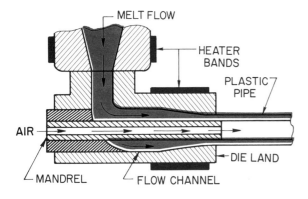

Fig. 5-15. Hollow Mandrel Pipe or Tubing Extrusion Die
This is a crosshead or 90° die.

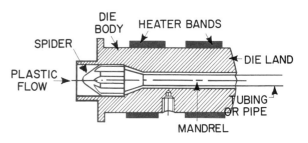

Fig. 5-16. Straight Pipe or Tubing Die

Tubing and Wire Coating

Tubing and wire coating require a die with a **mandrel,** or tapered guider, through the center to make the inside of the product hollow. Wire is pulled through the guider so that the plastic flows all around the wire. The wire is usually pulled through a **crosshead** or 90° angle die like that shown in Fig. 5-14. After the plastic forms a coating around the wire, it is cooled, tested, and rolled up.

Hollow plastic tubing is formed by a die through which a mandrel extends. Tubing dies may be either crosshead or straight. A crosshead tubing die is shown in Fig. 5-15 and a straight tubing die in Fig. 5-16. The plastic flows around the mandrel to form the tubing in the same way that it does in blown film extrusion and wire coating. Note that there is a flow channel or larger volume area behind the die opening. This flow channel helps the plastic to flow uniformily around the mandrel. If the flow is not uniform, the pipe or tubing will be weak or thinner on one side of the die.

Strand or Ribbon Extrusion

Strand and ribbon dies are used for compounding (mixing) plastic materials to make plastic pellets. The strands are fed into a cutter which cuts them into short lengths. This is often done right at the die. In some cases, it is done under water so the plastic will not stick to the die or be misshaped.

Importance of Extrusion

Extrusion is the number one plastics process in terms of the volume of plastics processed. More pounds of plastic materials are processed by extrusion each year than any other process. About 33% of all plastics materials (excluding blow molding) are now extruded. When blow molding is added, the volume is at about 36%. This high percentage of the total volume is expected to stay about the same for the next 10 years. Even with this great use of the process there are fewer extrusion machines in use than injection molding machines. One reason is that extruders process more pounds of plastic per hour, for the size of the machine, than injection molders. This is because extruders process materials continuously. An injection molder, by contrast, processes plastics with a start and stop action. Plastics extruders often have larger screws than injection molders. This also helps to reduce the number of machines needed.

Many plastics processes depend on the extrusion process. Most blow molding is built around the extruder. A large percentage of the thermoforming sheet is extruded. Most of the plastic pellets used for injection molding, blow molding, and extrusion are extruded into small rods and chopped into plastic pellets during their manufacture. The pellets are sold by the materials manufacturer to the processor to be made into useful products. These pellets are also extruded and then drawn (pulled) into smaller fibers and filaments for carpets, rope, and cloth.

The first screw extruder designed for thermoplastics was developed in Germany by Paul Troester in 1935. It was probably based on earlier pre-plasticizing (pre-melting) designs for injection molding. Earlier work on extruders was done in both England and the United States in the last half of the 19th century.

Advantages and Disadvantages

Extrusion, the most common way to produce long continuous plastic products, is a simple process in principle. However, it is a more difficult process to operate. It requires precise speed, feed, and temperature control to produce uniform products. Often the key to the success of an extrusion operation is in the take-off equipment, not the extruder. It is a process that should be started up and run continuously. Industrial extruders are run 24 hours a day during a work week and then shut down and cleaned over the weekend.

Industrial Extrusion Equipment

Industrial extruders range in size from 1½″ to 15″ (38 mm to 380 mm) in barrel diameter. Some experimental machines have been made with barrel diameters as large as 36″ (915 mm). In some cases, two, three, four and even five screws have been built into one machine. The most common multiple-screw machines have two screws. Multiple screws are used to increase the machine's output and mixing characteristics. Many claims and counter-claims have been made about single and multiple-screw machines. Both types will probably be used for many years to come. More single-screw machines are in use today than multiple-screw machines, however.

The **length** to **diameter** (L/D) ratio of the extruder is also important. The L/D ratio is the relationship between the length of the barrel to its diameter. A 1″ (25.4 mm) diameter barrel with a 20:1 ratio will be 20″ (508 mm) long. L/D ratios of from 8:1 to over 36:1 are used. A 24:1 L/D ratio is the most common.

TABLE 5-1
Average Hourly Extruder Through-Put

Extruder Barrel Diameter	Common Horsepower Needed*	Average Through-Put/Hr.*
1½" (38 mm)	15	60 pounds (27.2 kg)
2" (50 mm)	25	100 pounds (45.3 kg)
3" (76 mm)	50	300 pounds (136 kg)
4½" (115 mm)	150	800 pounds (363 kg)
10" (254 mm)	750-1000	5000-6000 pounds (2268-2721 kg)

*Depends on machine design and material processed.

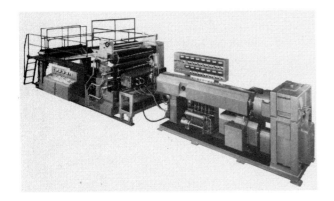

Fig. 5-17. Sheet extrusion line set up for use.

The higher the L/D ratio, the longer the material has to melt and mix in the barrel. Longer mixing is often needed for materials that are sensitive to high temperatures. They must be brought up to melt temperatures more slowly. Greater output is also claimed for higher L/D ratios. The output or through-put of commercial extruders is shown in Table 5-1. It is a measure of the efficiency of an extruder.

Materials Used in the Industry

Most thermoplastic materials may be extruded. Among the most important are polyethylene, polystyrene, polypropylene, polyvinyl chloride (PVC), ABS, and the cellulosics.

Plastic film and sheet are extruded from ABS, polyethylene, polystyrene, polypropylene, and cellulosics. Much of the extruded plastic sheet is used for thermoforming of numerous products. Plastic sheet is used to make refrigerator door liners, formed signs, advertising displays, and many other products. Examples of film products are food wrap, formed packages, and dry cleaner and sandwich bags.

Polyethylene, polypropylene, nylon, and acrylic fibers and filaments are used to make many types of carpet, fabrics, and rope. The plastic fibers in clothing are acetate, triacetate, acrylic, modacrylic, nylon, polyester, rayon, and others. Nylon monofilament fishing line is an extruded plastic.

Profile shapes and tubing are most often made from polyvinyl chloride (PVC), polypropylene, polyethylene, and cellulosics. A large amount of PVC and ABS pipe is used for carrying irrigation water, sewerage, hot and cold water, and as a conduit. ABS pipe does not weather as well as PVC, so it is not used where it will be exposed to continued sunlight. Plastic sewer and water pipes are used in both mobile and permanent homes. Many profile extrusions are used for gaskets and seals between metal parts. Examples are the seals on refrigerator and car doors.

Laboratory Extrusion Equipment

Laboratory extruders usually have a ¾" (19 mm) or 1" (25.4 mm) diameter barrel. The L/D ratio is usually either 20:1 or 24:1. A 20:1 ratio for school use is the most common. Through-put of such machines is about 6 to 10 pounds per hour (lbs./hr.) (2.72 to 4.54 kg/hr.) for ¾" (19 mm) models and up to 30 lbs./hr. (13.6 kg/hr) for 1" (25.4 mm) machines.

Because of the frictional heat produced in the barrel, automatic temperature controls are most desirable. At least three heating zones are needed:
1. Rear barrel,
2. Front barrel, and
3. Die.

The temperature controls should be able to hold at least a ± 10°F (5.5°C) variation for school equipment.

Safety Requirements

Certain precautions are necessary for the **safe** operation of an extruder. A shear pin should be provided in school laboratory extruders to keep the screw from breaking if it gets caught. Scrap materials will sometimes get into the material and cause the screw to stop. The shear pin, usually located between the screw and screw drive gear, also offers protection for the operator.

Chain guards and heat shields are needed to protect the operator from burns or injuries. A heat shield around the barrel and heater bands is absolutely necessary. These covers should be kept in place all the time the machine is running.

Hot metal parts are usually exposed even when all the guards are in place. **Safety zones** must be maintained around the **barrel and die** at all times. "Hot" signs should be placed over the **barrel and die** to warn others of HOT parts. Always protect people from hot parts so they will not get burned after the machine is shut down.

Operating Requirements

Air, electrical power, and plastic materials are needed for operation. Air-line pressure of about 100 psi (pounds per square inch) (690 kPa) should be available for hopper cooling. Water cooling may be used as a substitute for air. Either a water line and drain or a water circulation system may be used. A 230 volt single- or three-phase electrical power source sufficient to operate the drive motor and heating bands is needed. A few

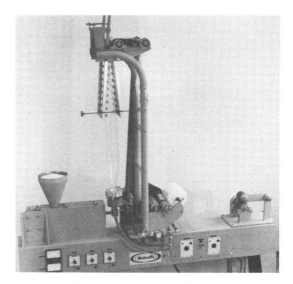

Fig. 5-18. Laboratory extruder making blown tubular film.

small extruders work on 115 volt single-phase electricity, but this is not most desirable. More efficient operation can be gained with 230 volt power. Plastic materials most often used for school laboratory extrusion are discussed later in this chapter.

Operation Sequence

Operation of the extruder is given in the following sequence list and in Table 5-2. The table of operations includes reasons for actions and troubleshooting. Follow the sequence carefully. Know the material you are working with. **Be sure that it is an extrusion grade plastic.** Know its melting point and flow characteristics (melt index or flow number). This information will help the operator run the extruder more efficiently. In most cases, two or three people should be assigned to the operation of the extruder system. This is important if separate take-off equipment is used with the extruder.

These are the five steps in the extrusion process:

1. Heat the barrel and die to operating temperature.
2. Start the screw drive motor and then fill the hopper with clean plastic.

3. Pull the extrudate (plastic) from the die and guide it through the cooling equipment.
4. Guide the plastic through the take-off equipment.
5. Adjust the screw speed, cooling rate, and take-off speed to produce the desired extrudate.

Table 5-3 is a guide to predicting the behavior of the extruder from material specifications. Table 5-4 lists start-up and operating temperature ranges for common thermoplastic materials. Always check the plastic manufacturer's instructions for exact information.

TABLE 5-2
Operation Sequence for Extrusion Molding

Sequence/Action	Reason	Troubleshooting
❶ Install the die wanted.	Each die is designed for a specific purpose.	**Die leakage:** The die has been installed wrong or too loosely. Tighten or readjust the die.
❷ Connect the air, water, and electrical power. Turn on the hopper coolant.	This gets the machine ready to start up. Coolant should be started before barrel heaters are turned on.	**Plastic sticks to the hopper:** Hopper coolant has not been turned on.
❸ Turn on the heaters and warm up the barrel and die of the machine. See Fig. 5-19. **Do not turn on the motor until the barrel and die are heated to operating temperature.** Bring the barrel temperature to about 75°F (41°C) **above** the melting point of the plastic. The die should be about 100°F (55°C) above the melting point at the start.	The barrel and die should be started well above the melting point of the plastic to be used. This will help prevent screw or shear pin breakage. The barrel must be hot before the frictional heat takes over.	**Screw or shear pin breaks:** (a) The barrel and die temperatures were too low. (b) Plastic was fed into the hopper before the machine was up to operating temperature. (c) The plastic pellet size was too large for the barrel and die temperatures.
❹ Set the screw drive for a **slow** screw speed. Turn on the drive motor. See Figs. 5-20 and 5-21. Check the screw rotation through the hopper opening.	The screw should run empty before any material is put in the hopper. This will protect the screw from breakage.	**Screw or shear pin breaks:** Screw started with old material in the barrel. This material may have been a higher melting point plastic.
❺ Fill the hopper with **clean** plastic material. See Fig. 5-22.	Dirt or metal particles in the material may clog the barrel or die or may damage the screw or shear pin.	**Streaks in the extrudate:** (a) Dirt or metal particles in the material are catching in the die. (b) Rust spots are in the die. Remove the die and clean at once. **Caution:** The die is very hot.

(continued on page 100)

Extrusion 99

❸

Fig. 5-19. Turning on the heaters.

❹
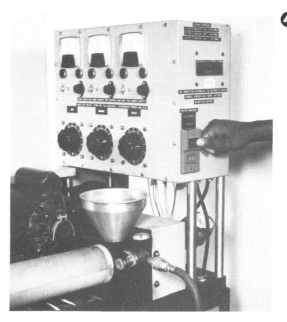
Fig. 5-21. Turning extruder motor switch on.

❹
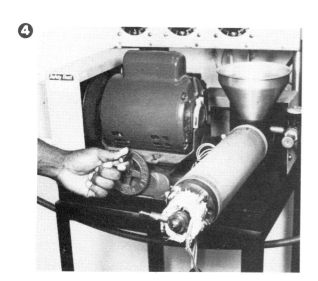
Fig. 5-20. Setting the screw speed.

❺

Fig. 5-22. Filling the hopper.

TABLE 5-2 (Cont.)

Sequence/Action	Reason	Troubleshooting
6 Using a stick, pick up the extrudate as it comes out of the die and pull it from the die. Feed it through the cooling and take-off units. See Figs. 5-23 and 5-24. (a) Adjust the take-off speed to match the extruder speed. (b) Adjust the barrel and die temperatures according to the need. See **troubleshooting** for information. (c) Adjust the cooling fans or cooling water temperature as needed.	Feeding gets the extrudate started through the system. (a) Matching speeds coordinate the extruder and take-off system speeds. (b) Adjusting the temperatures to the material gives a uniform extrudate. (c) Proper cooling gives extrudate the right surface finish.	**Extrudate thins out:** Take-off speed is too fast. Reduce take-off speed. **Extrudate sags at the die:** Take-off speed is too slow. Increase take-off speed. **Extrudate surface is coarse and rough:** Barrel and die temperatures are too low. Increase temperatures. **Extrudate is shiny and smoking:** Barrel and/or die temperatures are too high. Decrease temperatures. **Extrudate gets hard and cracks when cooling:** (a) Barrel and die temperatures are too high. (b) Cooling water is too cold and cools the plastic too fast. (c) Use cooling fans instead of water cooling. **Motor overloads when extruding:** (a) Barrel or die temperatures are too low. Increase them. (b) The screw speed is set too fast for the die and/or material being used.
7 Run the barrel empty when getting ready to stop the extruder. Dies are often removed to change type or clean at this time. See step 12.	An empty barrel will be less apt to damage the machine when restarted at too low a temperature. It will heat up faster when restarting.	
8 Turn off the screw drive motor.	Screw should be stopped while the barrel is still hot.	
9 Turn off the take-off unit.	Turn it off as soon as all the extrudate is run through it.	
10 (a) Turn off the barrel and die heating bands. (b) Put HOT signs on the barrel and die so others will not get burned.	This lets the machine cool off.	
11 Turn off the hopper coolant.	Turn it off after the barrel has started to cool.	
12 Remove the die while it is still hot. **Caution:** Use heat-resistant gloves. Clean the die and oil it.	Rust will accumulate if die is stored without cleaning. Oil surfaces to prevent rust.	**Material does not flow evenly from a die:** The die may be rusty or dirty inside.

Extrusion 101

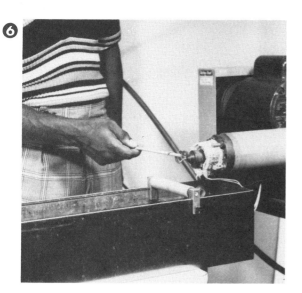

Fig. 5-23. Picking up the extrudate from the die.

Fig. 5-24. Running the extrudate through the cooling unit.

Checklist of Materials and Equipment Needed

Assemble the following materials and equipment for extrusion:

1. Extruder and die.
2. Cooling unit.
3. Take-off or pulling unit.
4. Plastic pellets or regrind (recycled) scrap plastic.
5. Insulated gloves.
6. Colorant (optional).
7. 6" - 12" (15.2 - 30.5 cm) wood stick.

TABLE 5-3
Extruder Operation Behavior

Material	Horsepower Required	Heat Required	Screw Speed
Polyethylene			
Low Density	L	L	H
High Density	H	H	L
Polystyrene			
General Purpose	M	M	M
High Impact	L	M	M
Polypropylene	M	M	M
Melt Index or Flow #			
High	L	L	H
Medium	M	M	M
Low	H	H	L
Density			
High	H	H	L
Medium	M	M	M
Low	L	L	H
Particle Size			
Large	H	H	L
Small	L	L	H

L = Low; M = Medium; H = High

TABLE 5-4
Common Extrusion Startup and Operating Temperatures for Selected Thermoplastics

Plastic Family	Minimum Melt Point* (Degrees F) (Degrees C)	Approximate Start-Up Temperatures (Degrees F) (Degrees C)			Approximate Operating Temperatures (Degrees F) (Degrees C)		
		Rear	Front	Die	Rear	Front	Die**
ABS	300°F	350-400°F	375-425°F	400-425°F	300-350°F	325-375°F	350-400°F
	149°C	177-205°C	190-218°C	205-218°C	149-177°C	163-190°C	177-205°C
Polyethylene, L.D.***	221°F	300-325°F	325-350°F	350-400°F	275-300°F	300-350°F	325-375°F
	105°C	149-163°C	163-177°C	177-205°C	135-149°C	149-177°C	163-190°C
Polyethylene, H.D.***	248°F	325-350°F	350-375°F	350-400°F	290-330°F	300-370°F	325-375°F
	120°C	163-177°C	177-190°C	177-205°C	143-166°C	149-188°C	163-190°C
Polypropylene	330°F	375-400°F	400-425°F	425-450°F	330-380°F	355-405°F	380-430°F
	166°C	190-205°C	205-218°C	218-232°C	166-193°C	180-207°C	193-221°C
Polystyrene, High Impact	374°F	425-450°F	450-475°F	475-500°F	375-425°F	400-450°F	425-475°F
	190°C	218-232°C	232-246°C	246-260°C	190-218°C	205-232°C	218-246°C

*Approximate melt temperature.
**Die temperatures should be approximately the same as the temperature of the plastic flowing through the die.
***Exact melt temperature depends on the density. Check this with the manufacturer's data sheet.

Materials for Laboratory Extrusion

As thermoplastic materials are extruded, they pass through a **physical** change. As they are heated, they are melted. When they go through the cooling device, another **physical** change occurs. As they are cooled, they solidify.

Thermoplastic materials most often used for school laboratory extrusion are polystyrene, polypropylene, polyethylene, and ABS because they are easy to extrude. Cellulose acetate and cellulose acetate butyrate may also be used.

In a school laboratory, three materials **should not** be used for extrusion:
1. PVC (polyvinyl chloride) because of the hydrochloric acid reaction,
2. Acetal because of its sharp degradation at about 400°F (204°C), or
3. Cellulose nitrate because of its flammability.

Refer to Chapter 4 for more information on this.

A plastic material should be chosen on the basis of how the resulting plastic product is to be used. **Polystyrene** may be used for sheet, profiles, rods, and tubing. Products extruded from this material will be quite rigid, especially if they are thick. High impact grades of polystyrene will be somewhat less rigid than general purpose polystyrene. Extrusion grade polystyrene is an easy material to extrude in the laboratory.

Polyethylene may also be extruded into rods, tubes, profiles and sheets. Polyethylene is slightly more difficult to extrude in the laboratory than polystyrene, however. It tends to be more sticky and takes longer to cool. Good temperature control and proper cooling methods are necessary for polyethylene. Low density polyethylene should be selected for products needing flexibility. High density polyethylene may be used for more rigid products.

Polypropylene may be used for products similar to those listed for polyethylene but requiring less flexibility. It is often chosen for a combination of stiffness, waxiness, and the ability to bend without breaking.

Extrusion requires special grades of plastics. Extrusion-grade plastics are usually low melt index (low flow rate) materials. The low melt index is needed so that the extrudate, or extruded material, will hold its shape as it leaves the die. Both powdered and pelletized plastics can be used. Pelletized plastics are used most often. The powdered materials tend to melt more quickly in a small extruder, however. The density of the material will depend upon the use of the product made. Low density materials will usually make softer, more flexible parts. High density materials are used for stiff products.

Operating Temperatures

Table 2-2 in Chapter 2 lists the softening points of several selected thermoplastics. A complete list of thermoplastic softening points is given in the thermoplastics materials chart in Appendix A. Barrel and die start-up temperatures for laboratory extruders range from 50°F to 100°F (27°C to 55°C) above the softening points of most thermoplastics. A good start-up point is 50°F (27°C) above the softening point for the rear zone, 75°F (42°C) above for the front zone and 100°F (55°C) above at the die. **These temperatures will usually need to be reduced by at least 25°F (14°C) to 50°F (27°C) at each zone once the extruder is in operation.** Table 5-4 lists several common thermoplastics which can be easily extruded in the school laboratory, their softening points, start-up, and operating temperatures. Consult the chart in Appendix A and the plastic material manufacturer's data sheet for further details.

After the extruder has been turned off, be sure to cover the hopper to keep dirt out. Remember that the extruder can get plugged up from dirt and chips falling into the hopper.

Use care in putting away so that other students will not get burned on the barrel or die while these parts are still hot. Place the HOT signs where they are needed.

Tooling Up for Production

Dies for laboratory extruders may be purchased from some of the extruder manufacturers to fit their machines. Specialized dies probably will need to be made in the school laboratory. In most cases the dies should be made of steel. Good results may be obtained by using cold rolled steel for laboratory extruder dies. In some cases aluminum dies may be made, although the use of such a die would be limited. It would probably wear more quickly on the land surfaces. Also, the opening of an aluminum die could be damaged more easily due to its softness.

Simple rod dies can be easily made by drilling and reaming a hole in a turned steel rod. The dies are fitted to the front of the extruder barrel by either screw threads or a turned flange. Note the turned **adapter flange** shown on the rod die, Fig. 5-25. Figure 5-26 shows typical rod dies, one having a heater band. Consult the machine manufacturer's instructions for exact adapter specifications.

Sheeting and tubing dies are more complex. A typical sheeting die made for a laboratory extruder is shown in Fig. 5-27. The

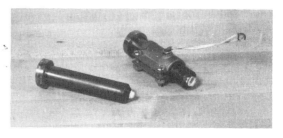

Fig. 5-26. Two typical rod dies for a laboratory extruder. One shown with a band heater installed.

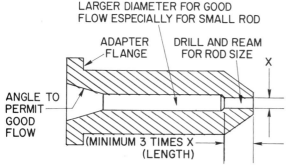

DIMENSIONS AS NEEDED TO FIT EXTRUDER

Fig. 5-25. Rod Extrusion Die

Fig. 5-28. Fan-Shaped Extrusion Die, Disassembled
The small part in the foreground is the cooled plastic removed from the inside of the die.

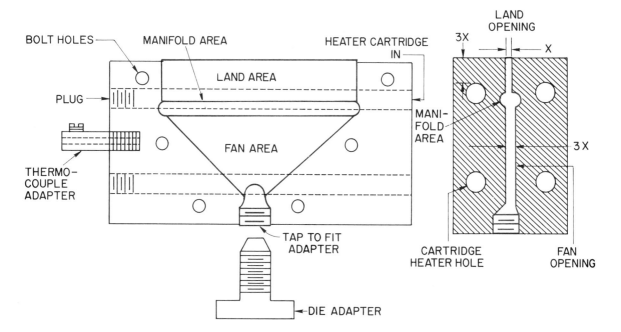

Fig. 5-27. Fan-shaped extrusion die with a manifold.

same die is shown in Fig. 5-28. Note that the depth of the **fan area** opening is three times the depth of the **land area** opening. The width of the land area should be about three times the thickness of the opening. This is to provide proper flow of the material in the die.

Recycling

Extrusion is an excellent process for the recycling of scrap thermoplastics. Cutoffs and scrap from thermoforming and injection molding may be reground and pelletized for use in the extruder or injection molder. Scrap thermoplastics from the home or school may also be recycled. This is an excellent way to help control pollution and reduce the cost of materials. Margarine tubs and lids, coffee can lids, plastic six-pack retainers, plastic spoons, detergent bottles, and other similar scrap plastic products may be used. Coffee can and margarine tub lids make excellent injection molded pop bottle caps, for instance.

Each type of plastic must be sorted out and run through the scrap granulator **separately**.

Safety Note

These plastics must be cleaned (labels also removed) and sorted before recycling.

The scrap granulator must be **well cleaned** between each color or type of material. Clean out the scrap granulator by taking out the screen and removing the hopper. Blow it out completely. During operation, connect a

Fig. 5-30. Examples of common recyclable plastic materials which can be used in the laboratory. **Each type of plastic must be recycled separately.**

Fig. 5-29. Recycling center with a typical scrap granulator. Granulated plastic is drawn out of the granulator with a shop vacuum. Drawers under the bench are used to store scrap plastic prior to recycling.

Fig. 5-31. Extrusion pelletizing line in action with cooling unit and extruder. Extruder is at right, cooling/take-off is center, and pelletizer is at left. Recycled plastic is extruded and pelletized for use in the injection molder or blow molder. Vacuum under the pelletizer collects the pellets.

shop vacuum to the scrap granulator to collect the material as it is being granulated. This keeps the laboratory clean. Empty the shop vacuum each time a color or type of plastic has been reground.

Recycled materials may be used for injection or extrusion as they come from the granulator without having to be processed further. They may also be extruded into 1/8" (3.17 mm) rods, cooled, and cut into pellets. This is the way that commercial plastic pellets are made. Pelletized materials are cleaner to use and more uniform in size.

STUDENT ACTIVITIES

1. Process through prepared dies as many extrusion materials as are available. Change dies to determine differences in processing characteristics for various resin formulations.
   a. Tubing die — polystyrene, polypropylene, polyethylene, etc.
   b. Rod die — polypropylene, polyethylene, polystyrene, etc.
   c. Flat sheet die — polyethylene, polypropylene, and polystyrene.
   d. Sheet, tubing or rod dies, and expandable polystyrene beads.
2. Experiment with materials by changing the following variables to determine end product relationship with specifications of resin:

   Note: Prepare a log sheet for experimentation with the extrusion process like the one shown in Appendix A, page 393. Save this record of your experimentation for later reference.
   a. Melt index — Use resins of the same family (i.e. PP, PS, or PE) and vary the melt index. Try formulations with several different melt indexes. Note the changes in the product.
   b. Particle size — Note how changes in particle size affect the melt temperatures and power requirements for extrusion.
   c. Density — Use different density resins and note the results — changes in rigidity, hardness, etc.
   d. Altering materials —
      (1) Mix two resins of the same family with different densities to obtain a density between the two parent resins.
      (2) Mix two resins of the same family with different melt indexes to obtain a melt index between the two parent resins.
   e. Mix dry color additives or color concentrates with the resins. Extrude as usual to obtain colored products.
   f. Mix dry foam concentrates with the resin and extrude a foamed plastic product. Consult manufacturer's directions for concentrations of foam additive.
3. Experiment with the process to understand and be able to work with these variables.
   a. Temperature — Process the materials at various temperatures at the rear, front, and die zones. Determine the lowest temperatures at which a given plastic can be successfully extruded.
   b. Speed of extrusion — Vary the extrusion speed to determine the fastest speed of extrusion at a given temperature without overloading the motor. **Motor overloading must be avoided.**
   c. Cooling methods and speeds — Experiment with different ways of cooling various types of extrudate — i.e. fans, water spray, water bath, and chill rolls. Change locations of coolers to determine changes in the end product.
4. Design various extruded products. Suggest designs for the following:
   a. Profile or rod shapes
   b. Tubing or hollow shapes
   c. Sheeting and film
5. Design and construct dies for various extrusion shapes:
   a. Straight dies
   b. "L" head dies
   c. Hollow shapes

6. Production organization and operation.
   a. Material supply
   b. Extrusion operation and control
   c. Cooling extrudate
   d. Material take-off
   e. Coiling or cutting to length
   f. Storage of extruded materials
   g. Safety organization
      1) Safety supervisor
      2) Safety zones and signs
   h. Record keeping
      1) Materials used
      2) Processing variables
         a. temperature
         b. speed
         c. cooling
      3) Labor used
      4) Rejected parts
      5) Assembly and finishing
   i. Quality control supervision — analysis of rejects — adjustment of process control
7. Build a model extruder similar to Fig. 5-32. Extrude Play-Doh®.
8. Experiment with Play-Doh extruder.
9. Research materials data in the **Modern Plastics Encyclopedia**.

Fig. 5-32. Model extruder that extrudes Play-Doh®.

Questions Relating to Materials and Process

1. What melt index and density specifications are best for the following products? Choose from the list at right.

	Melt Index	Density
a. Flexible PE tubing	6.0	0.924
b. Rigid PE tubing	0.2	0.958
c. Thermoforming PS sheet	1.9	0.919
d. Polystyrene rod	Hi Flow	1.06
e. Acrylic rod	Low Flow	1.06

   List the reasons you have chosen each material.
2. Which thermoplastics extrude best? Why?
3. In the following situations, choose from column B the material or materials you would suggest for each item in column A.

A	B
a. Refrigerator door liner — thermoforming sheet	Hi impact polystyrene
b. Plastic soda straws	L.D. polyethylene
c. Fiber optics rod	Extrusion grade PP
d. Garden hose	PVC
e. Monofilament fishing line	Acrylic
	Nylon
	ABS

4. Given specific materials to work with, how will you determine whether they are most effective for extrusion?
5. Using the criteria listed for question 4, what products would you produce by extrusion with the following materials?
   a. Polyvinyl chloride
   b. Polyethylene
   c. Polypropylene
   d. Polystyrene
   e. Cellulose acetate
6. What is the relationship between extruding plastics, ceramics, and concrete products? Is it possible to extrude these other materials? How could it be done?

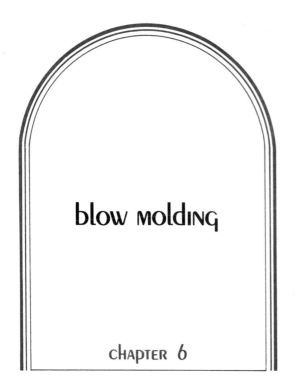

blow molding

chapter 6

Process Description

Blow molding is a process for forming hollow objects from thermoplastics. A hollow tube-like piece of plastic is placed between two open mold halves. The mold halves are closed on the plastic tube called a **parison**. Air pressure is used to blow the parison into the shape of the mold halves. Blow molding can be compared to blowing a balloon up in a bottle. The formed thermoplastic part holds its shape as it is cooled by the mold. The plastic hardens by a **physical** change. The mold halves are then opened and the part is removed and trimmed.

Three types of blow molding are in use today. They are the direct method, the indirect method, and the injection method.

Direct Method

The direct method of blow molding is often known as **extrusion blow molding**. It probably produces the largest amount of blow molded products. In this method a hollow plastic parison, or tube, is extruded between two open mold halves as shown in Fig. 6-2. (For a description of tubing extrusion, refer to Chapter 5.) When the plastic parison reaches the bottom of the mold, the mold closes. Air is blown into the center of the parison, and the plastic takes the shape of the mold. High-speed extrusion blow molding equipment produces detergent bottles, milk

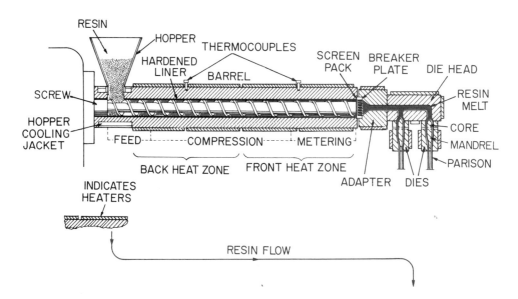

Fig. 6-1. Cross section of an extrusion blow molder with twin die heads.

Blow Molding

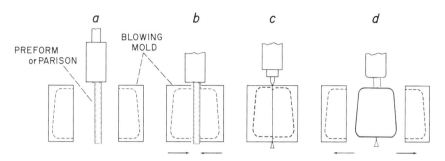

Fig. 6-2. Blow Molding — Direct Method
A. Parison extruded.
B. Molds close.
C. Parison blown by air to shape of mold.
D. Mold opened and part removed.

Fig. 6-3. Closeup of extrusion blow molding.

Fig. 6-4. Blow Molded Products

bottles, 5 gallon gas cans, tanks, and many other hollow containers with open necks.

Indirect Method

Products formed by the indirect method of blow molding may be made in either of two ways: (1) the tubing method and (2) the two-sheet method. The heated premade plastic tubing or sheets are clamped between the mold halves. Plastic tubing used in this process is extruded and cut to length before use. An air hose is placed in one end before the mold halves are closed on the plastic. Air is blown through the hose into the plastic, and the plastic takes the shape of the mold. It is cooled, removed, and trimmed. Long hollow objects, such as hosiery display legs, are often blown by the indirect tubing method. See Figs. 6-5 and 6-6.

In the two-sheet indirect method, the heat and pressure at the joint cause the two plastic sheets to weld together. Large hollow signs or tanks may be made by this method. See Figs. 6-7 and 6-8.

Injection Blow Molding

Injection blow molding is a method whereby a parison, or hollow plastic slug, is first injected over a mandrel. The parison mold halves open, and the parison is moved to the blow mold halves. The blow mold halves close over the parison which is still on the mandrel, and the object is blown. See Fig. 6-9.

Injection blow molding is the newest form of blow molding. It will make a completely finished part without flash and with no trimming required as in extrusion blow molding.

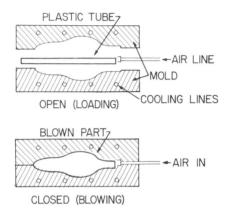

Fig. 6-5. Indirect tubing method of blow molding.

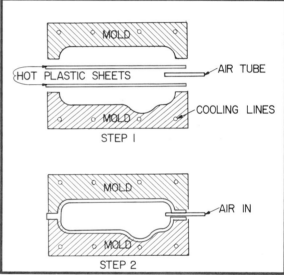

Fig. 6-7. Indirect two-sheet method of blow molding.

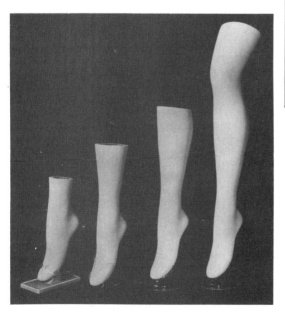

Fig. 6-6. Hosiery display legs formed by the indirect tubing method of blow molding.

Fig. 6-8. Toy boat formed by indirect two-sheet method of blow molding.

Blow Molding 111

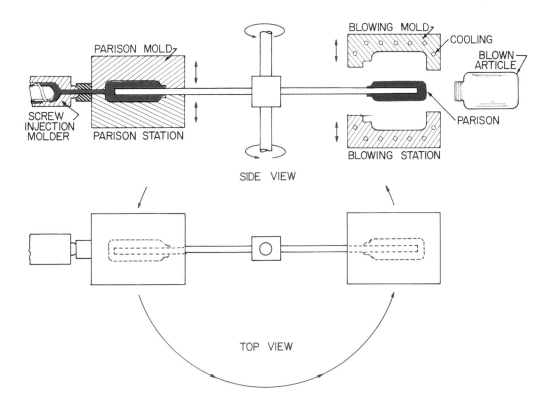

Fig. 6-9. Injection Blow Molding

Fig. 6-10. Injection Blow Molding Machine

Fig. 6-11. Closeup of injection blow molding machine head.

A part of very uniform wall thicknesses may be made by this method. Wall thicknesses in the parison can be varied so that more plastic can be applied where needed. This process is sometimes slower and more costly than extrusion blow molding, however. Tooling (molds) per cavity are more expensive for injection blow molding because two molds are needed to make each part. Injection blow molding does not allow production of products with handles or irregular shapes, such as toy wheels, heating ducts, flashlights, and similar parts.

Importance of Blow Molding

Blow molding now processes about 3% to 4% of all thermoplastic materials. This is expected to climb to 6% or more by 1980 as more products are packaged in plastics. As new plastics are developed that can be blow molded, new markets will be created for blow molded products. Blow molded plastics are rapidly replacing glass, metal, and paper containers.

Disposable packaging is on the increase. While plastic materials now make up only about 3% to 5% of the solid waste for disposal in the U.S., their increasing use alarms a number of ecology-minded groups. To solve the problem of disposal, new self-destructing plastics are being developed mainly for detergent and beverage bottles. The recovery of the energy in the plastic through incineration is one answer of the problem of disposal. Recycling is another.

Blow molded plastic containers save the customer money. Compared to glass, they break less easily in shipment and on the merchants' shelves. Plastic containers are also less costly to ship because they are lighter and take less space. These factors more than offset the slightly higher cost per unit for most blow molded plastic bottles. In some odd-shaped containers, there may be no actual cost difference.

Blow molding is becoming a more automatic process and the product costs are being reduced. This will help increase the use of blow molded products. Greater use of blow molded products should also help cut the cost of resins used in blow molding. With these developments, blow molding is rapidly becoming one of the major processes of the plastics industry.

Advantages

Blow molding provides a quick method of processing openended, bottlelike containers. Mold costs are low as compared with processes such as injection molding and compression molding. Most blow molds are either cast or machined aluminum and usually do not need to be hardened or chrome plated. Fine details and high mold polish are often not needed. Blow molded parts of many sizes from very small to quite large are possible. Also, the scrap plastic may be recycled. Blow molding can usually be adapted to common extrusion equipment. Newer blow molders, however, are built around specially designed extruders.

Disadvantages

Automatic, high-speed blow molding equipment is expensive. Prices can be compared to injection molding equipment. Extrusion blow molded parts tend to thin out in the corners. Some adjustment for thinning can be made in the parison die design, however. Injection blow molding reduces thin-

Fig. 6-12. Blow molded bottles that are light in weight and space saving.

Blow Molding 113

ning in corners. Blow molded parts often have built-in stresses that can cause early breakage of the part. Because of this, blow molding cannot be used for some parts. Rotational molding is usually used in place of blow molding in such cases. See Chapter 13 for a description of rotational molding.

Industrial Equipment

Blow molding machines are sold by the size of the blow molded product they produce. Product container sizes range from one ounce (29 cm^3) to 66 gallons (250 liters). The size of a blow molder is sometimes given by its extruder or injection screw size. It ranges from 1½ inches (38 mm) to 6 inches (152 mm) in diameter.

Often, blow molders have more than one die head. This is also considered when equipment is bought. From one to 12 die heads may be used on one machine.

The speed of the machine and the number of parts it can produce per hour is important. It is possible to make as many as 7200 small parts per hour in blow molding. The greater the wall thickness of the part, the longer the cycle time and therefore the fewer parts produced per hour. Larger parts usually have thicker walls and take longer cycle times.

Several types of extruder blow molder systems are used. Most are designed to constantly use the plastic parison coming from the extruder.

Rising Mold, Continuous Extrusion Method

The rising mold method is quite widely used on either single or multiple die (head) molders. The open mold halves rise up to the extruded parison, close over the parison, and drop back down. The plastic is blown inside the molds as they return to the "down" position. A new parison is extruded while the part cools in the mold. After the part cools, the mold halves open and the part is taken out. The machine is then ready to start a new cycle. See Fig. 6-14.

Fig. 6-13. Large Accumulator-Type Blow Molder
Note the large container it produces.

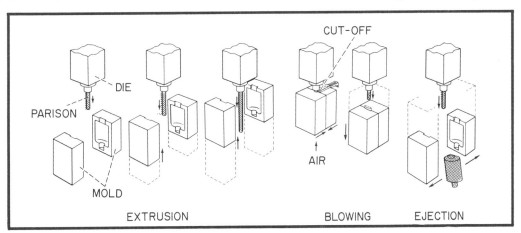

Fig. 6-14. Rising Mold Process

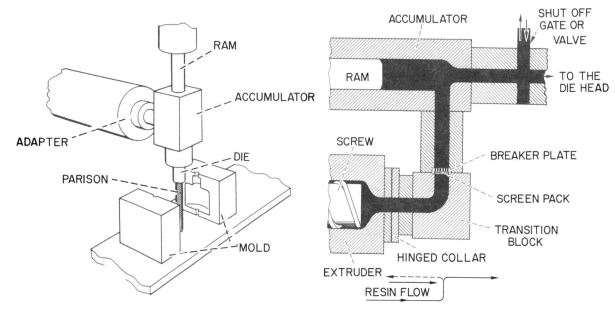

Fig. 6-15A. Fixed Mold Accumulator Blow Molding

Fig. 6-15B. Blow Molding Accumulator

Fig. 6-16. Garbage can molded by an accumulator type blow molder.

Fixed Mold, Accumulator Method

The fixed mold, accumulator method of blow molding works well for large parts. Large amounts of melted plastic can be stored up to extrude into a large parison. Plastic is first extruded into an accumulator cylinder above the die head. The accumulator cylinder then extrudes the plastic through the die down between the open mold halves. While the part is being blown and cooled, the extruder is refilling the accumulator cylinder. Accumulator systems cannot be used to process heat-sensitive plastics such as PVC and others. See Figs. 6-15 and 6-16.

Shuttle Valve Method

The shuttle valve, or alternate flow, system of blow molding is also very common. In this method, the molds open and close on a horizontal plane. The extruder has two or more die heads in two sections. One section of the die heads is extruding while the other is blowing and cooling. A shuttle valve on the extruder switches the flow of the plastic from one section of the die heads to the other. Figure 6-17 shows how it works.

Rotary Blow Method

The rotary blow molding method usually uses only one die head. Several molds are

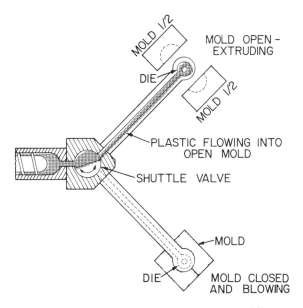

Fig. 6-17. Top view of shuttle valve blow molding.

Fig. 6-18. Large, four-head blow molder that makes gallon-size bottles.

rotated on a multiple-station indexing turntable, vertical rotary wheel, or an endless belt. Index positions for parison extrusion, mold close and pinch, blow and cool, and ejection are used. Figure 6-19 is a diagram of the vertical rotary wheel type of blow molding. Rotary blow molding is a high-speed production method for small to medium sized parts. Fig. 6-20 shows an endless belt system.

Continuous Tube Method

The continuous tube method is also designed for high-speed, high-volume production. In this process one to several molds are clamped over a continuous parison. The molds may be on a rotary table or may shuttle back and forth horizontally. The parison is usually extruded horizontally in this method. Air is blown by a needle into the parison at

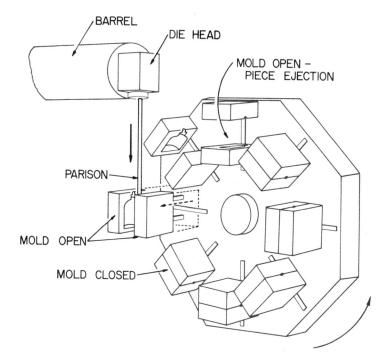

Fig. 6-19. Rotary Wheel Blow Molder

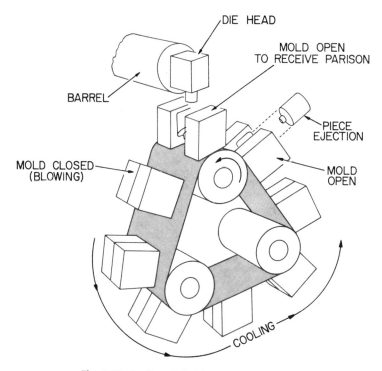

Fig. 6-20. Endless Belt Blow Molding System

each mold. The needle enters the parison in the scrap section of the part. Figure 6-22 shows this process.

Intermittent Extrusion with Fixed Molds

Either a plunger or screw type extruder may be used for intermittent (start and stop) extrusion. The plunger type blow molder is made like a plunger injection molder (see Chapter 4). The screw type blow molder is like a screw extruder (see Chapter 5). The extruder screw or plunger is started and stopped to extrude the parison when it is needed. The mold halves are clamped over the parison in the regular manner, blown, and cooled. See Fig. 6-23.

This method depends on external heat more than frictional shear for melting the plastic, especially in the plunger type. The melt temperature is more apt to vary, causing irregular parts.

Fig. 6-21. Rotary wheel blow molder that makes bottles.

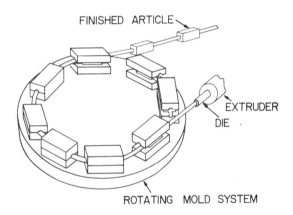

Fig. 6-22. Continuous Tube Blow Molding, Horizontal

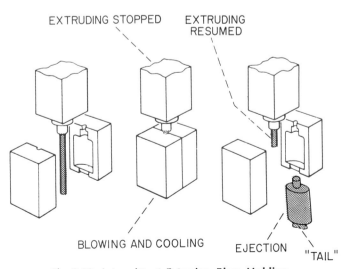

Fig. 6-23. Intermittent Extrusion Blow Molding

118 Section II MOLDING

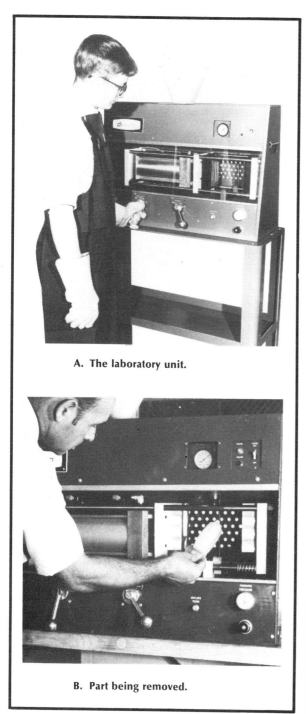

A. The laboratory unit.

B. Part being removed.

Fig. 6-24. Plunger Blow Molder/Injection Molder Combination

Materials Used in Industry

Almost any thermoplastic can be blow molded. The most common is polyethylene. Both high and low density polyethylene are used, with high density being the more popular. The polyethylene squeeze bottle is probably the most common blow molded product. High density polyethylene is used for more

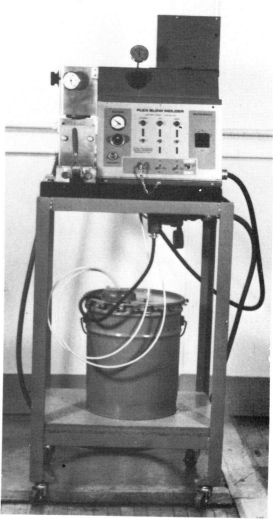

Fig. 6-25. Screw Extrusion Blow Molder

rigid containers like half-gallon and gallon milk or bleach bottles. Cellulosics, polyvinyl chloride, polycarbonate, acrylic, ionomer and polyacetal plastics are also used for blow molding. Recently nylon (polyamide) has been blow molded. Polypropylene is becoming a popular blow molding resin because it makes clearer bottles than polyethylene. These clear polypropylene bottles may be filled with hot liquids.

Laboratory Blow Molding Equipment

Some laboratory blow molders have appeared on the market recently. They produce parts of up to about 4 ounce (118 cm³) capacity. One is a plunger extrusion type machine which has the option of injection molding. Another is a screw type blow molder. Two such units are shown in Figs. 6-24 and 6-25. Both units are intermittent extrusion blow molders.

Operating Requirements

Laboratory blow molders operate on either 115 or 230 volt electrical current. They also require compressed air and may need water cooling units. Some blow molders are available with built-in air pumps and water coolers. Such blow molders need only be connected to an electrical source.

Electrical current is needed for the barrel and die heaters, extruder motor (if used), air pump, and water cooler. Compressed air is needed for blowing the parison in the molds. It may also be used to cool the molds. Water cooling is required on the hopper end of the extruder barrel in some units. The hopper end needs cooling to keep the plastic pellets from sticking to it. Water may also be used to cool the molds.

Molds for laboratory blow molders are usually cast or machined from aluminum. Water or air cooling channels can be built into the molds for more rapid cooling of the blown part. The molds must be fastened to the machine for proper operation. Guide pins in the molds are needed for correct mold alignment. They should not be cast into the molds but inserted after the casting is cured. Molds may also be cast from mass casting or metal filled epoxy.

Blow molding grades of plastic must be used, low density blow molding polyethylene being the most common. A density of from 0.915 to 0.930 and a melt index of 1.5 to 2.5 are recommended for most laboratory units. Blow molding polypropylene of 2.5 or less melt index may also be used. **Read the machine manufacturer's directions for exact recommendations and suggestions of other plastics which may be used.**

Operation Sequence

There are five steps in the blow molding process:

1. Heat the plastic, form the parison, and place it between the mold halves.
2. Close the mold halves over the soft plastic parison.
3. Blow the plastic parison against the mold walls.
4. Cool the plastic part.
5. Open the mold, remove the part, and trim it.

This sequence, the action taken, reason for the action, and troubleshooting suggestions are shown in Table 6-1. **Read the manufacturer's instruction manual for exact directions.**

Checklist of Materials and Equipment Needed

Assemble the following materials and equipment for blow molding:

1. Blow molder and molds.
2. Blow molding plastic, polyethylene between 0.915 and 0.930 density and 1.5 and 2.5 melt index.
3. Mold release — silicone or fluorocarbon.
4. Wrenches for adjusting the machine.
5. Color concentrates if desired.
6. Insulated gloves.

TABLE 6-1
Operation Sequence for Blow Molding

Sequence/Action	Reason	Troubleshooting
❶ (a) Bolt the mold halves onto the mold platens, Fig. 6-26. (b) Adjust the mold clamping pressure, Fig. 6-27. (c) Read the manufacturer's directions for their recommendations.	Molds need to be tightly fastened and properly aligned to close correctly and prevent flashing.	**Flashing:** (a) Platen pressure wrong. (b) Dirt on mating mold surfaces. (c) Too much blowing pressure. (d) Molds not properly aligned — mold flashes at die head.
❷ (a) Prepare the machine for use by checking oil levels in oilers and cleaning sediment bowls. (b) If the machine contains a cooling unit, check the cooling water level. (c) Make sure 2% rust inhibitor is present in the cooling water. (d) Oil moving parts where required.	(a) Cleaning prevents foreign matter from entering air lines. (b) Compressor or air cylinders must be properly oiled. (c) Inhibitor stops rust from gathering on interior parts. (d) Oiling keeps up proper operation of moving parts.	**Clogged die head:** Rust in the die head caused by water in the air.
❸ Turn on the barrel and die heaters, Figs. 6-28 and 6-29. Bring them up to operating temperature — 50°F to 100°F (27°C to 55°C) above the melting point of the plastic is needed. **Caution: Do not turn on extruder screw or plunger before barrel and die reach operating temperatures.** **Do Not Turn Heater Regulator Knobs to Highest Settings.** Select a medium setting and adjust temperature after heater indicator lights go out, Fig. 6-30. Check heater thermometers. Read the manufacturer's directions for exact preheating sequence.	(a) To melt the plastic in the heater barrel. (b) To prevent damage to the extruder screw or plunger. (c) Too much heat can cause damage to barrel, die, controls or heaters. Materials in the barrel or die may degrade (scorch).	**Barrel or die temperature too high:** Turn temperature control knob(s) to a lower setting. **Barrel or die temperature too low:** Turn temperature control knob(s) to a higher setting.
❹ Fill the hopper with material, Fig. 6-31. Use a low melt index (1.5 to 2.5) polyethylene to start with. Read machine manufacturer's directions for other materials.	Blow molding grade polyethylene is easiest to work with. Low melt index (high molecular weight) materials are necessary to prevent the parison from sagging out of shape.	**Uneven wall thickness on finished part:** Wrong material used — parison does not shape correctly. Change to correct grade material.

(continued on page 122)

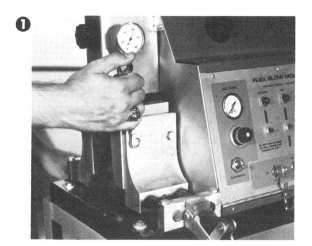

Fig. 6-26. Mount the mold.

Fig. 6-27. Adjust mold clamp pressure.

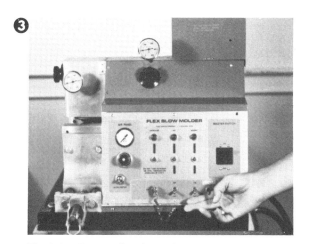

Fig. 6-28. Turn on heating units.

Fig. 6-29. Adjust barrel heater.

Fig. 6-30. Adjust die heater.

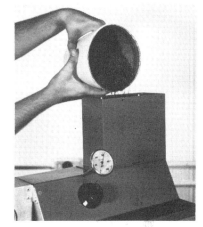

Fig. 6-31. Fill hopper.

TABLE 6-1 (Cont.)

Sequence/Action	Reason	Troubleshooting
5 (a) Connect air hose or turn on air compressor. (b) Adjust the parison blowing pressure regulator to about 10 psi (69 k Pa).	(a) To blow the molded part. (b) To regulate the blowing pressure.	**Part does not blow:** (a) Air not turned on. (b) Pressure set too low. **Flashing:** Too much blowing pressure. **Part splits:** Too much blowing pressure.
6 (a) Be sure barrel and die head are at operating temperatures. Extrude several lengths of molten plastic with both mold halves open. Check for uniform parison. (b) Remove trial parisons with a wood stick. **Do not touch the hot plastic. It will burn you.**	(a) Materials sometimes degrade in start-up. These materials should be removed. This will allow the parison to become uniform in color and temperature. (b) Wood will not scratch die face.	**Parison too thin:** Material too hot. Reduce barrel and die temperatures. **Parison too thick:** Material too cool. Increase barrel and die temperatures.
7 (a) Extrude a parison ¼" to ½" (6 mm to 13 mm) below the mold pinch-off, Fig. 6-32. (b) Close both mold halves fully, Fig. 6-33. (c) Turn on the blowing air for 10 to 15 seconds.	(a) The parison must be longer than the mold. (b) The parison must be tightly clamped in the mold. (c) The air blows the parison out to the mold walls.	**Parison does not extrude:** (a) Extrusion switch or valve not turned on. (b) Material hopper empty. (c) Material hopper gate closed. (d) Material is too cold. Machine not up to operating temperature. (e) Material "bridges" or blocks in the barrel feed throat.
8 Allow the part to cool in the mold for 20 to 30 seconds, Fig. 6-34. As the mold gets warm, allow a longer cooling time.	The part will deform (lose its shape) if it is removed before it cools properly. The thicker the part, the longer the cooling time. Cooling time may be shortened by air or water cooled molds.	**Part deforms during or after removal:** (a) Increase the blowing time. (b) Increase the cooling time. (c) Reduce the mold temperature.
9 (a) **Put on a glove to protect your hand during this step.** (b) Open the mold halves. (c) Remove the molded part, Fig. 6-35. (d) Inspect the finished part.	(a) The mold and part will be quite hot. (b) To remove the molded part.	**Parts defective:** Refer to the previous troubleshooting notes.
10 When parts are completed, shut the machine down. (a) Turn off the extruder motor switch. (b) Turn off the barrel and die heaters. (c) Turn off the blowing air. (d) When the barrel and die have cooled to the 150°F (65°C) shut off hopper cooling water or air (if used).	(a) To prevent cold extrusion. (b) To cool the extruder down. (c) Air not needed. (d) To cool the hopper while the machine cools off.	

Blow Molding 123

Fig. 6-32. Extrude the parison.

Fig. 6-34. Turn on the blowing air and cool the part.

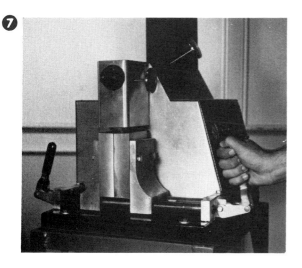

Fig. 6-33. Clamp the mold shut.

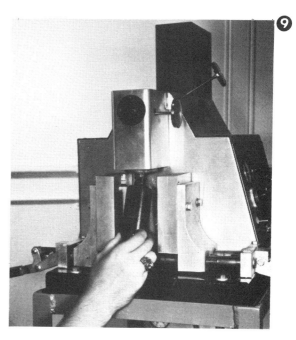

Fig. 6-35. Remove part.

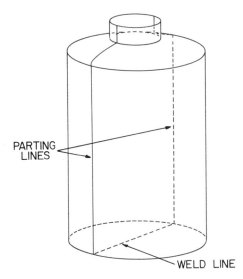

Fig. 6-36. Blow Molded Part

Fig. 6-37. Commercial blow mold with cooling lines.

Design of Blow Molded Products

Blow molded products require special design considerations, Fig. 6-36. Several factors must be studied when designing blow molded products and their molds.

Uniform Wall Thickness

Uniform wall thickness is usually hard to accomplish. The lower the blowup ratio, the more uniform the walls will be. The blowup ratio is the ratio between the maximum part diameter and the extruded parison diameter. Ratios of 2:1 to 5:1 are used in the plastics industry. A normal blowup ratio is about 3:1. Larger diameter blow molded parts need thicker parisons to make up for the wall thinning due to blowup.

Parison Size

Parison size is often limited by part design. The parison diameter should be no larger than the neck diameter for containers without handles. This will prevent pinch-off at the neck. The parison must be larger than the neck for containers with handles. The required wall thickness is controlled by the blowup ratio. The larger the blowup ratio needed, the thicker the parison wall needed.

Part Design

Sharp angles, edges, and corners should not be used when designing blow molded parts. Shapes should be as smooth as possible with round, eliptical, or oval shapes being the most desirable. Square parts should be avoided if possible. Changes in diameter or size should be gradual rather than abrupt.

Mold Cooling

If long production runs are expected, water cooling should be installed in the mold halves. Short runs of less than a one-hour class period usually do not need water cooled molds. Water cooling will, however, produce more uniform parts and more rapid cycles. Tap water may be used to cool the molds, with the used water going to a drain.

Tooling Up for Production

Molds for laboratory blow molders may be made from several materials. These may be machined or cast aluminum, cast metal filled or mass casting epoxy, cast high-strength gypsum cement or machined hardwood.

Machined Aluminum Molds

Machined aluminum molds are the most sturdy and usually produce parts with the best finish. Machined aluminum molds may be made similar to machined aluminum injection molds. The machining procedure is listed below. For further information see the mold making section in Chapter 4.

1. Select aluminum bar stock large enough to make two mold half blanks. Cut them off to length.

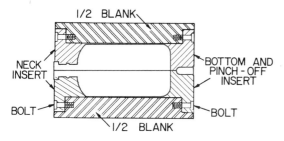

Fig. 6-38. Straight cavity mold for blow molding.

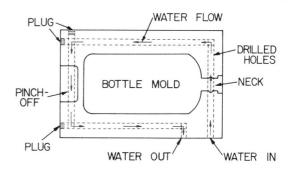

Fig. 6-39. Water Cooling Circuit
Cooling water should enter near the thickest and hottest area of the part.

2. Surface and square all six faces of each blank until it is the right size. Make sure the mold blanks will fit into the machine. Surfacing of the mold blanks may be done on either the metal lathe, milling machine, or shaper. See Figs. 4-48 and 4-50, Chapter 4.
3. Locate and drill the holes for the guide pins. Ream the holes to produce a press fit in one half of the blank and a slip fit in the other half of the blank. See Figs. 4-49 and 4-51 in Chapter 4.
4. With scribed lines, lay out the cavities on the mold blanks. Machine the cavities to size. A straight cavity may be machined from the solid blanks, Fig. 6-38. This may be done by clamping the two pinned mold halves together in a four-jaw chuck and drilling the center straight through. Use a large drill bit to drill the center out and finish the hole with a boring bar. End plugs may then be machined and fitted into the mold halves to form the neck and bottom of the container. A steel insert for the bottom pinch-off may be needed for better wearing qualities, although it is not absolutely necessary.
5. Polish all the surfaces as described in Chapter 4. If a soft surface finish is desired, sandblast the surface.
6. Locate and drill the water cooling channels if they are needed, Fig. 6-39. Assemble and mount the mold halves in the machine. Be sure to correctly align the mold under the die head.

Cast Aluminum Molds

Aluminum molds may also be made by the sand casting method. A simplified pattern-making procedure follows. A male model of the blow molded part is used to make the sand casting pattern.

1. Make a male model of the part to be blow molded. The model may be carved, machined, or formed from soap, wax, wood, plaster, or similar material. If both halves of the mold are to be the same, only one half of the model needs to be made. Split models may be turned out on the lathe. **Note: The model should be oversize because the aluminum will shrink when it is cast.** Aluminum shrinks 5/32" per foot (13 mm/m). A **shrink rule** should be used when the model or pattern is measured. If exact size is not important, this may not be necessary.

   In order to cast over wood, plaster, or another porous model, the surface of the model should be sealed. Sealing will prevent the gypsum cement (or plaster) from sticking. Seal the surface with shellac, lacquer, or other sealer. Soap or mold release should be put on over the sealer.
2. Place the sealed model on a base plate, Fig. 6-42. A Formica® or Micarta® sink

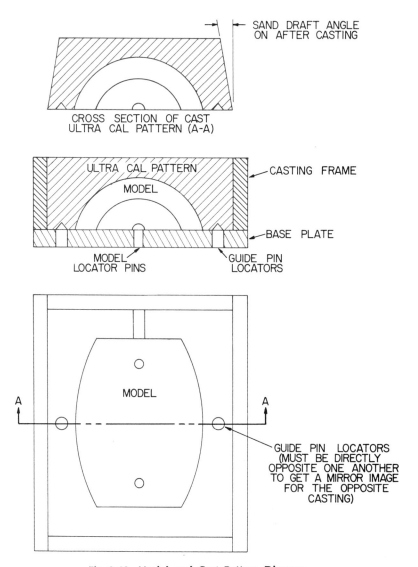

Fig. 6-40. Model and Cast Pattern Diagram

top cutout works well. Glue the model to the plate with rubber cement or silicone RTV bathtub sealer. Place two tapered locating pins in the base plate. Build a box around the model as shown in Fig. 6-43. Tape the edges of the box to keep it from leaking. Pour in a creamy mixture of high strength gypsum cement or plaster (U.S.G. Ultracal 30 or equal is recommended), Fig. 6-44. See Chapter 8 for Ultracal mixing instructions. After the gypsum cement is cured, remove it from the box. Sand a draft angle on the edges of the gypsum cement pattern. Use a belt or disk sander.

3. Ram up sand casting molds, using the cured gypsum cement patterns. Cast each half of the mold. Trim off the sprues and runners.

4. Machine the rear surfaces of both mold halves. Machine the parting line surface of each mold half parallel to the rear

Blow Molding 127

Fig. 6-41. Model on the lathe.

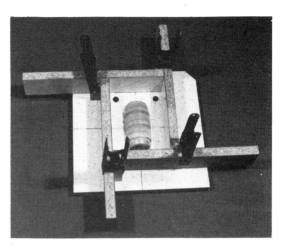

Fig. 6-43. Casting frame is in place on the mounted model.

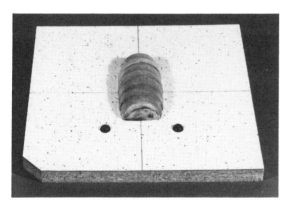

Fig. 6-42. Mounted Model

Fig. 6-44. The gypsum cement is poured into the mold. Note the clay dams that prevent leakage.

surface. Drill and ream the guide pin holes at the indentations caused by the tapered locating pins. See Chapter 4 for details. Align the mold cavities and clamp them together.

5. Press the guide pins into the interference (tight) fit holes in one half of the mold.
6. Clamp the pinned mold halves together in a machine vise and machine the other four edges. Locate, drill, and tap the mold mounting holes.

Cast Epoxy Molds

Cast metal filled or mass casting epoxy molds may be made following the same basic procedure as cast aluminum. The epoxy mold is cast directly over the model instead of taking a pattern from it.

1. Make and mount the model on a surface such as Formica® as described above. Build a box or place a metal casting frame around the model. The box will be removed as it is for the gypsum cement

Fig. 6-45. Cast gypsum cement pattern for blow mold ready to ram up for sand casting.

Fig. 6-47. Machining the face of the mold (parting line surface).

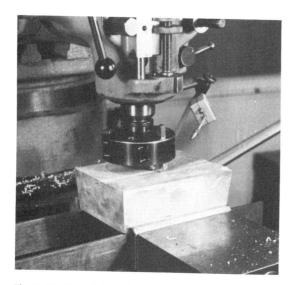

Fig. 6-46. Machining the rear surface of the cast aluminum blow mold half.

Fig. 6-48. Machining the edges of the mold.

sand casting pattern. The casting frame will, however, become a part of the mold half, if it is used. See Fig. 4-56 in Chapter 4 for details.
2. Coat the model, box, and mounting plate with fluorocarbon mold release. If a casting frame is used, do not put mold release on it. Apply the release only to the model and the mounting plate.
3. Mix and pour the epoxy compound, following the manufacturer's directions. Very thick epoxy compounds may be thinned by heating to 150°F (67°C) **before** adding the catalyst. Either vacuum or vibrate the epoxy to remove the trapped air. Let it cure.
4. Remove the model from the mold half. Align the cavities of the two mold halves

and then drill and ream the guide pin holes. See Chapter 4 for details on guide pin installation.
5. To achieve a mold that will withstand hard use, install a steel insert in the pinch-off area. Shape a piece of mild steel for each mold half. Drill and tap mounting holes in each mold half to hold them in place.

Cast Gypsum Cement Molds

Cast high strength gypsum cement (Ultracal 30 or equal) molds may be made by using the same procedure as cast epoxy molds. The gypsum cement can be poured over the prepared model in a metal casting frame. The metal casting frame must be used to keep the gypsum cement from breaking. Vibrate or vacuum the cement to remove trapped air. A metal pinch-off must also be used. Fasten it to the mold casting frame before the gypsum cement is cast.

Machined Hardwood

Temporary or experimental blow molds may be made from machined hardwood, preferably hard, fine-grained wood such as hard maple or birch. Such molds may be made by regular woodworking methods. A steel pinch-off plate should be screwed to the bottom of the mold and metal guide pins and guide pin inserts added.

Indirect Blow Molding in the Laboratory

A simple blow molding operation may be done with heated plastic tubing. Medium or high density polyethylene tubing seems to work best. A wooden, aluminum, epoxy, or gypsum cement mold may be used. A hinged gypsum cement mold is shown in Fig. 6-49. These are the steps in making the mold.

1. Make the mold halves using the procedures already given.
2. Hinge the two mold halves together with a piano hinge. Use a suitcase catch or latch to hold the mold closed.
3. Cut a length of plastic tubing an inch (25 mm) or so longer than the mold cavity.

Fig. 6-49. Gypsum cement blow mold with a hinged frame.

The tubing should have a 1/16" (1.587 mm) to 1/8" (3.175 mm) wall thickness. Attach a small wire to one end of the tube. Hang the tube in an oven at about 325°F to 350°F (162°C to 176°C) for about 3 to 5 minutes.

4. Remove the heated plastic tube from the oven. Quickly place it over one half of the mold. Close the mold and clamp it shut. Use an air gun to blow the heated plastic tube in the mold.

Caution: There must be an air regulator on the air line. Set the regulator at between 15 and 35 psi.

5. Hold the pressure on the tube for about 10 to 20 seconds. Let the plastic cool in the mold.
6. Open the mold and remove the part. Trim off the flash.

The pinch-off is the most critical area in blow molding. A metal pinch-off plate, filed to the right shape, should be attached or cast into the mold.

Materials and Products for Blow Molding

Laboratory blow molding requires a material which is easy to work with. For this reason, low density polyethylene is the most used laboratory blow molding resin. As stated before, it should have a melt index of between 1.5 and 2.5 and a density of between

Fig. 6-50. Products made on laboratory blow molders.

Fig. 6-51. Demonstration model of the blow molding principle.

0.915 and 0.930. Product suggestions for low density polyethylene follow.

Bottles
Coin banks
Chessmen
Balls
Toys
Practice golf balls
Salt and pepper shakers
Decorative objects

STUDENT ACTIVITIES

1. Process a variety of plastics in each available mold.
2. Experiment with plastics having different density and melt index.

Caution: Stay within the blow molding melt index range.

Note: Prepare a log sheet for experimentation with the blow molding process like the one shown in Appendix A, page 393. Save this record of your experimentation for later reference.

   a. Try several different melt index resins for a given mold.
   b. Mix two **blow molding grade** polyethylene resins together and blow mold them. Note how the blend averages to a density between the two parent materials.

3. Experiment with the blow molding process. Change **only one** variable at a time.
   a. Starting with normal established settings for a given material and machine, do the following:
      1) Increase the blowing time one second with each part.
      2) Decrease the blowing time one second with each part.
      3) Blow several parts, increasing the blowing pressure 1 psi (6.9 kPa) for each part.
      4) Blow several parts decreasing the blowing pressure 1 psi (6.9 kPa) for each part.
      5) Increase the barrel and die temperatures slightly (10°F or 5.5°C), mold several parts, and note what changes occur in the parts. Increase another 10°F (5.5°C) and repeat the above. Do not go more than 50°F (27.5°C) above the normal settings in this experiment.

4. Mix dry color pigments with the plastic pellets. Produce a variety of colored products. Various colors may be mixed with each other also.

5. Design small products that can be blow molded. Consider the following for good design:
   a. Blowup ratio
   b. Pinch-off
   c. Parison wall thickness
   d. Part shape
   e. Mold finish
   f. Variation in part size
   g. Good design principles

6. Construct molds for the products designed in item 5. Consider the following:
   a. Type of mold material to be used
   b. Procedure for mold making
   c. Pattern type
   d. Die to mold fit

7. Organize a company and mass produce blow molded products. Offer the products for sale or combine them with other products.
   a. Organization
      1) Operations sequence
      2) Equipment locations
      3) Supplies
      4) Safety
   b. Record keeping
      1) Materials used
      2) Labor required
      3) Processing variables for quality control
         a) temperature
         b) speed
         c) time
   c. Assembly of parts, if needed
   d. Finishing

8. Look up materials data in the **Modern Plastics Encyclopedia**.

Questions Relating to Materials and Process

1. How are the material specifications determined for blow molding?
   a. Melt index?
   b. Specific gravity?
   c. Molecular weight?

2. What density of materials (PE) would be the best for the following? Choose from these ranges: (a) low density = 0.910 to 0.925, (b) medium density = 0.926 to 0.940, and (c) high density = 0.941 to 0.965.
   a. Gallon milk containers
   b. Soft squeeze bottle
   c. Detergent bottle
   d. Shampoo squeeze tube

3. From the following materials, choose those which might be suitable for blow molding.
   a. Polyethylene, density 0.924, melt index 2.0
   b. Polypropylene, density 0.850, melt index 2.5
   c. Polyethylene, density 0.960, melt index 20
   d. Polypropylene, density 0.956, melt index 1.6
   e. Polyethylene, density 0.950, melt index 0.7

4. What is the most effective way to find out whether a plastic will work for a given product under certain conditions?

5. Record the code numbers from the bags of various blow molding materials and look up the specifications in the data books available from the resin manufacturers.

6. Use the **Modern Plastics Encyclopedia** as a basic reference for the materials, processes, and suppliers of the plastics industry.

COMPRESSION AND TRANSFER MOLDING

CHAPTER 7

Process Description

Compression and transfer molding are processes in which plastic materials are compressed into shape and cured or hardened in a metal mold. Although thermoplastic and thermosetting plastics may be molded in this way, thermosetting plastics are usually used. Injection molding is much faster and more economical for thermoplastics and is also often used now for thermosetting plastics. See Chapter 4 for a description of injection molding of thermosets.

Products made by the compression molding process include electric coffee pot bases, electric toaster handles, light-switch cover plates, and plastic dishes. Common transfer molded parts include pot and pan handles, auto distributor caps, and electric switch parts.

Fig. 7-1. Transfer and compression molded heat-resistant parts for automobile transmission.

Fig. 7-2. Hand-operated compression press used during the Civil War.

Compression Molding

Compression molding is probably the oldest molding method. It dates back to prehistoric people who pressed clay into pots by hand. At the start of the 15th century, the printing press was developed. The principle of this press was later applied to molding. During the middle 1800's, natural material such as wood flour-filled shellac was compressed into decorative articles with a screw press. See Fig. 7-3. Compression molding was one of the first processes used for molding plastics. Phenolic plastic became the main compression molding material after it was developed by Dr. Leo H. Baekeland in 1909.

Compression molding machines operate in the following manner. Plastic materials are measured by weight or volume and placed in the open **mold cavity** (1), Fig. 7-4. The mold is usually heated to from 290° to 400°F (143°C to 205°C). The mold cavity is usually bolted to the **lower platen** (2) of the compression press. The **mold force** (3), or plunger half of the mold, is bolted to the **upper platen** (4) of the press. One of the platens forces (compresses) the two mold halves together. Compression pressures of 1000 to 12,000 pounds per square inch (psi) (6900 kPa to 82 800 kPa) are common. The plastic melts, flows, and then cures into shape in the mold cavity. When the thermosetting material has **polymerized** (cured), the mold is opened and the part is removed. A **chemical change** takes place in the mold when thermosetting plastics are compression molded. It usually takes from 1 to 20 minutes for this chemical change to occur. The cure time depends on the material, mold temperature, and thickness of the part.

Fig. 7-3. This molded daguerreotype case was compression molded of wood flour-filled shellac compound about 1855. Some of the very finest mold work was done during that period with primitive mold making tools.

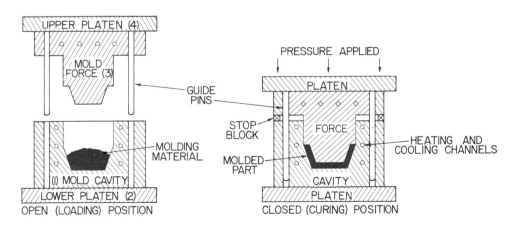

Fig. 7-4. Compression Molding (Fully Positive Mold)

Fig. 7-5. Compression mold used for thermoset molding compounds.

Fig. 7-6. Pellets and Granular Thermosetting Plastics

The thermosetting plastic which is placed in the mold cavity is called a **prepolymer.** A prepolymer is a plastic which has been partly polymerized during its manufacture. The long polymer chains have started to form, but cross-linking has not begun. The cross-linking takes place in the mold during the molding process. Once it is polymerized, or cross-linked, the thermosetting material cannot be reformed. Chapter 3 describes polymerization and cross-linking in detail.

The plastic materials placed in the mold cavity may be in granular or pellet form (Fig. 7-6), or preformed into tablets called **preforms** (Fig. 7-7) Sometimes called "pills," these preforms are a measured amount of granular thermosetting plastic that is pressed into tablet form at room temperature. Preforming provides a more uniform way to measure the plastic and a cleaner method of handling than the granular plastic does. Phenolic, melamine, urea, and DAP (diallyl phthalate) plastics are often premeasured this way.

Liquid thermosetting plastics, such as polyester and epoxy, may also be measured and poured into the mold cavity. They are usually mixed with glass fibers (fiberglass) for added strength or mixed and formed into dough and shaped like logs. This process is discussed in more detail in Chapter 10. Recently, glass-filled polyesters and epoxies have become available in powdered and granular form. These may be handled much

Fig. 7-7. Compression and Transfer Molding Preforms

the same as phenolics in compression and transfer molding.

Cold molded plastics are formed by compression molding, too. Inorganic materials like cement, talc, asbestos, and gypsum cement with a plastic binder are usually used. This plastic is compressed in an unheated mold, removed from the mold, and then baked in an oven to cure, Fig. 7-8. This process is used to produce parts which require high temperature and/or electrical properties under severe conditions. Cold molded parts are used primarily for heat resistant and electrical parts in switches and appliances. Cold molding is a very small but highly specialized segment of the plastics industry.

When **thermoplastics** are compression molded, the heated mold must be cooled before the part is removed. Cooling the mold is necessary to harden the thermoplastic, a **physical change.** The mold must be heated again at the start of the next cycle. Vinyl phonograph records are molded in this manner. Preheated "slugs" of vinyl may also be used with a constantly cooled compression mold for phonograph records. Thermoplastic sheets may also be laminated together by compression molding presses. This is discussed in Chapter 10.

Transfer Molding

Transfer molding is a thermosetting molding process similar to plunger injection molding. It is associated with compression molding because it uses similar equipment and materials and produces products similar to those typically compression molded. The process was developed in 1926.

A diagram of the transfer molding process is shown in Fig. 7-9. Thermosetting plastic

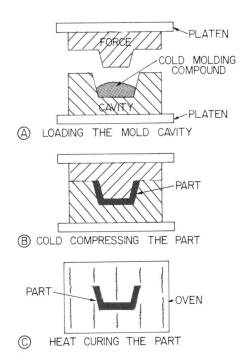

Fig. 7-8. Cold Molding

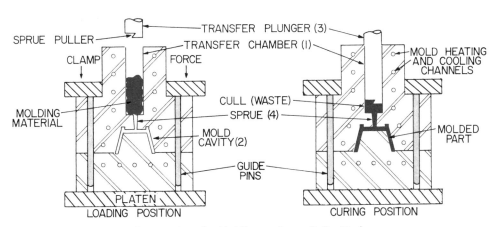

Fig. 7-9. Transfer Molding — Sprue Puller Design

136   Section II   MOLDING

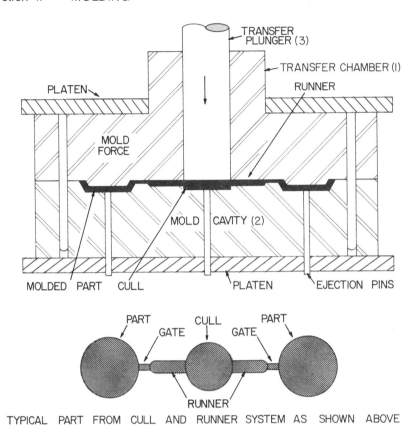

Fig. 7-10. Transfer Molding — Cull and Runner Design
The entire system — cull, runners, and parts — is removed from the mold at once.

is placed into a heated **transfer chamber** (1). After it is heated, the plastic is transferred (forced) into a heated, closed **mold** (2) by the **transfer plunger** (3). Transfer molding pressure (transfer plunger pressure) is usually twice that of compression molding. A pressure drop between the **sprue** (4) and the mold cavity makes this necessary. Actual pressure in the mold is similar to compression molding. Mold temperatures of 320° to 350°F (160° to 177°C) are common. Polymerization (cross-linking, a chemical reaction) takes place in the mold cavity. After polymerization, the mold is opened and the cured part is removed.

Thermosetting materials are also weighed or measured for transfer molding. They may be in a preform, pellet or granular form. A 10% to 20% overfill of the transfer chamber is usually necessary to make sure the mold fills. Often several cavities are filled from a single transfer chamber. A runner system similar to injection molding is used to carry the plastic to each mold cavity. This process is often more economical than multiple-cavity compression molding.

Multiple-cavity compression molding needs exact weighing of plastic charges for **each** cavity. Each cavity must also be filled separately. This takes more time than multiple-cavity transfer molding where all cavities are filled equally from one plunger.

Transfer molding works better for parts which have molded-in inserts, holes, or vari-

Compression and Transfer Molding 137

Fig. 7-11. A typical transfer mold used for thermosetting compounds. Note the transfer plunger hole in the center of the lower half of the mold with runners leading to the various cavities.

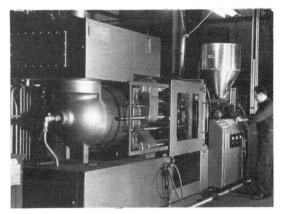

Fig. 7-12. A 375-ton (340 Mt) in-line screw injection press configured for molding thermoset materials.

ations in wall thickness. Compression molding tends to move the inserts around and bend the pins which form the holes. For this reason, transfer molding is often used for parts such as distributor caps, ignition rotors, and switch parts.

Importance of Compression and Transfer Molding

Compression and transfer molding combined are about fifth in importance in the plastics industry. About 4% to 5% of the plastics materials are processed by these methods yearly. Output by these processes may decrease to about 3% to 4% of the market by 1980. Even with this small part of the market, compression and transfer molding processes will still be important for years to come. They still provide an economical way in which to produce large and small thermosetting parts and are very important to the production of glass-reinforced plastics.

Many compression molded parts have been replaced by injection molded thermoplastic and thermosetting parts. Telephone parts are a good example. Most of the plastic telephone parts were originally made by compression molding. When color telephones were introduced in the early 1950's, they were injection molded from thermoplastics. Many small radio and television knobs, now thermoset injection molded, were compression or transfer molded earlier.

Parts such as household electrical switches, switch plates, electrical outlets, and light fixture parts will probably be compression or transfer molded for years. Plastic dishes and certain other plastic housewares require hard, scratch-resistant surfaces which compression molded thermosetting plastics provide. Also, designs are easily molded into plastic products by compression or transfer molding, and colors run all the way through these products.

Advantages

Compression molding offers a method of processing thermosetting plastics into various sizes and thicknesses, including parts

Fig. 7-13. An assortment of parts molded from black phenolic thermoset molding compounds, with heat- and impact-resistant properties.

Fig. 7-14. Typical Transfer Molded Parts

with heavy walls. These parts may be made with little or no evidence of shrinkage. Very dense, hard-to-scratch products are possible. Fiber-reinforced thermosetting plastics develop maximum strength when compression molded. Transfer molding and injection molding tend to break up the fibers, reducing the strength of the product.

Transfer molding offers many of the advantages of compression molding. In addition to these, holes and inserts can be molded in with ease. Multiple-cavity molds may be economically operated in transfer molding. Transfer molding materials as well as those for compression molding are less costly than comparable thermoplastics.

Disadvantages

Compression and transfer molding need longer molding cycles than thermoset injection molding. With longer cycles, machine time and operator labor are more costly. These added costs often offset the lower material cost of compression and transfer molding. However, thermoset injection molding equipment is currently unable to make as large parts as compression molding can, so compression and transfer molding will still be needed.

Compression and transfer molding materials are quite often dirtier to handle than thermoplastic. They tend to contaminate other materials near them, and for this reason these methods of operations are usually located in separate rooms. Recently, new "contamination free" thermosetting materials that come in pellet form have been introduced.

Materials Used in Industry

Most thermosetting plastics may be compression or transfer molded. Common thermosetting plastics, compression and transfer molded, are phenolic, melamine, urea, alkyd (polyester), diallyl phthalate (DAP), and epoxy. Phenolic plastics are used for auto distributor caps, electron tube bases, and pan handles. These plastics are cheaper than other thermosets but darker in color. Melamines are used for colorful dishes, buttons, and laminated table tops (Formica®, Micarta®, etc.). Urea plastics are compression and transfer molded into screw-type bottle caps, lamp reflectors, buttons, and housings for small home appliances. Melamines and ureas are more expensive than

Compression and Transfer Molding 139

Fig. 7-15. An assortment of engine ignition parts available in black phenolic or black and colored alkyd compounds.

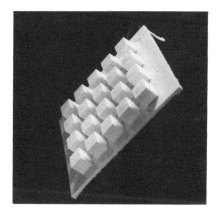

Fig. 7-17. Dice molded on the Hull 200-ton (181 Mt) automatic compression molder. These products are not easy to release from the mold.

Fig. 7-16. A 200-Ton (181 Mt) Automatic Compression Molder

phenolic, but they will accept almost any color or shade. Polyester plastics are popular for fiberglass reinforced products such as auto taillight housings, car bodies, and boat hulls. Epoxy plastics are used with glass fibers and as bonding agents for ceramics and laminated products. DAP plastics are compression molded into electrical switch parts and other electrical and chemical resistant parts.

Fig. 7-18. A vertical, 220-ton (199 Mt) molding press for compression or transfer molding of thermoset materials.

Industrial Equipment

Compression press sizes are usually listed by tonnage, daylight opening, and platen size. Tonnage is the **total** amount of force, in tons, that can be applied to the mold by the movable platen. Daylight opening is the **maximum** amount of distance between the two platens when they are completely open. The platen size is the **length and width** of the platens, generally between the tie rods.

Industrial compression molding machines range from 10 tons (9 metric tons) to over 4000 tons (3628 metric tons) clamp force. Daylight openings range from about 12" (300 mm) to 80" (2000 mm). Platen sizes are from 7" x 9" (175 mm x 225 mm) to 70" x 90" (1775 mm x 2300 mm). In a transfer molding press the transfer pressure will also be listed in addition to the above. It usually ranges from 20 to 30% of the clamp tonnage of the main ram.

Compression Molds

Three general types of compression molds are used: fully positive, semipositive, and flash. Each has its advantages and disadvantages. What is an advantage of one often becomes a disadvantage of another.

A **fully positive mold**, Fig. 7-19, usually makes parts with the most uniform density. It gives the plastic positive pressure (packs it) during the whole compression stroke. As soon as the two mold halves start to close, pressure is applied to the plastic material. The two positive mold halves usually telescope (slide one inside the other) a long distance. Very little flash (material escaping) occurs with the use of a fully positive mold. Very accurate material measurement is needed for uniform sized parts. Too much plastic in the mold will make the part too big, and too little will cause a short shot (unfilled part). A wide range of **bulk factor** materials may be used with a fully positive mold. Bulk factors are like compression ratios. The larger the bulk factor, the more it can be compressed. Higher bulk factor materials are often cheaper. The longer the telescoping of the mold, the higher the bulk factor material that can be used. Bulk factors are discussed in more detail later in this chapter.

A **flash mold**, Fig. 7-20, does not telescope. The two halves just come together,

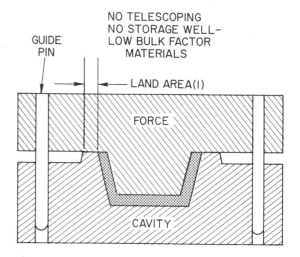

Fig. 7-20. Flash Mold

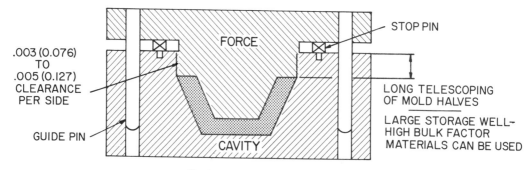

Fig. 7-19. Fully Positive Mold

the **land area** (1) at the opening keeping the mold halves from closing too far. Flash molds do not need accurate material measurement. In fact, extra material is necessary to properly mold a part. This extra material is wasted, but this waste may be cheaper than the labor needed to accurately weigh out the plastic for a positive or semipositive mold. Flash molds do not have the storage capacity which telescoping gives. Very low bulk factor materials are necessary for maximum part density from flash molds. Flash molds usually do not need to be hardened or made from more expensive mold steel for short production runs.

A **semipositive mold,** Fig. 7-21, is a compromise between fully positive and flash molds. The semipositive mold halves telescope, but only about 5/16" (8 mm). The short telescoping distance lets extra molding material escape until the two halves start to close. It is similar to a flash mold until the last 5/16" (8 mm) of closing. Positive pressure is then put onto the material in the mold. Parts may be less dense near the top or opening of the mold. Low to medium bulk factor materials are usually used in semipositive molds and need not be measured as accurately as for positive molds.

Both positive and semipositive molds should be hardened because the mold halves rub together as they telescope (close) inside each other. The metal in the mold halves scrapes and tends to score, or gall, as the halves rub together. Hardened tool steel or carburized mild steel (1018 or 1020) is necessary for positive and semipositive molds. Either H-13 or P-20 mold steel is an excellent choice for a long wearing mold. These steels (H-13 or P-20) are most often used for industrial production compression and transfer molds.

Laboratory Compression and Transfer Molding Equipment

Laboratory compression molding machines are from about 10 to 75 tons (9 to 68 metric

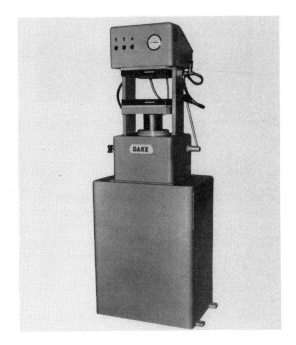

Fig. 7-22. 25-Ton (23 Mt) Laboratory Compression Molding Press

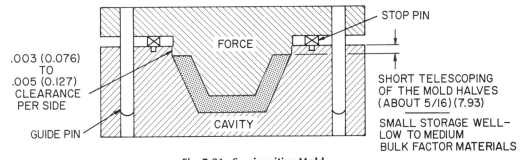

Fig. 7-21. Semipositive Mold

Fig. 7-23. 25-Ton (23 Mt) Laboratory Transfer Molding Press

tons) clamp capacity. Platen sizes range from about 6" x 6" (150 mm x 150 mm) to 12" x 18" (300 mm x 450 mm) or larger. Daylight openings on laboratory presses of 6" (150 mm) or more are common, and often are adjustable. Most laboratory compression molding presses have electrically heated platens.

Laboratory transfer molding presses are of similar sizes. Due to the more complex nature of this press, transfer molding is not often done in the school laboratory. Also, these presses are more expensive than compression molding machines.

Laboratory Compression Molds

Laboratory compression molds are usually **hand molds** which are put in and taken out of the press by hand for each cycle. Also, the parts are usually removed from these molds by hand. These molds are not fastened to the platens. The advantage of hand molds is that a different mold may be put in the press for each cycle. Some laboratory hand molds may be bought from equipment manufacturers and other suppliers, but most must presently be made in the school laboratory. Information on their construction is given under "Tooling Up for Production" in this chapter.

Operating Requirements

Laboratory compression molding machines need 115 or 230 volt electrical power for heating. In addition, many laboratory compression presses are designed with water cooling channels in the platens. This allows the platens to be cooled for thermoplastic laminating. Therefore tap water and a water drain must be provided if the platens are to be cooled.

Materials generally used for laboratory compression molding are phenolic, melamine, and polyester. These are discussed later in this chapter.

Operation Sequence

There are six steps in the compression molding process:

1. Preheat the mold to the right temperature and coat it with mold release.
2. Weigh out the amount of plastic needed.
3. Fill and close the mold. Put it in the press.
4. Apply enough force (pressure) to the mold to make a compact part with good surface finish.
5. Release compression pressure to allow gases to escape. Reapply molding force.
6. Let the part cure in the mold.
7. Open the mold and remove the part.

Table 7-2 shows the operation sequence. Follow it carefully. **Read the manufacturer's instruction manual for exact specifications and instructions.**

Checklist of Materials and Equipment Needed

Assemble the following materials and equipment for compression molding:

1. Compression press, heated platens.
2. Compression mold(s).

3. Mold release — carnauba wax, such as Trewax.
4. Mold ejection frame or device and/or arbor press.
5. Accurate scale, gram increments.
6. Paper or plastic measuring cups.
7. Bench top covered with transite board or a steel top bench.
8. Plastic molding material — phenolic general purpose, medium flow, low to medium bulk factor.
9. Heavy insulated gloves.

Operating Pressures and Temperatures

Operating pressures vary according to the material and mold used. A rule of thumb which may be followed for pressures is listed in Table 7-1.

Often the compression temperatures are held constant while the pressures are varied. Many compression molding presses have

Fig. 7-24. Materials and equipment for compression molding.

temperature controls on each platen. For good results, both platens should be kept at the same temperature. For this reason they are often left at one temperature once they are adjusted, as adjusting takes time. If one thermoswitch controls both platens, this is not a problem. Temperatures of about 300°

TABLE 7-1
Operating Pressure Rule of Thumb for Compression Molding

Material	PSI (kg/cm²) for Surface Area[1]	P/I (kg/cm) for Sides[2]
Phenolic	2000 (140 kg/cm²)	1000 (178 kg/cm)
Urea	3000 (210 kg/cm²)	1500 (268 kg/cm)
Melamine	4000 (280 kg/cm²)	2000 (357 kg/cm)

[1] Pounds (kilograms) of force for each square inch (kg/6.45 cm²) of flat molding surface area.

[2] Additional pounds of force for each inch (kg/cm) of side height (pounds per inch of side height or kilograms per centimeter side height).

Example:

A dishlike part has 4 square inches (25.8 cm²) flat surface area and is 1 inch high (2.54 cm), as shown below. It is to be made of phenolic.

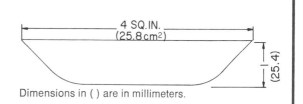

Dimensions in ( ) are in millimeters.

Answer:

2000 psi × 4 sq. inches surface = 8000 lbs.
(140 kg/cm² x 25.8 cm² = 3612 kg)
1000 P/I × 1 inch high sides = 1000 lbs.
(178 kg/cm x 2.54 = 452 kg)
—————
9000 lbs.
(4064 kg)

9000 pounds (4064 kg) total force should be applied to the ram of the press.

to 350°F (149°C to 177°C) may be used. Lower temperatures can be used, but they increase the cure time.

Shorter curing times will usually result if higher curing temperatures are used, but this increases the risk of overcuring. Undercuring may result from lower curing temperatures. The fact that thick parts need more cure time than thin parts must also be considered. A curing temperature of 350°F (177°C) has been found best for general laboratory use.

TABLE 7-2
Operation Sequence for Compression Molding

Sequence/Action	Reason	Troubleshooting
❶ Adjust the daylight opening of the press for the mold used. Allow enough room for a filled mold. Place the empty mold in the press and close the press platens.	Molds are of varying heights. The mold must be able to slip into platen opening when filled.	**Mold does not fit the opening:** Some presses have adjustable strain rods. Others have adjustable platen push rods. Adjust the press as necessary.
❷ Plug in the electrical cord, turn on both platens, and allow the **mold and platens** to heat up together, Fig. 7-25. 300°F to 350°F (149°C to 177°C) is normal.	The mold and platens must come to operating temperature before being loaded with material.	**Platens do not warm up:** Check fuses and plugs to be sure the power is on. Be sure the platen switches and pilot lights are on.
❸ Weigh out the proper amount of material for the mold, Fig. 7-26. Experience with a mold will determine the exact amount of charge.	The mold must have enough material to produce a good part. Do not waste material with overcharges.	**Part suddenly seems to be undercharged:** (a) A material change may require a different amount of material. (b) Moisture in the air is absorbed by the material. A sudden change to damp weather may require additional material weights.
❹ (a) Remove the mold from the platen. **Caution: Hot. Use gloves.** (b) Wax the heated mold with carnauba wax, Fig. 7-27. (c) Be sure to have a heat-resistant (transite) cover on wood bench tops for compression molding or use steel-top benches.	(a) To put plastic in the mold. (b) To prevent sticking. (c) To prevent hot molds from burning bench tops.	**Part sticks to the mold:** Wax the mold with carnauba wax mold release. Mold release is usually not added between each part molded, but only when needed.
❺ (a) Pour the charge of material into the mold, Fig. 7-28. (b) Distribute it evenly in the mold.	(a) To fill the mold. (b) To prevent the mold from twisting.	**The mold twists:** The charge may be uneven in the mold.

(continued on page 146)

Compression and Transfer Molding

Fig. 7-25. Heat the platens and mold.

Fig. 7-27. Wax the mold.

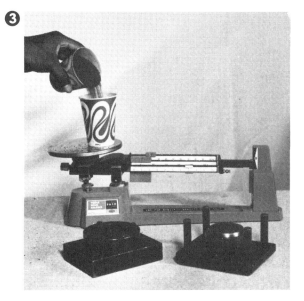

Fig. 7-26. Weigh the material charge. (Mold is in the foreground.)

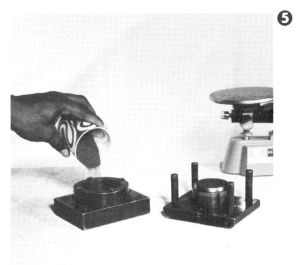

Fig. 7-28. Fill the mold.

Section II MOLDING

TABLE 7-2 (Cont.)

Sequence/Action	Reason	Troubleshooting
6 Close the mold, Fig. 7-29. It should close only one way. Match up the index marks for proper closing.	Mold may be damaged if closed the wrong way.	Mold scores or pins out of alignment: Mold put together wrong. Mold should have one offset or odd-sized guide pin to prevent this.
7 Place the mold on the lower platen **in the center, both front to back and side to side,** Fig. 7-30.	The mold must be centered to prevent twisting of the mold and the press.	Mold closes crooked or twists: (a) Mold off center in press. (b) Material not well distributed in the mold.
8 (a) Compress the material, Fig. 7-31. Apply pressure as figured by the rule of thumb, Table 7-1. (b) "Breathe" the mold cavity by releasing the molding pressure. Build the molding pressure back up again until the part is cured. (c) Hold this pressure for several minutes, Fig. 7-32. Start with 5 minutes at 350°F (177°C). Some parts may be cured in a shorter time. (d) Note: Pressure will not build up rapidly at first.	(a) Each square inch of mold surface requires a definite amount of pressure. **See operating pressures, Table 7-1.** (b) Breathing will allow internal gases to escape. (c) The pressure needs to be held until the material melts, flows, and cures. The cure time depends on the part thickness. The thicker the part, the longer the cycle. (d) Material is melting and flowing in the mold. Pressure will build up when the flow stops.	Part is dull and particles are not compacted: (a) Pressure too low. (b) Pressure not held long enough. (c) Mold not hot enough. Part surface blisters: (a) Pressure held too long. (b) Mold to hot.
9 Open the press and remove the mold, Fig. 7-33. Remove the part from the mold. Allow the part to cool.	The part should be cooled before being handled.	The part pops when taken out: The cure time was too short.
10 Inspect the part for defects.	To make adjustments on the next cycle.	Dull finish: (a) Cycle too short. (b) Mold too cold. (c) Mold rough. Blistered finish: (a) Cycle too long. (b) Mold too hot.
11 Trim the flash from the edges with a file or sandpaper.	Flash is sharp and can cut your hands.	

Fig. 7-29. Close the mold.

Fig. 7-31. Compress the material.

Fig. 7-30. Place the mold on the lower platen.

Fig. 7-32. Set the timer.

Fig. 7-33. Eject the part.

Safety Precautions

Hot platens and hot molds must be guarded so accidental burns will not happen. Safety zones around the machines and **hot** signs are necessary. When machines are shut down, **hot** signs should be hung in front of the platens, Fig. 7-34.

Hot molds should be waxed with **carnauba** wax and placed inside of a bench or cabinet or back out of the way. **Do not place on a wooden shelf as a fire may result.** Bench tops and shelves around compression molders should be metal or covered with asbestos-cement (Transite) board.

Molds should be left open when cooling to keep them from rusting.

Tooling Up for Production

Although compression mold design is similar in some ways to injection molds, there are, however, a number of differences. Only one cavity should be built into a laboratory compression mold. Compression molds may be made from mild steel (1018 or 1020), tool steel, or special mold steels. H-13 or P-20 mold steels may be used for long wearing molds. Positive and semipositive molds should be hardened. Flash molds do not need to be hardened, but will wear longer if they are. Although Figs. 7-19, 7-20, and 7-21 show solid construction, the molds are usually made in several parts. Typical construction details are shown in Fig. 7-36. Care should be taken to make the mold strong enough to withstand high compression pressures.

Hand-operated compression molds should be designed so they can be put together only one way. A mold can be designed with either one **leader pin** offset or with one or two oversize leader pins. The leader pins must be long enough to enter the opposite half of the mold ¼″ to ½″ (6.35 mm to 12.7 mm) **before** the force enters the cavity. This will prevent the force from damaging the cavity. The leader pins should be pressed into one half of the mold so they will stay. Leader pins and leader pin bushing holes must be drilled **before** the mold is hardened. Pins and bushings may be made or purchased ready-made.

Molds may be machined by turning, milling, electrical discharge machining, and grinding. Mating, telescoping surfaces of positive and semi-positive molds should be **ground after** they are hardened. There should be a clearance between the telescoping mold halves of .003″ to .005″ (0.076 mm to 0.127 mm) **per side.** All parts of the mold should be built, polished, and assem-

Fig. 7-34. Hot Sign on the Platens

Fig. 7-35. Compression Mold under Construction

bled before any parts are hardened. The mold must then be **taken apart** and the pieces hardened **separately.** Hardening may warp the parts or cause growth or shrinkage. Some grinding or polishing may be needed after hardening to produce a proper fit between the parts.

Procedure for Making a Mold

The following procedure is for a mold similar to that shown in Fig. 7-35.

1. Turn the force to the desired shape. The **locator** is used as a part to chuck the piece as well as an extension through the force plate for proper alignment.
2. Bore the locator hole in the center of the force plate about .001" to .002" (0.025 mm to 0.051 mm) larger than the **force locator.** Clamp the force into the force plate and drill the mounting bolt holes. Drill and tap the mounting bolt holes in the **force.** Counterbore the force plate for the bolt heads.
3. Turn the cavity inside diameter (I.D.) to .002" (0.051 mm) larger than the force outside diameter (O.D.).

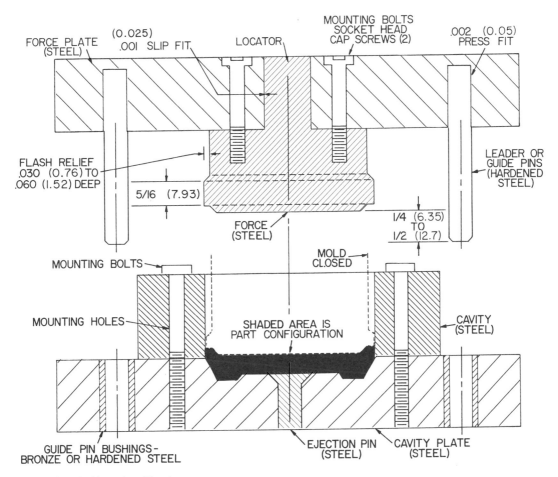

Dimensions in ( ) are in millimeters.

Fig. 7-36. Fully positive (relieved) mold for a coaster.

4. Turn the cavity in the cavity plate. Bore the ejection pin hole. Center and clamp the cavity to the cavity plate and drill the mounting bolt holes in the cavity plate (tap drill size). Tap the mounting holes into the cavity plate.
5. Bolt the force to the force plate. Bolt the cavity to the cavity plate.
6. Place the force inside the cavity and align the force and cavity plates. Turn the unit upside down and clamp it to a vertical milling machine or sturdy drill press table in the inverted position.
7. Drill the leader pin holes through the cavity plate and into the force plate. Do not drill clear through the force plate. The leader pin holes should be staggered or one leader pin should be larger than the others so the mold can be put together **only one way**. Ream the leader pin holes in the **force plate** for a .002″ (0.051 mm) **press fit**.
8. Enlarge the **cavity plate** leader pin holes so that bronze bushings may be pressed in. Be sure to complete the drilling, reaming, and boring of each hole before moving to the next hole location. The leader pins and pin bushings **must** be in perfect alignment. Do not press the leader pin bushings in yet.
9. Take the whole unit apart and harden the force, cavity, ejection pin, and cavity plate. The force plate need not be hardened. After hardening, it will probably be necessary to grind the inside of the cavity and the outside of the force where the two surfaces mate. A .003″ to .005″ (0.076 mm to 0.127 mm) clearance **per side** is best (.006″ to .010″ or 0.152 mm to 0.254 mm) total diameter difference.
10. Reassemble the mold and check the leader pin alignment. Press in the leader pins and bushings.
11. Be sure the outside of both the top and bottom plates of the assembled mold are smooth with no projections. It is well to surface-grind the assembled plates. They need to be flat and smooth in order to conduct heat from the platens of the press and not damage the press platen surfaces.

Material Specifications

General purpose phenolic thermosetting material is most commonly used for laboratory compression molding. A good choice is a medium flow (8-12) wood flour-filled phenolic with low to medium bulk factor (2.0-3.5).

The **bulk factor** is the compression ratio of the material. It is the ratio of the volume of loose material (before molding) to the volume of the finished part (after molding). It may be compared to the compression ratio of a gasoline engine.

The **flow** (or plasticity) **number** refers to how long the plastic takes to cure (polymerize). The higher the flow number, the longer the cure time. Flow numbers range from 1 to 20 or 10 to 200 depending on the manufacturer. A comparison of scales is shown in Table 7-3. Most manufacturers now use the low scale. Melamine and urea flow characteristics are expressed in words: low flow, medium flow, medium high flow, and high flow.

Several different fillers are used in phenolic, melamine, urea, and DAP materials. Fillers are put in to reduce the cost, provide body, reduce shrinkage, add strength and improve chemical, electrical and mechanical properties of the plastic. See Table 7-4 for list of fillers and their effects on phenolic plastics. Similar results are obtained in melamine and urea materials.

Colorants and lubricants are also added to these materials. Colorants produce the color desired in the product. Lubricants help the materials to flow more easily in the mold.

TABLE 7-3
Phenolic Flow Number Scales

	Quick Cure		Medium Cure		Long Cure
Low scale	1	5	10	15	20
High scale	10	50	100	150	200

TABLE 7-4
Fillers Used in Phenolic Molding Materials

Filler	Effect
Wood flour	General purpose
Cotton flock and wood flour	Improved shock resistance
Cotton flock	Better shock resistance
Cord	High shock resistance
Fiberglass	Very high shock resistance and high strength
Asbestos	Heat resistance
Asbestos and cotton flock	Shock and heat resistance
Rubber and wood flour	Rubber modified
Mica	High frequency insulation

Adapted from *Plastics*, J. Harry DuBois and Fredrick W. John, Reinhold Publishing Co., New York, N. Y., 1967, p. 23. Original chart courtesy Union Carbide Corp., Bound Brook, N. J.

Fig. 7-37. Compression molded parts made in the laboratory.

Colors that can be applied to phenolic materials are limited, and most are very dark. Because natural phenolic plastic is orange in color, only dyes that will cover the natural color can be used. Black, brown, dark red, and some dark greens which will hide the orange color are used as color pigments.

A number of products may be compression molded from phenolic materials. Among them are ash trays, serving trays, windshield scrapers, coasters, desk pen bases, threaded bottle caps, knobs, buttons, poker chips, and checkers.

Powdered pure phenolic plastics are available from most phenolic manufacturers. These plastics can be mixed at a ratio of about 1 part phenolic to 3 parts wood chips, by volume, and then molded. A ratio of up to 1 part phenolic powder to 6 parts wood is sometimes used commercially. Either pure black walnut jointer or planer shavings and pure powdered phenolic plastic make attractive molded ash trays, coasters, and other objects. They have a woodlike appearance. Care must be taken to keep the mold well waxed during processing. Molding cycles are much longer than for regular phenolic, usually being double the time.

General purpose, low to medium bulk factor, medium flow melamine or urea compression molding materials may also be used for laboratory compression molding. These materials are usually more colorful than phenolics. They are often used for dishes, buttons, and small pan handles as well as many of the products listed for phenolics.

Two new phenolic materials have recently been introduced to the market. One is **Genal®**, made by the General Electric Company, and the other is **Durez Si®**, made by Hooker Chemical Corporation. Both are designed for high-speed thermosetting injection molding. They come in pellet form and may also be used for compression molding.

Polyester-fiberglass materials may also be compression molded. This is discussed in Chapter 10.

Phenolic, melamine, and urea plastics absorb moisture easily when in the unmolded state. Moisture in the material changes the bulk density so that more weight of material is needed to mold a given part. Drying these materials before molding gives a more uniform bulk density. Thus, more accurate weighing of the material may be done. Drying may be done in a regular oven at about 125° to 150°F (52°C to 66°C) for 45 to 60 minutes. This drying also preheats the plastic which cuts down the molding time. Preheating may also be done in an electronic preheater which works like an electronic oven.

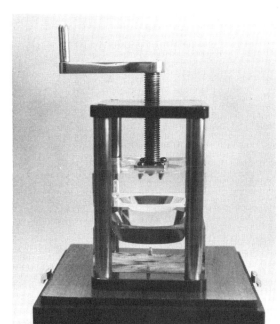

Fig. 7-38. Model Compression Press

Fig. 7-39. Model Transfer Press

Student Activities

1. Process a variety of thermosetting materials with prepared molds:
   a. Phenolics.
   b. Melamine.
   c. Urea.
   d. Genal® or Durez Si®.
   e. Polyester/fiberglass.
   f. Powdered pure phenolic and wood chips.
2. Experiment with materials in prepared molds. Make the following variations:

   Note: Prepare a log sheet for experimentation with the compression molding process like the one shown in Appendix A, page 394. Save this record of your experimentation for later reference.

   a. Material types.
   b. Colors.
   c. Mix colors of the same material type.
   d. Change bulk factors.
   e. Mix in filler such as powdered wood (sanding dust) or ground glass fibers (⅛" or 3.175 mm long). Compare with products made of unmodified resin.
   f. Change flow numbers.
   g. Mix two materials of the same type but with different flow numbers. The flow characteristics should be somewhat between the two parent materials.
3. Experiment with molding process variables such as time, temperature, or pressure. Change only one variable at a time so that you can identify the cause-effect relationship. Make the following changes:
   a. Molding time — shorter and longer.
      1) Find the shortest cycle time that will produce consistently good parts with the other variables constant.
      2) Find the undercure time for a given temperature/pressure setting.
      3) Find the overcure time for a given temperature/pressure setting.

b. Molding pressure — low and high.
   1) Find the lowest possible molding pressure that will produce good parts for a given time/temperature setting.
   2) Determine the molding pressure in pounds per square inch (kPa) of flat surface from the results of item **b-1**. Divide the total force on the mold by the square inches of flat surface on the part.

$$\text{psi (kPa)} = \frac{\text{total force}}{\text{sq. in. (cm}^2\text{) flat surface}}$$

c. Molding temperature — colder and hotter.
   1) Find the lowest molding temperature that will produce good parts at a given time/pressure setting.
   2) Find the undercure temperature for a given time/pressure setting.
   3) Find the overcure temperature for a given time/pressure setting.
4. Design products which can be produced by compression molding:
   a. Round shapes.
   b. Square and rectangular shapes.
   c. Irregular shapes.
5. Substitute uncured tire retread rubber for plastic resins and produce rubber parts.
6. Production organization and operation:
   a. Obtain material supply.
   b. Dry the material.
   c. Preheat the mold.
   d. Wax the mold.
   e. Preweigh material charges.
   f. Fill and close the mold.
   g. Place in press and close press. Apply pressure.
   h. Time the cycle.
   i. Open press and remove mold.
   j. Eject the part.
   k. Perform deflashing and finishing of part.
   l. Assemble.
   m. Package.
   n. Record keeping
      1) Materials used.
      2) Processing variables.
      3) Assembly and finishing.
      4) Rejects.
      5) Labor.
7. Build a model compression press with clear acrylic molds. Mold Play-Doh®. See Fig. 7-38.
8. Build a model transfer press with clear acrylic molds. Mold Play-Doh®. See Fig. 7-39.
9. Look up materials data in the **Modern Plastics Encyclopedia.**

Questions Relating to Materials and Process

1. What flow characteristics (low, medium, or high) would be best suited for the following situations and why?
   a. 1/8" (3.175 mm) wall section, 4" (101.6 mm) diameter ash tray.
   b. 1/2" (12.7 mm) wall section, 2½" (62.15 mm) diameter flat disk.
   c. 1/16" (1.587 mm) wall section, 4" (101.6 mm) tall tumbler.
   d. Transfer molded distributor cap.
   e. A telephone hand set.
2. What bulk factor would be best suited for the following situations and why?
   a. Positive mold with a small storage well.
   b. Positive mold with a deep storage well.
   c. Shallow flash mold.
   d. Shallow semipositive mold.
3. What fillers would be best suited for the following situations in phenolic moldings and why?
   a. Electrical insulation value.
   b. Heat insulation value.
   c. Least expensive product.
   d. Greatest bulk factor.
4. How do time, temperature, and pressure relate to each other in compression molding?
   a. How does a reduction in one variable affect the others?
   b. How does an increase in one variable affect the others?

5. In compression molding, what would happen to phenolic, melamine, and urea molding compounds if only pressure and no heat were applied? Why?
6. Of the three materials — phenolic, melamine, and urea — which is the least expensive to mold and why? Which takes the least amount of compression pressure?
7. Why does compression molding pressure build up slowly at first and rapidly later? What causes a pressure drop during the start of molding?
8. What three stages do thermosetting compression molding compounds go through after heat and pressure are applied? In what order do they happen?

Process Description

Thermoforming is a plastics process in which heated plastic film or sheet is formed over or into a mold by vacuum, compressed air, or mechanical pressure. Thermoplastic sheet may also be formed by a combination (mixture) of two or all three of these methods. It may also be free-blown. When the plastic sheet has cooled on the mold, it will hold its shape. It hardens by a **physical change.**

Thermoforming is divided into four categories, the first of which can be subdivided into two types:
1. Vacuum forming
    a. Drape vacuum forming
    b. Cavity vacuum forming
2. Pressure forming
3. Free blowing
4. Mechanical stretch forming

Vacuum Forming

In vacuum forming, a vacuum (removal of atmospheric pressure) is drawn on one side of the heated plastic sheet after it is **sealed** against the mold. Atmospheric pressure on the other side of the sheet forces the plastic against the mold. The difference in air pressure between the two sides of the sheet provides the force. With a good vacuum on one side, the maximum force applied to the other side of the sheet is the atmospheric pressure that is available. Atmospheric pressure, which is 14.7 psi (101.4 kPa) at sea level, is lower at higher altitudes. Thus, vacuum forming becomes less effective at higher altitudes. In order to get the most benefit from the atmospheric pressure, a tight vacuum seal must be maintained between the mold and plastic sheet.

Vacuum forming is probably the most used thermoforming method. It is divided into two kinds: **drape** vacuum forming and **cavity** vacuum forming.

Drape vacuum forming is the forming of a heated thermoplastic sheet **over** a male mold. The plastic stays on the mold until it is cooled and hardened. The formed plastic part is then taken off the mold and the extra plastic trimmed off. The cycle is then re-

THERMOFORMING

CHAPTER 8

peated. Drape vacuum forming is probably the most used school laboratory thermoforming method. It is also the simplest.

Cavity vacuum forming is the forming of a heated thermoplastic sheet or film **into** a female or cavity mold. It is cooled and taken off the mold in the same way as drape forming. It is also trimmed before use.

In most thermoforming, the thermoplastic sheet must be held tightly on all four sides

Fig. 8-1. Automatic Thermoforming and trimming line in action.

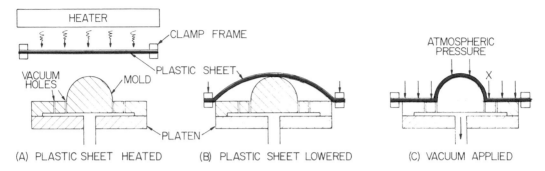

Fig. 8-2. Drape Vacuum Forming Sequence

with a clamp frame. The clamp frame keeps the plastic from shrinking out of shape when it is heated. It also helps the plastic form a tight vacuum seal on the mold.

In thermoforming, the plastic sheet thins out in the places where it stretches most. In drape vacuum forming, the plastic sheet usually thins out around the outside edges of the mold, indicated by the "X" in Fig. 8-2. In cavity vacuum forming, the plastic sheet stretches and thins out in the bottom of the cavity, indicated by the "O" in Fig. 8-4. The choice between cavity and drape vacuum forming often depends upon where thinning out of the plastic will hurt the product least. Thicker plastic sheets tend to stretch more evenly. They do not thin out as easily in the corners or bottoms as thin sheets do.

Mark-offs sometimes occur during processing. These are the imperfections that show in the formed plastic sheet where it touches rough spots in the mold or dirt on the mold surface. Choice of the molding method often depends on whether surface detail or smoothness as well as the strength of the part is important.

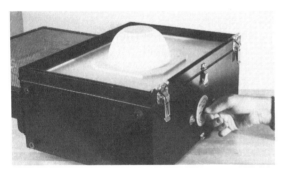

Fig. 8-3. Drape Vacuum Forming

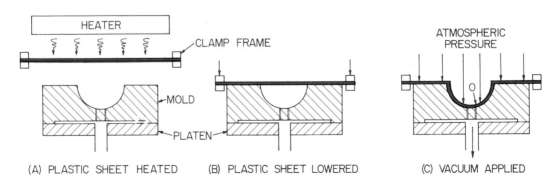

Fig. 8-4. Cavity Vacuum Forming Sequence

Fig. 8-5. Cavity Vacuum Forming

Pressure Forming

In pressure forming, the thermoplastic sheet is "blown," by air pressure, onto or into the mold. A pressure box or mold is usually lowered over the heated thermoplastic sheet. Air pressure is added to the box to force the plastic sheet against the mold surface.

Pressure forming is often used for thick or stiff thermoplastic sheet. It is also used to speed up the production of thin thermoformed parts. More pressure is put onto the surface of the plastic sheet than in the vacuum forming method. Fifty to 100 pounds per square inch (345 to 690 k Pa) may be used in pressure forming. In vacuum forming, pressure on the surface of the sheet is limited to atmospheric pressure (14.7 psi or 10.4 k Pa at sea level). This is not enough pressure for some stiff or thick thermoplastic sheets.

Free Blowing

Free blowing is a method of thermoforming used to form optically clear parts. No mold is used. Instead, the hot plastic sheet is clamped between a ring and platen. Air is blown through a hole in the platen and the plastic blows up like a dome. This is the way in which acrylic astrodome panes, serving tray covers, and other similar parts are formed.

Simple equipment for free blow forming may be built in the school laboratory for use in forming acrylic plastic sheet. See Fig. 8-9. Free blowing may also be done on laboratory and commercial thermoforming equipment.

Mechanical Stretch Forming

Mechanical stretch forming (also called plug forming) is a thermoforming method in which mechanical pressure is used to form the sheet. Usually, air or hydraulic cylinders

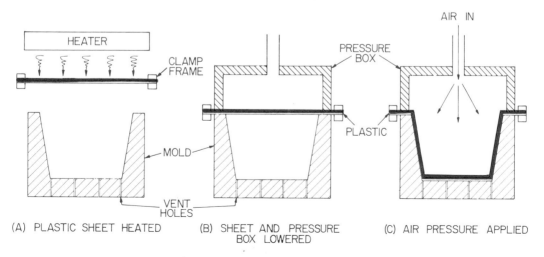

Fig. 8-6. Pressure Forming Sequence

move the mold(s) up and down to stretch or form the thermoplastic sheet. The mold(s) could also be gear, lever, or screw operated.

Mechanical stretch forming serves other purposes as well. It is used to form thick or stiff plastic sheet. It may also be used as a helper for vacuum or pressure forming. The helper, or plug, stretches the thermoplastic sheet before the vacuum or air pressure is applied. In both of these methods, mechan-

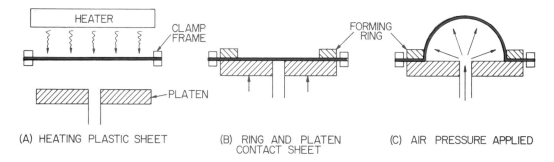

Fig. 8-7. Free Blowing Sequence

Fig. 8-8. Free Blown Parts

Fig. 8-9. Simple Shop-Made Free Blowing Equipment

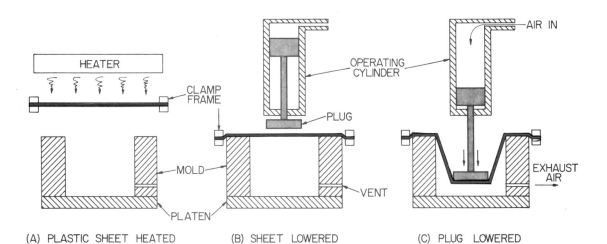

Fig. 8-10. Mechanical Stretch Forming Sequence

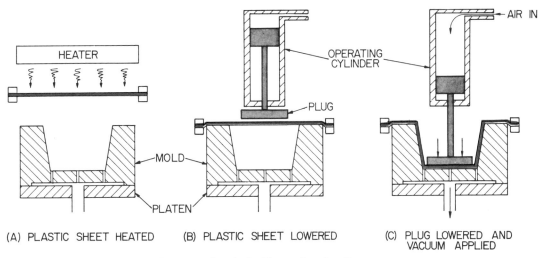

Fig. 8-11. Plug Assist Thermoforming Sequence

ical stretch forming helps achieve a more uniform thickness of parts.

Process Combinations

Often, combinations (mixtures) of two or three thermoforming processes are used together. Pressure forming and vacuum forming may be combined. Mechanical stretch forming may be added to either of these processes or their combination to stretch the sheet before the vacuum or pressure is applied. The plastic may also be stretched before being formed by compressed air. It can then be vacuum formed in the same mold after a plug (mechanical stretch forming) pushes it down into the mold. There are a number of combinations of thermoforming and new ones are being developed. Important to remember is the fact that any two or all of the thermoforming processes may be combined.

Thermoforming Techniques for Packaging

Many different kinds of products reach the market enclosed in see-through plastic packaging that is inexpensive and permits easy consumer selection. Methods of packaging using the thermoforming technique are described here.

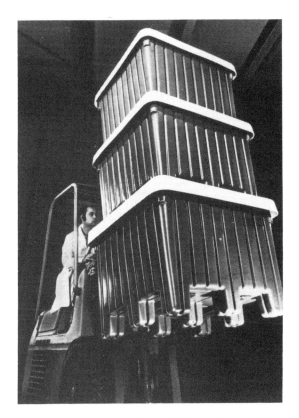

Fig. 8-12. Vacuum and plug assist thermoformed tote boxes and covers.

Blister Forming. Blister forming is a vacuum forming technique used to make package covers for products. Clear plastic sheet or film is usually drape or cavity vacuum formed to a shape slightly larger than the product to be enclosed. The plastic blisters or domes are cut out of the sheet and attached to printed card stock with the product enclosed between. Products such as razor blades, tools, toys, automobile parts, and other consumer items are often packaged in this way, Fig. 8-13.

Skin Packaging. In skin packaging, a product is placed on a special perforated card in the thermoformer, Fig. 8-14. A thin ionomer or polyethylene film is vacuum formed over the product. The card stock is perforated so the air can pass through it. The plastic film sticks to the card stock and holds the product in place. The card stock used for polyethylene film is specially treated to help the film stick. Special treating is not needed for the ionomer film. No mold is used, as the product forms the shape of the plastic. See Fig. 8-15.

Shrink Packaging. In shrink packaging, a thin plastic film is placed around a product and shrunk by heat. Although technically not thermoforming, it is included here because (1) it is similar to thermoforming and (2) it is used for packaging. The plastic material used is a flat or tubular film in which the molecules are oriented during extrusion. The orientation sets up strains in the material which are released when it is heated. After the plastic is placed around the product, it

Fig. 8-13. Display of blister packaged batteries.

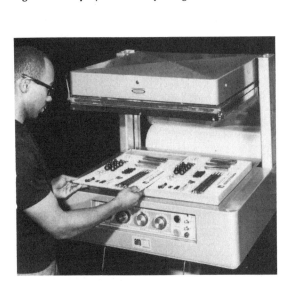

Fig. 8-14. Skin Packaging

Fig. 8-15. Skin Packaged Dishes

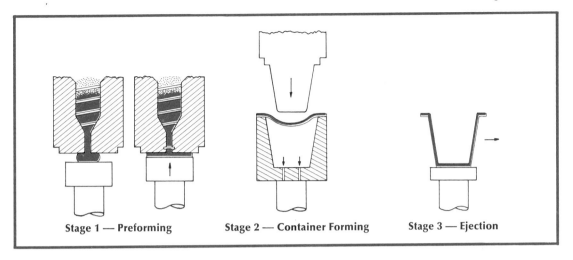

Fig. 8-16. Three stages of monoforming.

is heated by a heat gun or in a heat tunnel. An electric hair dryer may be used for small shrink packaging jobs. The heat shrinks the plastic film tightly around the product.

Monoforming — a Combination Process

Monoforming is a new, patented process which combines screw extrusion, compression molding, and thermoforming. It is used for high speed production of containers. Monoforming has three stages:
1. Preforming which employs both extrusion and compression molding techniques,
2. Container forming which is basically plug assisted thermoforming, and
3. Ejection of the part.

The three stages are shown in Fig. 8-16.

The main advantages of monoforming are low tooling costs, high quality parts at a lower cost, thicker walls than conventional thermoformed parts, and lower material costs. The material cost is low because of the direct use of resin rather than prior sheet extrusion and the absence of trim or scrap to recycle. This process is economical for short production runs according to the machine manufacturer.

Importance of Thermoforming

One of the top users of thermoplastics is thermoforming. It will likely be one of the most important plastics processes of the future. One of the main reasons for a bright future for this process is its low machine and tooling cost.

Fig. 8-17. Polystyrene cups are formed and ejected.

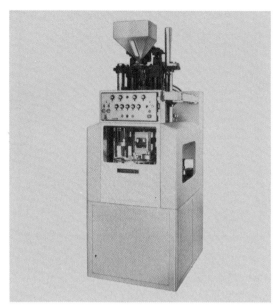

Fig. 8-18. Monoformer Vertical Extruder Thermoformer

Fig. 8-20. Building with thermoformed windows.

Fig. 8-19. Thermoformed World War II Bomber Nose

Thermoforming was invented in the 1930's, and developed slowly. One reason was that cellulose acetate was the only available plastic that could be successfully thermoformed. The first fully automatic thermoforming machine was built in 1938. It was used to make Christmas tree spires and stars, cigarette tips, and ice cube trays from cellulose acetate sheet. Not until World War II were other plastics invented that could be thermoformed. Airplane windshields and bomber noses were made from clear acrylic plastic sheet, Fig. 8-19. Army maps were vacuum formed to show ground contour details better.

Thermoforming has become a very important plastics process. Many packages for consumer products are made this way. Vacuum formed individual serving packages for jelly, cream, and butter are common in many restaurants. Two-piece thermoformed and spin-welded milk, kitchen cleanser, and fruit juice cans are found in grocery stores. Many large advertisements and signs, such as lighted gasoline station signs, are now thermoformed. It is an inexpensive process for many products. The automobile body of the future, it is thought, will also be thermo-

Fig. 8-21. ABS-polycarbonate alloy sheet thermoformed car is called "Formacar."

formed. One such car body has been developed by the Borg-Warner Corporation, Fig. 8-21.

Advantages

Thermoforming of plastic sheet is economical when large thin-walled parts are needed. It is also economical for short production runs. Machines and molds for thermoforming are less expensive, for their size, than those for many plastics processes. Production thermoforming machines cost less than one-third that of an injection molder to make the same size part. Thermoforming molds cost about one-quarter that of the same size injection mold.

Molds for thermoforming do not have to take high pressures. They may be made from inexpensive materials such as wood, plaster, fiberglass, and cast aluminum. The low cost of thermoforming machines and molds cut the total production cost of thermoformed products.

Parts as large as 25 feet long, 12 feet wide, and 5 feet deep (7.62 m x 3.65 m x 1.52 m) can be made on thermoforming machines available today. Compression or injection molded parts that large cannot be made with the machines now available. Fiberglass lamination is the only other plastics process that can now make parts that large. One of the world's largest thermoformed products is shown in Fig. 8-22. This combination houseboat/travel trailer has a 90 horsepower motor and can go 20 miles per hour (32 km/hr) on water.

Fig. 8-22. Ship-A-Shore Combo Cruiser

Fig. 8-23. Artist's conception of a small multipurpose plastic car for grass cutting, snow plowing, lazy golfing, and commuting short distances. This car could be one answer to the energy crisis.

Scrap thermoforming plastic may be saved and recycled, or chopped up for reuse, as described in Chapter 5. Recycling makes thermoforming even more economical. The recycled material may be reused in an extruder or injection molder, but this **requires the separation** of each family of plastics. Families must not be mixed during granulation (chopping up). Colors from the **same** plastic family, however, may be mixed during recycling.

Disadvantages

Sheet plastics for thermoforming are higher priced **per pound** than injection molding pellets. The reason is that thermoforming sheet must first be extruded or calendered before it can be thermoformed. Making the plastic sheet adds another process. This raises the cost per pound of the plastic sheet. In small quantities, the cost of thermoforming sheet is about twice the price of injection molding plastic. In large production quantities the difference is much less. Injection or compression molding may be more economical for long production runs of similar parts. Lighter weight parts may be made by thermoforming, however. This tends to reduce the per part cost to a comparable level.

Another disadvantage of thermoforming is that parts thin out in places during processing. More uniform wall thicknesses can be made by many other processes. In addition, stresses are often built into thermoformed parts during forming. The advantages and disadvantages must be carefully studied before any process is chosen.

Materials Used in Industry

Any thermoplastic material that can be formed into sheet or film probably may be thermoformed. Plastic as thin as .0005" (5/10,000" or 0.0127 mm) to thicker than .500" (½" or 12.7 mm) can be successfully thermoformed.

High-impact polystyrene sheet is one of the most popular thermoforming plastics. It is thermoformed into refrigerator and freezer door liners, refrigerator inner liners, wall plaques, three-dimensional signs, cigarette packages, and food containers. Polyethylene film and sheet are thermoformed into many types of package covers. Polyethylene may also be thermoformed into molds for casting thermosetting plastics. Ionomer (DuPont Surlyn "A") film is used to make very tough packages. Both ionomer and polyethylene are used for skin packaging, a type of vacuum forming. They are formed right over the package card and product. Tools are often sold this way. Cellulosics, such as cellulose acetate and cellulose acetate-butyrate, are frequently used to make clear package

Fig. 8-24. A 72" x 96" (1828 mm x 2438 mm) Industrial Thermoformer

Fig. 8-25. Thermoformed liner for upright refrigerator and freezer.

covers. Polyvinyl chloride (PVC or vinyl) film and sheet are thermoformed into clear and colored packages and other products. Many products, especially thick and/or strong ones, are thermoformed from ABS.

Laboratory Thermoforming Equipment

Laboratory thermoformers range in size from about 10″ x 14″ (250 mm x 350 mm) to 24″ x 24″ (600 mm x 600 mm) sheet size. A number of inexpensive machines from about 10″ x 14″ (250 mm x 350 mm) to 18″ x 21″ (450 mm x 530 mm) sheet size can be bought which only perform the vacuum forming process. See Figs. 8-28 and 8-29. Of these, many will only do drape vacuum forming. These machines are usually called **vacuum formers.** The smaller vacuum formers may be used with a vacuum cleaner or shop vacuum as a vacuum source. Many now have vacuum motors built into them.

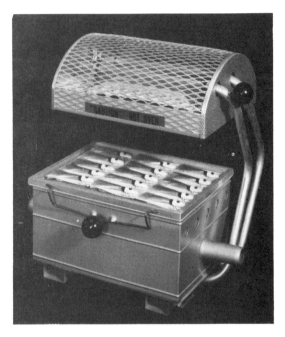

Fig. 8-28. Vacuum former, 10″ x 14″ (250 mm x 350 mm).

Fig. 8-26. Thermoformed Plastic Parts

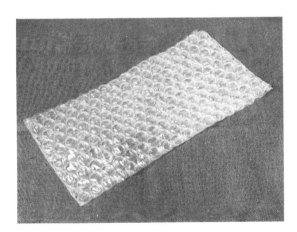

Fig. 8-27. Thermoformed and Heat-Sealed Plastic Packing Material

Fig. 8-29. Vacuum former, 14″ x 20″ (350 mm x 500 mm).

Fig. 8-30. Thermoformer 24″ x 24″ (600 mm x 600 mm) capable of all types of thermoforming operations.

Fig. 8-31. Manual thermoformer 18″ x 18″ (460 mm x 460 mm) capable of all thermoforming processes.

Larger, heavy duty, more expensive machines from about 16″ x 16″ (400 mm x 400 mm) to over 24″ x 24″ (600 mm x 600 mm) sheet size can usually do vacuum, pressure, and mechanical stretch forming processes. These machines are called **thermoformers** since they can do a mixture of all the thermoforming operations.

Operating Requirements

The only operating needs for small vacuum formers are an adequate electrical circuit, a mold, and the plastic sheet. Small vacuum formers often need a 15 to 20 ampere electrical circuit. If the circuit is too small or overloaded, the vacuum former will blow the fuse.

Larger machines that do all the thermoforming operations need compressed air, a heavy duty electrical circuit, mold(s), and the plastic sheet. Many large thermoformers use 230 volt single-phase electricity. Compressed air is used to move the mold platens up and down on some machines. In pressure forming, compressed air is also needed for forming the sheet. A fan is sometimes used to cool the plastic quickly after it is formed.

Thermoforming 167

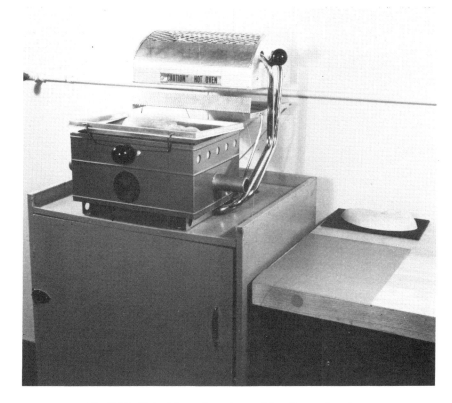

Fig. 8-32. Materials and equipment for thermoforming.

Plastics most often used for laboratory thermoforming in schools are high-impact polystyrene, cellulose acetate, ABS, polyethylene, and ionomer. These plastics will be discussed later in the chapter along with project suggestions.

Operation Sequence

There are six simple steps in performing the thermoforming process:

1. Install the mold on the platen.
2. Clamp the plastic sheet or film into the thermoformer.
3. Heat the plastic until it begins to sag.
4. Bring the plastic in contact with the mold.
5. Apply the vacuum or pressure, or both, as needed.
6. Cool the formed plastic part.
7. Remove the part and trim it if needed.

Refer to Table 8-1 for the operation sequence in thermoforming. **Read the machine manufacturer's instruction manual for exact directions.**

Checklist of Materials and Equipment Needed

Assemble the following materials and equipment for thermoforming:

1. Plastic sheet.
2. Thermoformer.
3. Paper cutter, shears, or scissors to cut the plastic sheet.
4. Mold(s).
5. Mold release wax (carnauba works well).

TABLE 8-1
Operation Sequence for Thermoforming

Sequence/Action	Reason	Troubleshooting
1 Put the mold(s) on the platen(s) of the thermoforming machine, Fig. 8-33. Adjust the mold(s) so the mold(s) center under the clamp frame. Fasten the mold(s) if necessary.	Enough clearance must be kept between the molds and clamp frames so the part will form well. Do not crowd the molds against the clamp frames.	**Webbing between the mold and clamp frame:** Move the mold away from the clamp frame.
2 Cut the plastic sheet to the right size for the clamp frame, Fig. 8-34. Sheet should be slightly larger than the clamp frame in most cases. This operation is usually not needed for roll feed machines.	Proper cutting assures plastic sheet will fit the machine. Also saves plastic.	**Sheet pulls out of the frame during forming:** (a) Sheet cut too small. (b) Clamp pressure too light.
3 Dry the plastic in an oven. **This operation is only for high moisture absorbing plastics such as polycarbonate.** Skip this operation for other plastics unless they bubble.	Plastic sheet such as polycarbonate absorbs water from the air. The water comes out of the sheet in the form of bubbles when sheet is heated.	**Bubbles show up in the surface of the sheet when heated:** (a) Dry the sheet to manufacturer's specifications. (b) Heat the sheet slower in the thermoformer. (c) Do not get the sheet too hot.
4 Put the plastic sheet in the clamp frame, Fig. 8-35. Center it well. Narrow frames usually need at least ¼" to ½" (6.35 mm to 12.7 mm) extra plastic around the frame. **Example:** For a 10" x 14" (25.4 cm x 35.5 cm) thermoformer, cut the sheet about 11" x 15" (279 mm x 381 mm).	The plastic often pulls in when thermoforming. Extra plastic prevents the sheet from pulling out of the frame.	**Plastic sheet pulls in on the edges:** (a) Cut the sheet larger. (b) Center the sheet more carefully.
5 Heat the plastic sheet until it sags slightly, Fig. 8-36. The amount of sag can be determined by experience. Light smoking of polystyrene is normal. Polyethylene and polypropylene sheet turn clear when they are ready to form.	Plastic sheet must be heated enough to form easily. Some sag is needed to stretch the sheet before forming. Different plastics require different amounts of sag before forming.	**Sheet does not pull down tight on corners and edges:** (a) Heat the sheet longer. (b) Sheet too thick. (c) Loss of vacuum or pressure.
6 Bring the mold(s) and the heated plastic together, Fig. 8-37. Either lower the plastic or raise the mold as required. **Important: a tight seal between the plastic and the mold is needed.**	The plastic **must** seal well against the mold to keep from loosing vacuum or air pressure during forming.	**Sheet does not pull down right:** (a) Poor vacuum or air-seal on the mold. (b) Vacuum holes needed in the mold.

(continued on page 170)

Fig. 8-33. Mount mold in press.

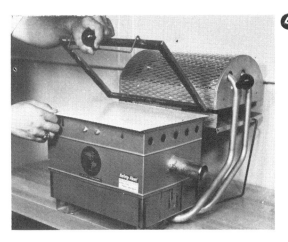

Fig. 8-35. Put plastic in clamp frame.

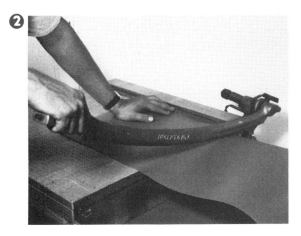

Fig. 8-34. Cut plastic to fit frame. Thicker sheet can be cut on a sheet metal shear or table saw.

Fig. 8-36. Heat the plastic until it sags.

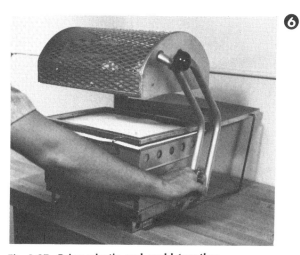

Fig. 8-37. Bring plastic and mold together.

TABLE 8-1 (Cont.)

Sequence/Action	Reason	Troubleshooting
7 Apply the vacuum, air pressure, or mechanical pressure as needed.	This pulls or pushes the plastic over or into the mold.	**Vacuum seal is lost:** Move the plastic and mold tighter together.
8 (a) Remove the heat from the plastic and let it cool, Fig. 8-38. (b) Use cooling fan or compressed air to cool the plastic (**optional**).	(a) This lets the plastic cool. (b) This speeds up the cooling rate and keeps the molds from getting too hot.	**Plastic cools too fast on the mold:** (a) Keep the heat on the plastic longer. (b) Do not use a fan or compressed air.
9 Move the mold and plastic apart. Open the clamp frame, Fig. 8-39.	To remove the plastic.	**Plastic warps after being taken off the mold:** Cool longer or faster.
10 Take the plastic off the mold, Fig. 8-40. Remove the plastic sheet from the clamping frame.	To remove the part from the machine.	**Plastic sticks to the molds:** (a) Use mold release such as carnauba wax, grease, or fluorocarbon spray. (b) Look for undercuts on the mold.
11 Trim the extra plastic from the part, Fig. 8-41. Use a tin snips, paper cutter, sharp scissors, or a die cutter. The type of trimmer will depend on the part, the materials used, and the equipment available.	Trimming gives the part its proper outside shape.	**Plastic sheet webbs:** (a) Sheet was too hot. (b) The sheet was too thin. (c) Molds too close to each other or the frame. (d) Corners on the molds too sharp.

Fig. 8-38. Remove heat and cool.

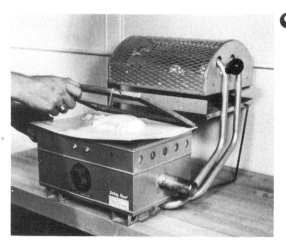

Fig. 8-39. Open the clamp frame.

Thermoforming 171

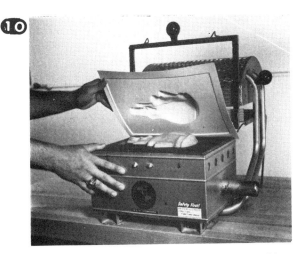

Fig. 8-40. Remove the plastic part from the mold.

Fig. 8-42. Note how webbing has formed at the corners of the vacuum formed part.

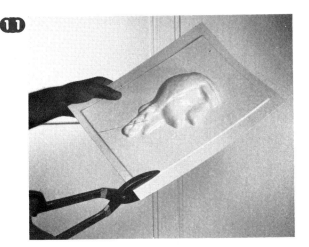

Fig. 8-41. Trim the part.

General Operating Hints

Webbing, a finishing defect, often occurs when thin sheet is vacuum formed. See Fig. 8-42. This problem may be overcome by using thicker sheet, forming at cooler temperatures, or spacing the molds farther apart. Webbing will form between a mold and the frame if the mold is too close to it. Deep draw drape molds will web more easily than shallow draw drape molds. Female molds (cavity forming) or plug assist forming should be used for deep draw parts wherever possible. Webbing is not common in pressure or mechanical stretch forming.

Thin sheet draws down more tightly to the mold than thicker sheet does during the vacuum forming process. To help the sheet draw down better, small vacuum holes must be drilled in the mold. Number 60 drill holes are the best size to use as they will not show on the finished product. A slightly larger vacuum hole made with a number 55 drill may be used for sheets over .060" (1.52 mm) thick. Larger holes tend to mark the surface of the sheet more easily. The vacuum holes should be drilled in low spots or corners of the mold — places where the plastic will draw down last. They can also be drilled where the plastic will stretch the most. Vacuum holes should not be drilled until after the mold has been tried. Drill the holes as needed in spots where the atmospheric pressure is trapped and the sheet does not form properly.

> **Safety Notes**
>
> The heating unit on a thermoformer should always be protected. Guards around the unit will prevent burns. If it is not well guarded, place a HOT sign on the heating element.
>
> Be sure the heating element is turned off when not in use.
>
> Thick plastic sheet holds heat a long time. Cool it long enough before handling. Use a glove.

Tooling Up for Production

Molds for thermoforming are simple, easy, and inexpensive to make. Of all the plastics processes, thermoforming gives students the best chance to make molds. Wood, plaster, metal, thermosetting plastics, or ceramics may be used for molds.

Wooden Molds

Wooden molds, made by cutting, turning, or carving, are simple to construct. Choose a smooth, close grained wood such as poplar, soft maple, birch, or hard maple to make a mold. Clear pine and basswood can also be used but dent very easily. Choose a piece of wood large enough or glue pieces together to achieve the right size. Shape the mold, and then seal its surface. Use a heat-resistant paint — urethane varnish or polyester resin. Allow the sealer to cure or dry well before using the mold. Several coats of carnauba wax should be applied to the finish. For less permanent molds, simply applying a few coats of carnauba wax over the bare wood is sufficient.

Double-tempered hardboard (Masonite®) may be used for shallow molds. A coat of polyester resin will keep the hardboard from flaking during use. Tempered hardboard may also be used for baseboards for wood and plaster molds. Molds should be glued to the hardboard with a silicone bathtub seal. Other glues may fail in the heat. Be sure that the bottom of the mold is flat.

Drill vacuum holes only as needed. Try the mold first to find out where holes should be drilled, if at all.

Fig. 8-43. Wooden Mold

Fig. 8-44. Turned Wooden Mold

Plaster Molds

Molds can be made from plaster of paris or gypsum cement with equal ease. Ultracal 30, made by the United States Gypsum Company, is an example of a good high-strength gypsum cement. Other plaster manufacturers make similar mold making gypsum cements. They are preferred over plaster of paris because they are much stronger, but cost about the same.

Master patterns for plaster and gypsum cement molds are available in stores or can be made in the laboratory. For example, a plaster wall plaque was purchased from a local variety store. See **A** in Fig. 8-45. A **master pattern (B)** was vacuum formed from the purchased wall plaque used as the model. An Ultracal 30 mold (C) was cast from the master pattern. The cast mold was sanded smooth and flat on the bottom, glued to a baseboard, and used for vacuum forming. The vacuum formed master pattern (B)

Thermoforming

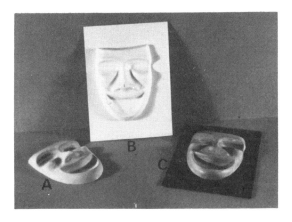

Fig. 8-45. Three steps in making a thermoforming mold from gypsum cement.
  A. Original
  B. Thermoformed Master
  C. Mounted Thermoforming Mold

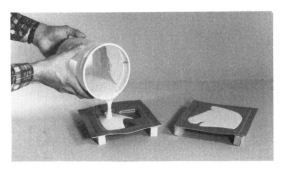

Fig. 8-46. Pouring an Ultracal 30 gypsum cement mold.

and the original wall plaque used as a model (A) were stored for future use. A broken mold could then be replaced by a new one cast from the master pattern.

Similar molds can be made from aluminum kitchen gelatin molds by using the same method. **Caution:** Do not pour Ultracal 30 or similar high-strength gypsum cements directly into the aluminum mold. A chemical reaction will cause the cement to stick to this mold permanently. Vacuum forming may be done directly over the mold, however. Sometimes vacuum holes will need to be drilled into the aluminum when it is used in this way.

High-strength gypsum cements like Ultracal 30 should be mixed to a thick, creamy state. They should never be watery or thin. Pour them just as soon as they are well mixed.

Steps for Mixing Ultracal 30 Gypsum Cement

1. Weigh an empty container such as a pail, jar, or paper cup. Write this down as weight **A**.
2. Fill the master mold with water. Pour the water into the container just weighed.
3. Weigh the water and the container. Write this down as weight **B**. Subtract the weight of the container (A) from the weight of the container and water (B). This will give the weight of the water alone (B − A = C).
  C equals the weight of the water alone.
4. Throw away 1/2 of the water **C**. Write down the weight of 1/2 of **C**. Call this weight **D** (1/2 of **C** = **D**).
5. Add 3 parts of Ultracal 30 to 1 part of the water that is left, **by weight.** If D weighed 2 pounds, multiply 3 × 2. The resulting 6 pounds of Ultracal 30 in this example are to be added to the 2 pounds of water. Slightly more plaster than is needed will be made by this method, but this will allow for waste.
6. Add the Ultracal 30 powder to the water slowly. **Never add water to the plaster** as it will become lumpy. Let the plaster stand in the water for about 2 or 3 minutes before stirring. Mix the Ultracal without whipping air into it. Power mixing with a hand drill and a paint mixing blade works well for large amounts. Small amounts should be mixed with a flat stick.
7. First, brush a coat of the plaster mixture over the pattern. Clean the brush out in water right away. **Slowly** pour the rest of the plaster mixture into the mold. Pour it into one corner to push the air out ahead of the plaster. Jiggle the mold to get any trapped air out.

8. Let the Ultracal set up over night before use.
9. **Do not pour waste plaster or gypsum cement down the sink. It will plug the drain when it hardens.** Use disposable mixing containers and throw them away after use.

Metal Molds

Metal molds may be made from stamped sheet metal (like the gelatin mold discussed earlier), cast metal, or machined metal. Sand casting patterns from the school foundry area often make good vacuum forming molds. Molds may also be made by common sand casting methods. Regular casting aluminum works well for this.

Aluminum or steel molds may be milled or turned. Care must be taken to allow enough draft angle, or taper, on the sides of all thermoforming molds for easy separation of the part after processing. Two to five degrees of draft is common.

Plastic Molds

Thermosetting plastic parts such as phenolic ash trays, metal-filled epoxy parts, and phenolic/paper laminate board make long-wearing thermoforming molds. Phenolic/paper laminate board can be cut, sanded, milled, and drilled like wood or metal. It is

Fig. 8-49. Turned Metal Thermoforming Molds
This mold is used to make disposable plastic vending machine cups by plug assist vacuum forming.

Fig. 8-47. Cast aluminum molds for ice cubes or egg cartons.

Fig. 8-48. Machined aluminum mold and thermoformed part on which letters are added later.

Fig. 8-50. Richlite® Paper/Phenolic Machined Molds

much harder than wood and needs no mold release. Richlite® and General Electric laminate board are examples of those that machine very easily.

Ceramic Molds

Ceramic molds for thermoforming may be made by **slip casting**. Make a master slip casting plaster mold. Pour clay slip in the master mold. Let the slip stand in the mold until a layer builds up. Pour out the extra slip. Let the clay harden enough to remove it from the mold. After removal, dry the clay completely. Bisque-fire the clay in a kiln. The ceramic mold may be used without glazing.

Materials and Products for Laboratory Thermoforming

Plastics commonly used for thermoforming were listed earlier. Probably the most used thermoplastic sheet in the laboratory is **high-impact polystyrene**. It makes opaque objects, is inexpensive, and molds easily and quickly. **Polyethylene** forms easily but shrinks a lot. **Cellulose acetate** sheet, often known as acetate, is used for clear objects and forms very well. **Ionomer** (DuPont Surlyn "A") **film** makes very tough skin packages. It will stick to common cardboard when formed. **Polyethylene film** may also be used for skin packaging, but requires a special coated cardboard. **Acrylic** plastic is used for thick walled parts, either clear or opaque. It is stiff and forms better by pressure or mechanical stretch forming.

Several other thermoforming plastics are not often used in the schools, but may be. They are either higher priced, harder to thermoform, or not as easy to find. These include polypropylene, ABS, and polycarbonate. Polyvinyl chloride sheets should be avoided for thermoforming in the school. PVC may release styrene monomer and/or chlorine gases if it gets too hot.

Following are some suggested product ideas for several different kinds of thermoplastic sheet:

1. **High-Impact Polystyrene**
   Wall plaques
   Model boat hulls
   Margarine or butter dishes
   Coasters
   Candy dishes
   Decorative license plates
   Chip and dip dishes
   Thermoforming mold masters
2. **Cellulose Acetate**
   Clear covers for products
   Blister packages for mass production
   Painted wall plaques
3. **Polyethylene Sheet and Film**
   Polyester casting molds (sheet)
   Fiberglass forming molds (sheet)
   Skin packaging (film)
4. **Acrylic Sheet**
   Bowls
   Dishes
   Tray covers
   Desk sets
   Novelties
   Boat hulls
   Chip and dip trays
   Penholder bases

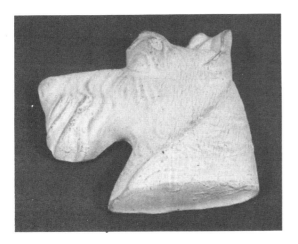

Fig. 8-51. Bisque-Fired Ceramic Thermoforming Mold

STUDENT ACTIVITIES

1. Thermoform several different plastic materials on each mold.

   Note: Prepare a log sheet for experimentation with the thermoforming process like the one shown in Appendix A, page 394. Save this record of your experimentation for later reference.

2. Materials experimentation:
   a. Vary the kind of materials on a given mold. Discover the differences in moldability and technique.
   b. Vary the material thickness. Notice the changes in the product made.

3. Process experimentation:
   a. Vary the heating time. Notice the changes in product quality. Notice the different sag in the plastic.
   b. Vary the heating speed by placing the heater at different distances from the material. Notice the changes in the product.
   c. Vary the speed of advance of the mold or plastic to the mold. Notice any change in the product.
   d. Vary the length of vacuum time. Notice what happens.

4. Design products that can be thermoformed. Consider the draft angle, undercuts, parting line, cavity size, material thickness, processing methods, and depth of draw.

   a. Drape formed products
   b. Cavity formed products
   c. Mechanical stretch forming
   d. Combination processes
   e. Free blown products

5. Design and construct molds for various products. Consider different types of mold making materials. Choose the mold material which is best suited for the situation.

   a. Wood
   b. Metal
   c. Gypsum cement
   d. Plastic
   e. Composition materials
   f. Ceramics
   g. Ready-made products
      1) Glass products
      2) Metal parts
      3) Plastic products
      4) Wood products
      5) Ceramics

6. Production organization and operation
   a. **Material supply:** consider —
      1) Sheet stock, standard sizes
      2) Roll stock, widths available
      3) Size of thermoformer
   b. **Cutting the stock to size**
      1) Paper cutter
      2) Circular saw
      3) Squaring shear
   c. **Clamping and heating:** heating time
   d. **Forming:** forming time
   e. **Cooling:** cooling time
   f. **Trimming**
      1) Method
      2) Time
   g. **Safety**
      1) Safety zones
      2) Safety guards
      3) Safety signs
   h. **Record-keeping**
      1) Materials used
      2) Process variables
         a) heating time
         b) cooling time
      3) Rejects
      4) Assembly and finishing
      5) Labor per unit

7. Build a small thermoformer from wood or metal. The heating element may be taken from a small home appliance.

8. Experiment with toy thermoformers. Compare the results with commercial laboratory thermoformers.

9. Look up material data in the **Modern Plastics Encyclopedia**.

Questions Relating to Materials and Process

1. Pick out thermoforming materials for the following products. Tell why you would choose the material. Is it price, strength, weathering, ease of processing appearance, etc. List the name of the plastic for each product to be thermoformed.
   a. Toy car body
   b. Fishing boat hull
   c. Model boat hull
   d. Desk organizer

e. Ice cube tray
   f. Telephone organizer
   g. Clear punch bowl
   h. Snack tray
   i. Small piece of luggage
   j. Briefcase
2. What thickness of material would you select as best for thermoforming the following products? Use three size ranges: less than .030" (0.76 mm)
   .030" to .080" (0.76 mm to 2.03 mm)
   more than .080" (2.03 mm)
   a. Plastic drinking glass (tumbler), tall
   b. Wall plaque
   c. Toy boat hull
   d. Chip and dip tray, shallow draw
   e. Snack tray
   f. Duck decoy sides, center to be filled with urethane foam.
   g. A desk organizer
   h. An ice cube tray

Tell why you chose a particular thickness for each item. Was price a factor in your choice of sheet thickness?

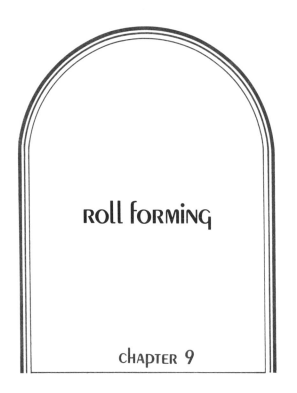

Roll Forming

Chapter 9

Roll forming is a group of processes in which the plastic is formed with the aid of rollers. The plastic is either formed into or onto a sheet. This group includes calendering, calender coating, knife (spread) coating, and extrusion coating.

Calendering — Process Description

Calendering is a process that is used to form plastic sheet or film from about .002" (0.05 mm) to .050" (1.27 mm) thick. It may also be used to put a plastic coating on textile fabrics, paper, synthetic leather, fiberglass fabrics, and foamed plastics. The finished plastic sheet or film surface may be either smooth or textured (embossed). The surface finish is controlled by the finish on the final rolls just before cooling.

Calendering of plastic is similar to the calendering of rubber and paper and the rolling of metal. It is a process of squeezing the material between two or more rollers to produce the desired thickness of product. It is similar to the old fashioned wringer on a clothes washing machine. The main difference between the calendering of plastics and paper and the rolling of metal is the type and position of the work rolls.

In a metal rolling mill, two small rolls are backed up by larger rolls. The metal is passed back and forth between the rolls until it is the right thickness, Fig. 9-1. The small work rolls reduce the rolling loads. The backup rolls make the mill more rigid.

In paper calendering, a series of rolls are used which are loaded from the bottom roll.

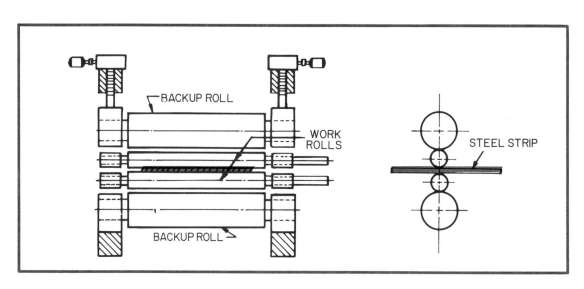

Fig. 9-1. Metal Rolling Mill

Roll Forming 179

The calender rolls put the final thickness and finish on the paper as it passes between them, Fig. 9-2.

Plastic calendering is similar to continuous extrusion except that rolls replace the barrel and screw of the extruder. It is an outgrowth of developments that originated in the rubber industry. In calendering, hot molten plastic is run between several large heated rotating rolls. The rolls squeeze the plastic into sheet or film. The thickness of the plastic is controlled by the distance between the rolls.

Unsupported plastic film and sheet are often calendered on an inverted "L" four-roll calender. See Figs. 9-3 and 9-4. The plastic is fed between the top pair of rolls (rolls 1 and 2). It follows roll 2, then is picked up by roll 3, and finally, is picked up by roll 4. The plastic is then removed from roll 4, cooled, and wound up on a spool. The plastic may also be **embossed** (textured) after it leaves roll 4. This is usually done between an engraved steel roll and a rubber backup roll.

The flow of plastic from roll to roll is controlled by either speed or temperature. The plastic will flow from the slower to the faster roll. It will also flow from the cooler to the hotter roll. In this way, the plastic can be made to flow as desired. This is shown in Figs. 9-4 and 9-5. The percentages shown

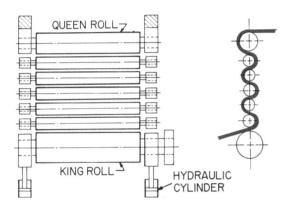

Fig. 9-2. Paper calender with hydraulic loading of bottom roll.

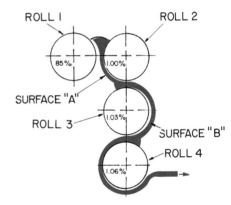

Fig. 9-4. Plastic web on a four-roll inverted "L" calender.

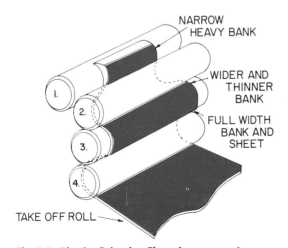

Fig. 9-3. Plastic Calender Flow Arrangement

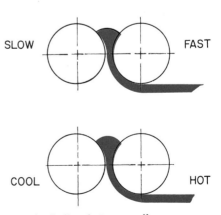

Fig. 9-5. Plastic flow between rolls.

on each roll in Fig. 9-4 indicate how speed might be varied to control the flow.

The differences in speed between rolls, as shown in Fig. 9-4, causes frictional heat in the plastic. The various rolls are internally temperature-controlled to maintain the proper processing temperatures. Excess heat is removed by the rolls. The heat, friction, and mixing action of the rolls blends and forms the plastic as it goes between them. The result is a uniform sheet of plastic.

Calendering plastic often requires more than just a calender, or system of rollers. The ingredients in the plastic mixture must first be mixed by a ribbon blender-type mixer. This is called **blending**. The plastic resin, plasticizer, stabilizer, fire retardants, pigments, and lubricants are well mixed in this first step. The temperature of the materials is not raised to any degree by this type of blending.

The blended plastic mixture is then placed in an intensive mixer where it is fluxed, or completely combined. Fluxing raises the temperature of the materials by frictional heat. The heat and pressure drive the materials into each other.

Both the blender and intensive mixer are batch mix units. The calender needs a continuous flow of materials to work. An extruder/strainer is often put between the batch mixers and the calender. This gives the calender a steady feed of material. A typical calendering system is shown in Fig. 9-7.

Importance of Calendering

Calendering blends, mixes, and forms plastic into sheet and film with less chance of heat degradation than in the extrusion process. This is one reason why calendering is the most widely used method of forming polyvinyl chloride (PVC). Because of its heat sensitivity, PVC degrades more easily

Fig. 9-6. Intensive mixer for fluxing.

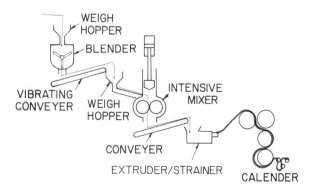

Fig. 9-7. A Calendering System

Fig. 9-8. Four-roll Inverted "L" Calender

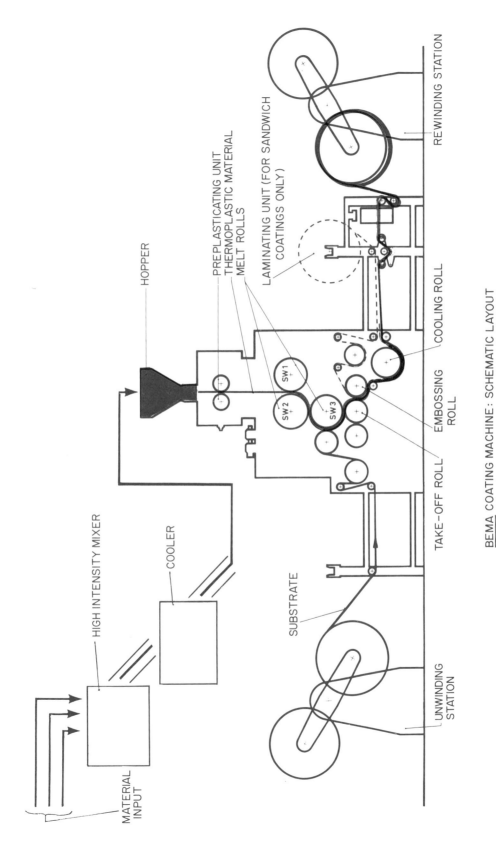

Fig. 9-9. Calender Coating System

in processing than many other thermoplastics. Using the calendering method of forming PVC relieves this problem. Other thermoplastic sheet and film are sometimes formed by the calendering method, but more often are extruded into unsupported film and sheet.

Advantages

Calendering provides a method by which heat-sensitive thermoplastics may be formed into sheet or film. Designs or patterns may be embossed directly into the sheet as it is being formed. Calendering permits the use of lower cost PVC materials than are used for extrusion. Both supported (backed) and unsupported (unbacked) sheet or film may be made by this process.

Disadvantages

A calendering system, such as the four-roll "L" calender, which requires the blender mixer and intensive mixer, should be used for long production runs of a given color and type of compound. Clean up between colors or compounds is time consuming. The range of thickness of sheet and film, compared to extrusion, is limited. Equipment costs for calendering are usually greater than for a similar capacity extruder. Also, calendering is limited to flat sheet or film products.

Materials Used for Calendering

Almost any thermoplastic can be calendered into film and sheet, but, as already stated, the most common is PVC. PVC requires more blending and mixing for particular applications so it is well suited for calendering. PVC is often custom-blended for each job right at the processing plant rather than purchased as preblended materials.

Industrial Equipment

Industrial calendering machines range in size from 18" (457 mm) diameter by 54" (1371 mm) wide rolls to 36" (914 mm) diameter by 120" (3048 mm) wide. Power required to drive the calender rolls varies from 75 horsepower (hp) on the small machines to 1000 hp on the largest. An intensive mixer for fluxing such as shown in Fig. 9-6 has a main drive motor of 600 hp. It drives the mixer up to 125 revolutions per minute (rpm).

Calender Coating — Process Description

Calender coating is a process for putting plastics, latex, and synthetic rubber materials on a backing material. The backing material is called a **substrate**. The final product is called a **supported material.** It is similar to calendering except that the plastic is applied to the substrate during the calendering operation.

A substrate such as cotton sheeting, woven fiberglass, burlap, paper, and nonwoven fabric is used. It is pulled through the calender as the plastic or other material is calendered down onto it. The plastic is melted in the calender rolls before it meets the substrate. Following the calendering operation, the material may be run between an embossing roll and backing roll. This will provide a textured surface if desired. An example of this process is shown in the diagram in Fig. 9-9, page 181.

A high intensity mixer and cooler, such as shown in Fig. 9-10, is often needed to blend the materials before placing them on the preplasticizing rolls. In the unit shown in Fig. 9-9, the preplasticizing rolls replace the intensive mixer and extruder/strainer often used with the inverted "L" calender unit. This makes a more compact production unit. It takes less floor space. Calender coating may also, however, be done on the conventional inverted "L" calender.

Importance of Calender Coating

Calender coating may be used to single-coat, double-coat, or laminate practically any substrate (backing material) that can be wound on a roll. It is one of the leading methods for coating such materials. Examples of products that are calender coated are crinkle-finish imitation leather, handbag fabrics, shoe tops, upholstery and cushion fabrics, lampshades, wall coverings, floor coverings, packaging materials, and credit card stock.

Roll Forming 183

Fig. 9-10. High Intensity Mixer and Cooler

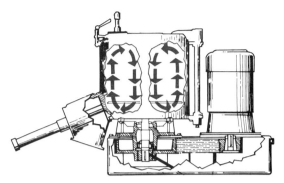

Fig. 9-11. Cross section of a high intensity mixer.

Fig. 9-12. Calender coated fabrics decorate the wall and cover the folding door and chair.

Advantages

Advantages of the preplasticizing calender coating system are shorter production runs, less cleanup, less costly equipment, and less floor space needed. It uses lower cost materials than plastisol knife coating. Less labor is needed because less equipment and material handling are necessary.

Disadvantages

The calender coating system shown will not produce as uniform a product as the "L" calender line. It needs longer production runs than the plastisol knife coating system, but can be shorter than the inverted "L" calender. Calender coating is limited in thickness from about .002" (0.05 mm) to .040" (1.02 mm) in one pass. Successive passes may be used to build up thicker sheets. Actual coating thicknesses depend upon the coating material used.

Materials Used for Calender Coating

The main material used for calender coating is dry blend rigid polyvinyl chloride (PVC). Plasticized PVC, low density polyethylene and high density polyethylene, polystyrene, polyurethane, polypropylene, and polyisobutylene may also be used. Latex and synthetic thermoplastic rubbers can be calender coated on a variety of materials. They include substrates such as cotton sheeting, woven fiberglass, burlap, paper, and nonwoven fabrics.

Industrial Equipment

Industrial calender coaters range in width from about 16" (400 mm) to over 86" (2200 mm). Large horsepower motors are needed to drive the rolls. They range from 12 to 100 hp. Electrical power for heating the rolls ranges from 25 kilowatts (kw) for 16" (400 mm) calenders to 380 kw for 86" (2184 mm) machines. Larger machines require motors and heaters proportionately higher in horsepower and kilowatts. Speeds of from 59' per minute (18 m/min.) to 160' per minute (50 m/min.) may be achieved.

Fig. 9-13. Calender Coating Machine

Knife Coating — Process Description

Knife coating is a method for coating paper or fabric substrates with liquid plastic materials. Vinyl plastisols, vinyl organisols, or plastic/water emulsions are usually used.

The substrate is unwound by pull rolls, tensioned and pulled under a doctor blade or knife which controls coating thickness. The coated substrate is then pulled through a heater unit to cure the coating. After the heat cure, an embossing roll may be used to put on a textured finish, or the product may also be left smooth. See Fig. 9-14.

The coated substrate is cooled by several cooling rolls. An accumulator takes up the slack in the material when a roll change on the winder is needed. Slack is taken care of by moving the bottom accumulator rolls up and down to make a longer or shorter distance between the winder and cooling rolls.

Importance of Knife Coating

Knife coating is one of the leading methods of making coated decorative wall coverings and coated fabrics for the home, business, schools, and industry. It produces a durable, washable, flexible material that will last for years. Folding door coverings, auto-

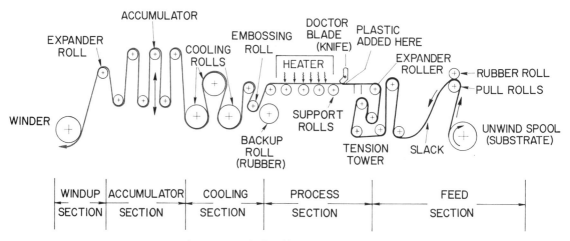

Fig. 9-14. Typical Knife Coating System

mobile and aircraft interior fabrics, and upholstery are typical examples.

Advantages

Short runs of different colors and formulations of coatings and substrates may be made in knife coating. Startup, shutdown, and cleanup time for this equipment takes little time. Coating thicknesses are limited to the .002" (0.05 mm) to .010" (0.25 mm) range.

Disadvantages

The thickness of coating is much more limited and raw materials are more expensive than for calender coating operations. Because knife coating is less automatic, labor costs are higher than for some other coating operations. Knife coating is more likely to generate scrap during startup and operation. The backing materials which can be used are limited because of the intense curing heat needed. Thermoplastic sheets cannot be used as a substrate as they are in calender coating. Also, the process is limited to the available widths of substrates which is often 60" (1524 mm) or less.

Industrial Equipment

Industrial knife coating machines usually produce coated fabrics from about 36" (914 mm) to 60" (1524 mm) wide. Roll lengths may vary from 500 yards (457 m) of heavyweight to 2000 yards (1830 m) of light-weight material. These machines do not require large motors to drive them. They do, however, need large or rapid curing ovens. These ovens are either gas or electrically heated and require substantial energy.

Materials Used for Knife Coating

The knife coating process is generally limited to the applications of vinyl plastisols, vinyl organisols, and latex paint type emulsions. Substrates are usually limited to paper, cotton, and burlap fibers.

Vinyl Fabric Casting

Vinyl fabrics may be made in the school laboratory from vinyl plastisols. They may be unsupported (no backing), supported (backed), or supported and foam backed. The same plastisol that is used for dip casting of coin purses in Chapter 13 on thermofusion may be used for this process. The plastisol may be colored as described in that chapter.

A special embossed release paper is available for producing a variety of patterns on the vinyl fabric. The release paper may be reused a number of times. Follow the pro-

Fig. 9-15. Typical Knife Coating Line

cedure below for producing a cast embossed vinyl fabric which may be used to cover chairs, foot stools, and other objects.

Operation Sequence for Vinyl Fabrics

A cast vinyl fabric may be made in one of three ways:

Unsupported Fabric

1. Place the special embossed release paper on a cookie sheet or on a special holder which is available from the release paper supplier.

2. Pour a thin coat of vinyl plastisol on the embossed release paper, Fig. 9-16. Spread it out to a uniform layer with a straight edge.
3. Place the sheet with the vinyl coat in an oven at 210°F (99°C) for two minutes to pre-gel the vinyl.

---**Safety Note**---
Use insulated gloves in steps 3, 4, and 5.

4. Place the sheet in an oven at 375°F (190°C) for two minutes to cure the vinyl.
5. Remove the cured material from the oven. The cured vinyl may be cooled and removed from the release paper. Do not remove the release paper if cloth is to be added.

Supported Fabric (Steps 1-7)

6. A backing may be added if increased strength is desired. Cotton cloth may be placed over the cured plastisol and a thin layer of plastisol added. Place the material back in the oven for three to four minutes at 375°F (190°C).
7. If you do not wish to add a foam base layer, cool and peel the cured plastisol from release paper, Fig. 9-17. If a foam base is desired, do not peel from the release paper yet.

Fig. 9-16. Pouring vinyl plastisol on embossed release paper for vinyl fabric casting.

Fig. 9-17. Removing cast vinyl fabric from release paper.

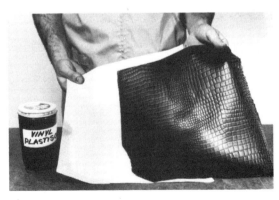

Fig. 9-18. Vinyl Plastisol, Release Paper, and Cast Embossed Vinyl Fabric.

Foam-Backed Fabric (Steps 1-10)

8. To add a foam base, mix the vinyl foam and blowing agent according to the manufacturer's instructions. Spread the vinyl foam material over the cured vinyl sheet.
9. Put the sheet back in the oven for two to four minutes at 375°F (190°C). The vinyl foam backing will expand about four times (400%) its original volume.
10. Cool and peel from the release paper, Fig. 9-18. Don't throw the release paper away; it can be used over and over again.

Extrusion Coating — Process Description

Extrusion coating is a combination of extrusion and roll forming operations. In this process, a thin extruded film of molten plastic is pressed onto or into a substrate. The plastic film is extruded downward from a flat film die. It is basically the same type of film die described in "Sheet and Film Extrusion," Chapter 5. A diagram of the extrusion coating process is shown in Fig. 9-19.

The substrate is unrolled and run through a series of rolls. It is brought between the pressure roll and chill roll directly below the film die. The plastic resin joins the substrate between these rolls. The substrate may either be heated or unheated before the extruded plastic is applied. The coated product is then cooled (if necessary), treated, slit, and wound up.

Extrusion may also be used to laminate two or more substrates together such as paper and aluminum foil. In this related process, called **extrusion laminating**, the substrates are fed into the extrusion coater and laminated together with the plastic film in between them.

Importance of Extrusion Coating

Extrusion coating has made a wide range of packaging and other sheet materials avail-

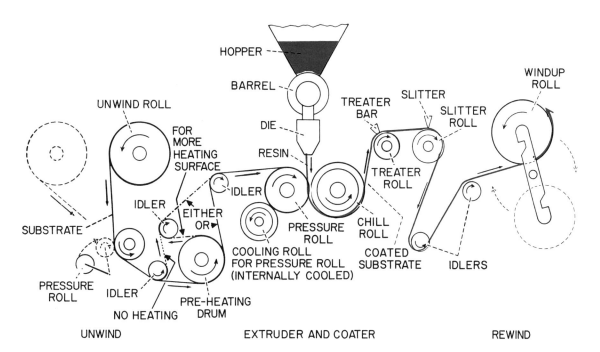

Fig. 9-19. Schematic cross section of the extrusion coating setup, with unwind and rewind equipment.

able. One such package is the polyethylene coated milk carton which originally was coated with wax, Fig. 9-20. Other applications for extrusion coating are bags for shipping food, fertilizer, chemicals, and plastic resins. Pouches for liquid and frozen foods, vacuum forming materials for blister packaging, and inexpensive luggage are other uses.

Advantages

Extrusion coating combines the best properties of both the substrate and the coating material. Polyethylene coated substrates are easy to heat-seal, stronger, and capable of keeping out oil, grease, and moisture. They remain flexible at refrigerator temperatures, can be printed, and have good surface gloss. Extrusion coating is a low cost method of coating. One pound of polyethylene, for instance, will coat about 30,000 square inches (193 548 cm^2) of substrate at a .001″ (0.0254 mm) thickness.

Disadvantages

Some adhesion problems exist in extrusion coating. The plastic sometimes does not stick to the substrate as well as it should. This is caused by oxidation of the hot plastic web between the die and its contact with the substrate. Air may also be trapped between the plastic and the substrate in extrusion coating. When these problems occur, they result in a poor quality product.

Also, extrusion coating equipment costs are high. To make best use of the equipment, extrusion coating should be done in long production runs at high speeds of 500 to 1500 feet/ minute (152 to 458 m/min).

Fig. 9-20. A paper milk carton is a typical extrusion coated product. Millions are used each year.

Fig. 9-21. Extrusion Coating Line

Materials Used for Extrusion Coating

Polyethylene is probably the most common plastic being extrusion coated today. Others include ethylene-vinyl acetate, copolymers, ionomers, and polypropylene. Substrates such as paper, paperboard, glassine (a thin, translucent paper), cellophane, polyester film, metal foil, glass fiber, and others are being used.

Industrial Equipment

Equipment necessary for an extrusion coating line includes a regular extruder, film die, and an extrusion coater attachment with wind and unwind equipment. Common extruder sizes and coating die widths are shown in Table 9-1.

TABLE 9-1
Common U.S. Extruder Sizes and Die Widths for Extrusion Coating

Extruder Size	Extrusion Coating Die Width
3½″ (89 mm)	24″ to 48″ (610 to 1220 mm)
4½″ (114 mm)	36″ to 90″ (915 to 2290 mm)
6″ (152 mm)	54″ to 140″ (1370 to 3555 mm)
8″ (203 mm)	to 160″ (4065 mm)

Fig. 9-22. Closeup of the film meeting the substrate in the extrusion coating process.

STUDENT ACTIVITIES

1. Collect samples of calendered, calender coated, knife coated, and extrusion coated materials.
2. Compare the samples of calendered, calender coated, knife coated, and extrusion coated materials.
3. Roll Play-Doh between two rollers to simulate calendering. Run a piece of paper or other substrate (backing material) through with the Play-Doh to simulate calender coating.
4. Make fabrics using vinyl plastisols and embossed release paper.

Questions Relating to Materials and Process

1. How is calendering related to extrusion?
2. What is the most used plastic material in calendering and calender coating?
3. Of the two, calendering and extrusion, which processes the greater amount of plastic material into sheet and film?
4. Why is an extruder/strainer necessary in many calendering systems?
5. There are four roll-forming systems in this chapter. Which system requires the smallest drive motors?
6. What is the difference in the plastic materials used in extrusion coating as compared to those used in calendering?

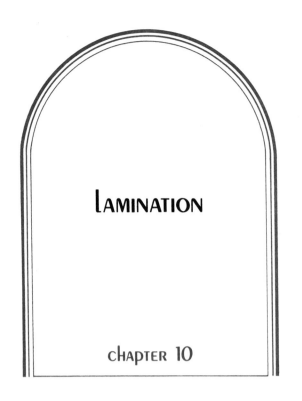

LAMINATION

CHAPTER 10

Fiberglass chair seats and car body parts which have been compressed at over 1000 psi (6900 kPa) are examples of high pressure fiberglass lamination. Formica® and Micarta® table-top materials are high pressure laminates. Colorful nameplate engraving materials similar to Formica are also high pressure laminates. Other examples are paper/phenolic and fabric/phenolic laminates which are used for circuit boards, automobile engine timing gears, and bearing spacers. They are available in sheets, rods, and tubes as well as made into parts.

Three types of low pressure lamination — the kinds most likely to be used in the school laboratory — are discussed in this chapter. They are (1) fiberglass reinforced plastics, (2) thermoplastic lamination, and (3) compreg lamination.

Lamination is a group of processes in which layers of materials are formed into sheets or parts. Often they are pressed together under pressure. Plywood is an example of wood laminated material. Plastic lamination involves the bonding of layers of plastics or plastics that are combined with other materials. Lamination is divided into two groups:

1. Low pressure lamination and
2. High pressure lamination.

Low pressure laminates are formed at from 0 to 1000 psi (6900 kPa) and high pressure laminates at pressures over 1000 psi.

Low pressure lamination includes most fiberglass reinforced plastic products and thermoplastic sheet lamination. Contact-molded fiberglass parts, such as boat hulls and large containers, are examples of low pressure fiberglass lamination. Thermoplastic lamination of photos and identification cards is another. Compreg (plastic filled wood) molding is also a low pressure laminating technique.

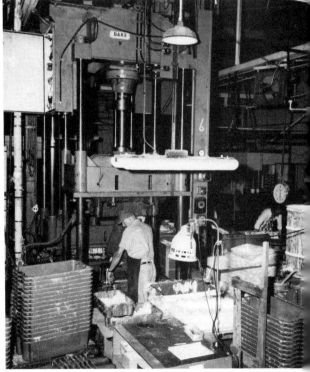

Fig. 10-1. A 100-ton (90 Mt) fiberglass molding press that produces tote boxes.

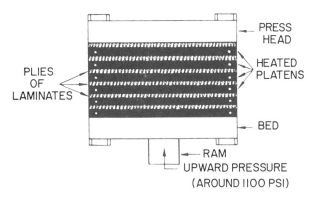

Fig. 10-2. High Pressure Lamination

Reinforced Plastics — Process Description

Reinforcing is a lamination process in which fibers are added to plastic materials to form a stronger product. Glass fibers are usually used as the reinforcing fibers. Polyester plastic resin is the most common plastic component. Together, the glass fibers and polyester resin form what is known as fiberglass reinforced plastic laminates or fiberglass reinforced plastics (FRP). Actually, glass fibers can be combined with a number of thermoplastic and thermosetting plastics. Also, fibers other than glass may be used as reinforcements. Boron filament and nylon strands are used in some cases.

The glass fibers reinforce (strengthen) the plastic materials much the same as steel rods strengthen concrete. The plastic material binds the strong, tough fibers together. It makes the product rigid, acting like the cement in concrete. The best qualities of the two are combined — the strength of the glass and the rigidity of the plastic. Together, they form a new material.

Glass fibers are added to plastic when the plastic is in a liquid state. The fibers may also be added to thermoplastics during their manufacture. Glass-filled thermoplastic pellets may then be injection molded into reinforced products. Parts such as automobile fans, fan shrouds, heater ducts, and under fender liners are often reinforced. Nylon, polycarbonate, and polypropylene are the thermoplastics most often reinforced.

Fig. 10-3. High Pressure Lamination Press

Fig. 10-4. Laminated Kayak

Glass fibers are also mixed with many thermosetting plastics during their manufacture. They are sold to molders as glass-filled thermosetting plastics for high strength and electrical insulation uses. Phenolic, melamine, urea, and DAP materials are among those processed this way. They are compression molded or transfer molded into many useful products.

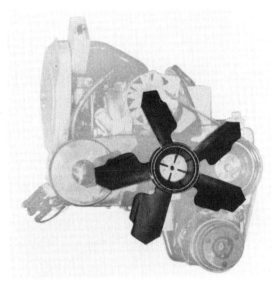

Fig. 10-5. Glass reinforced nylon fan blade for Chevrolet Vega.

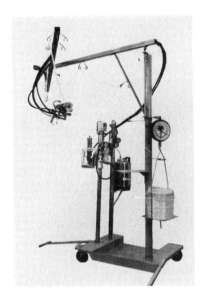

Fig. 10-7. Fiberglass Chopper Outfit

Fig. 10-6. Glass-filled phenolic part for a vacuum cleaner.

Fiberglass Reinforced Plastic Lamination

Fiberglass reinforced laminates are made by adding the glass fibers to the liquid thermosetting plastic in or on the mold. They may be combined by **hand lay-up** or **machine lay-up**. In hand lay-up, the glass materials and polyester resin are put on the mold by hand. The glass materials are most often cut from prepared mat or cloth to the size needed. The plastic resin is brushed on with a paintbrush. Hand lay-up is used both in industry and in the school laboratory. It will be explained in more detail later in this chapter.

Machine lay-up requires a fiberglass chopper/sprayer similar to the one shown in Fig. 10-7. It chops long continuous strands of fiberglass, called **rovings**, into short glass fibers. The machine also sprays the polyester resin onto the glass fibers as the fibers are being sprayed onto the mold, Fig. 10-8. The polyester resin is catalyzed as it is sprayed. The glass fibers and polyester resin are built up on the mold until the right thickness is reached. Heat caused by the catalyst cures the polyester resin, and the product becomes rigid. After it is cured, it is removed from the mold. Boat hulls and one piece bathtub and shower units are made this way.

Polyester resins may also be poured over preformed glass fiber mats on the mold.

Lamination 193

Fig. 10-8. Closeup of fiberglass chopper/resin sprayer head.

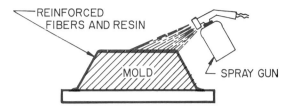

Fig. 10-9. Spray-Up Process

Fig. 10-10. The hull of this fishing boat is fiberglass.

Premixing is another method of combining glass fibers and resin for molding. Both are compression molded. They are explained under "Industrial Equipment" in this chapter.

Epoxy plastic resins are combined with glass fibers in much the same way as described earlier. They form very high-strength laminates. Epoxy/glass reinforcements are more expensive than polyester/glass materials. Epoxy/glass laminates are used for high-strength aircraft parts and industrial stamping dies. New boron fiber/epoxy combinations are extremely light and strong, but they are very expensive. They are used in space rocket and jet engine parts.

Importance of Reinforced Plastics Process

How the fiberglass reinforced plastics process ranks in the industry is difficult to determine. However, fiberglass reinforced products, such as boats and shower stalls, when considered alone, probably make up 5% or more of the plastic industry. This lamination process promises to continue to grow as new applications in the construction and transportation industry are developed.

Fiberglass provides rigid, hard-surfaced, strong, lightweight products for the consumer and military markets. Few plastics processes can provide another way of making some of the things now being produced with fiberglass.

Advantages

Fiberglass reinforced plastics make a much stronger product than many plastics without this reinforcement. The products are very durable and are resistant to weather, chemicals, and electricity. Usually, they require less expensive molds than plastics in other processes. The molds may be made from fiberglass, plaster, wood, or metal. Although no pressure is required for molding, better products can be made in most cases by pressure bag or matched die compression molding. There seems to be no limit to the size of the fiberglass reinforced plastic parts. See Fig. 10-10. Also, color can be permanently molded into fiberglass products, and a very smooth, hard, shiny surface is possible.

Disadvantages

Production of fiberglass products may be slower than other plastic processing methods. Sometimes only one part can be made in a mold in a day. Therefore, a number of molds are needed to speed up production of such large parts as boat hulls. Higher speed processing of small parts is done with matched die compression molding. Compression molding cycle times for fiberglass are similar to those for phenolic molding.

Fiberglass materials are messy to handle and present a cleanup problem in the plant. The glass fibers are very fine and fly around the plant. Because the polyester resin is very sticky, workers find it hard to clean off themselves and the tools.

Industrial Equipment

Several types of processing equipment are described here. Each type is often used alone to produce a product, but in some cases may be used in combination with another.

A fiberglass **chopper/spray gun combination** is widely used in the industry. It is low in cost as compared to equipment used in other fiberglass production methods. See Fig. 10-7. It meters and sprays the right amount of plastic resin and catalyst as it chops and sprays the glass fibers. Resin, catalyst, and fibers all converge (meet) on the mold at one time. The chopper/spray gun provides a fast way of building up fiberglass and polyester resin on the mold. It works well on large or small curved, flat, or uneven surfaces.

Another method, called **premix,** uses a mixing machine like the one in Fig. 10-11. The doughlike mixture of glass fibers and polyester resin is later placed in a compression mold and processed under heat and pressure to make the products. A catalyst is usually not needed in this process. Premixing is used for mass production of small to medium-sized parts like front-end assemblies and taillight housings for cars.

Performing is a way in which formed fiberglass mats are made to fit a given mold. They are made in a plenum chamber like the one shown in Fig. 10-14. Glass fibers and a resin binder are sprayed over a screenlike form. A fan pulls air through the

Fig. 10-11. Polyester Glass Premixer and Log Extruder

Fig. 10-12. Polyester/Glass Premixer

screen, and the glass fibers take the shape of the screen form. The preform is often put in an oven to dry the resin binder which holds the fibers together. The preform is then placed in a compression mold where the plastic resin is added. It is cured under heat and pressure, Figs. 10-14 and 10-15. Preforms are used for fast production of chair seats, auto body parts, suitcase sides, and similar products.

Filament winding is a method of making super strong fiberglass products such as high-pressure space rocket fuel tanks, light-

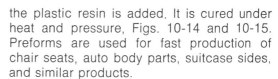

Fig. 10-13. Premix logs ready for compression molding.

Fig. 10-15. Fiberglass preform being made in plenum chamber.

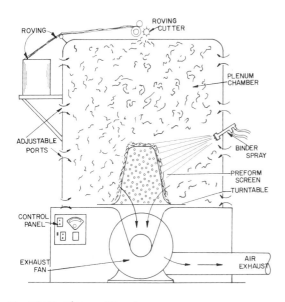

Fig. 10-14. Plenum Chamber

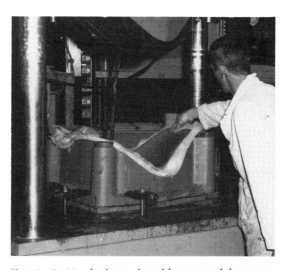

Fig. 10-16. Matched metal molds are used for compression lamination of fiberglass chairs in a 150-ton (136 Mt) press.

weight pipe, and barrels. Glass filaments, which have been saturated with polyester or epoxy plastic resin, are wound on a spool or core. The winder moves the filaments back and forth rapidly over the length of the spool or core, causing a crisscross pattern, Figs. 10-17 and 18. When the resin hardens, a very strong part is formed. The spool or core may or may not be removed from the cured part, depending on its shape. Shapes from straight pipe to ball-shaped tanks may be made by this method.

Materials Used in Industry

Glass fibers and polyester resin are the two basic materials used in fiberglass reinforced plastics. Several other items such as color pigments, fillers, thickeners, surfacing agents and catalysts are also used. Glass fibers, color pigments, plastic resins, and catalysts will be considered here.

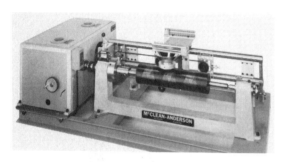

Fig. 10-17. Laboratory FRP Filament Winding Machine

Fig. 10-18. Filament Wound Tanks

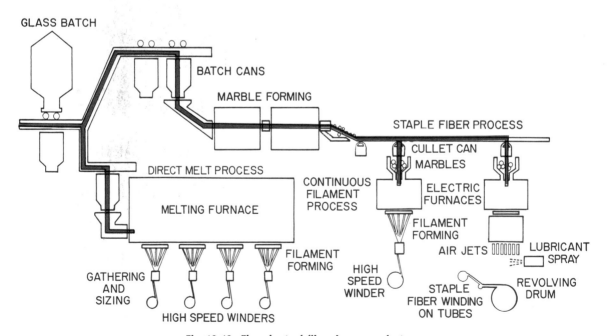

Fig. 10-19. Flowchart of fiberglass manufacture.

The **glass fibers** are manufactured from melted glass. The glass is pulled through very small holes at the bottom of the glass melting furnace. They are stretched as they are pulled out, making them finer. Ninety-three miles (150 km) of glass fiber measuring .00023" (0.0058 mm) in diameter can be pulled from the glass in a 5/8" (15.8 mm) marble. The fibers are wound on spools by a high-speed winder. A number of these fine glass fibers are twisted together to form glass threads which are made into glass yarn, cloth, roving, or mat. These forms of glass materials are molded into many useful fiberglass products. A flowchart on fiberglass production is shown in Fig. 10-19.

A. The complete unit.

B. Closeup of the top of autoclave.

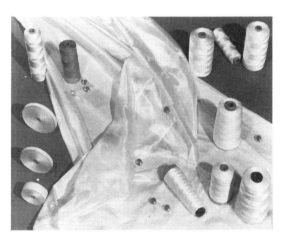

Fig. 10-20. Several glass materials used for laminating.

Fig. 10-21. Batch-type autoclave for the manufacture of polyester resins.

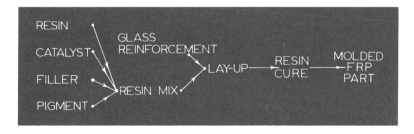

Fig. 10-22. Flowchart of fiberglass lay-up.

The polyester plastic resins are thermosetting **prepolymers**. The polyester resins are made in large **autoclaves**, or "pressure cookers," Fig. 10-21. The blend of basic materials is cooked and stirred under heat and pressure until it begins to thicken. The polymerization process is stopped at this point, and the polyester is sold as a half-cooked resin called a **prepolymer**. An **inhibitor** (chemical that stops a chemical reaction) is added to keep the prepolymer from setting up by itself. The resin is then in the form of parallel polymer chains that are not crosslinked. For the resin to become a crosslinked thermosetting plastic, a catalyst is added just before the resin is molded. The catalyst reacts with a promotor in the resin to cause internal (exothermic) heat which finishes the polymerization.

The **catalyst** that is usually used for polyester resins is MEK peroxide (methyl-ethyl-ketone peroxide). It contains about 11% active or "free" oxygen. The oxygen **oxidizes** (combines with) the small amount of highly reactive **cobalt** metal flakes in the cobalt naphthenate **promotor**. When the cobalt combines with the oxygen in the MEK peroxide, a lot of heat is given off by the chemical reaction (oxidation). The cobalt naphthenate promotor is usually put in the polyester resin at the factory. It is activated by the addition of the catalyst just before molding takes place. This promotor gives the plastic resin a purple or pink color.

Two types of polyester resins are used: (1) air-inhibited and (2) non-air-inhibited. Air-inhibited resins do not normally cure on the surface open to air. Contact with the air keeps the surface from curing, or polymerizing. Gel coat resins are usually air-inhibited so that the laminating layers will stick to them. Gel coats keep the glass fibers away from the surface of the laminate. Non-air-inhibited resins contain a liquid wax additive. The wax rises to the surface of the resin during cure and keeps air away from the resin. The resin then cures all the way through. Laminating resins are usually non-air-inhibited. They are designed to hold the glass fibers together.

If non-air-inhibited laminating resins are allowed to cure, the wax must be taken off **before another laminating layer can be put on.** The wax will keep another layer from sticking. The wax may be taken off with coarse steel wool, acetone, or lacquer thinner. Steel wool and acetone work well.

Many polyester laminating resins have **thixotropic** materials added to them. Thixotropic materials are those that raise the viscosity of, or thicken, the resin so it will not flow as easily. This helps prevent the resin from sagging easily on vertical surfaces. Resins without thixotropic materials in them are called **nonthixotropic**.

Polyester resins and MEK catalysts should be refrigerated for longer shelf life.

Safety Note

Do not store MEK peroxide at a temperature below 40°F (4.5°C) as it may crystallize.

Do not store food in the same refrigerator. It will cause the food to taste of the peroxide and may cause illness.

Color pigments are colored pastelike materials that are made for use only with polyester resins. Care should be taken in selecting only the right kinds of colorants for polyester resins as others may keep the resins from curing.

Laboratory Fiberglass Laminating Equipment

Very little special equipment is needed for fiberglass work in the school laboratory. Most of the equipment needed is usually available in other processing areas, mainly woodworking. Equipment such as sanders, saws, lathes, and other power tools may be needed for moldmaking. A ½ hp, 1725 rpm double-shaft buffer with 6" to 8" (152 to 203 mm) loose-sewn buffing wheels should be used for buffing fiberglass products. Acrylic buffing compounds may be used.

Laboratory fiberglass choppers are available. They are either electric or air motor driven, and do not have resin spray guns

Lamination 199

attached to them. If a lot of large fiberglass products such as boats or canoes are made, a commercial fiberglass chopper/resin spray system may be needed. These choppers will all work well on molds that have sharp curves.

Fiberglass work should be done in a well ventilated room. If spray equipment is used, a special spray booth should be installed to exhaust the overspray.

---- **Safety Note** ----
Be sure to have adequate ventilation whenever you are using polyester resins.

Fiberglass Laminating in the Laboratory

Much of the fiberglass laminating that is done in the school laboratory is by hand lay-up contact molding. A simple contact molded product that is useful to many people is a hot pad or trivet, which protects tables and other surfaces. A clipboard may be made by the same procedure.

A flat mold for these products is very simple to make. Cut one or two pieces of high pressure laminate (Formica) from a sink top cutout or other source to about the size of the product wanted, Fig. 10-24. A

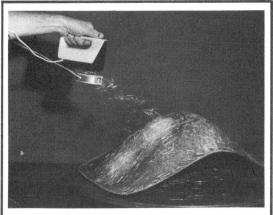

A. Laboratory FRP Chopper.

B. Air-powered industrial grade.

C. Electric-powered industrial grade.

Fig. 10-23. Fiberglass Choppers

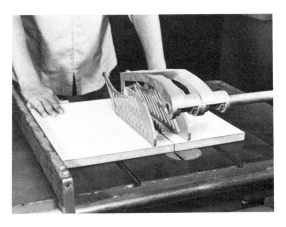

Fig. 10-24. Cutting Formica® molds for flat laminated FRP products.

single mold, which will produce one smooth surface, may be used for contact molding. If both surfaces are to be smooth, cut two molds. Use the second mold for a top mold, and press the two together with the laminate in between. Use some cement blocks, weights, or a press for pressure.

Coat the Formica mold surfaces with several coats of carnauba (Trewax® is one) paste wax or a paste-type fiberglass mold release. Polish surfaces lightly to a high lustre with a soft clean cloth after each coat of wax.

Other molds for fiberglass may be made in the laboratory or purchased from several suppliers. Molds for such products as chair seats, tables, cycle helmets, and trays are available. Additional information about fiberglass work and mold making may be found in **Fiberglass: Projects and Procedures.**[1]

Molds for square tanks, sinks, and similar objects may be built up from pieces of mounted Formica laminate. The parts of the mold may be put together with wood screws and the corners sealed with a small fillet of

---

[1] Authored by Gerald L. Steele and published by McKnight Publishing Company, 1962.

Fig. 10-25. Several molds available for FRP lamination.

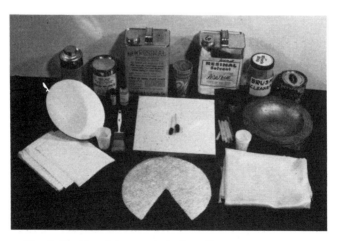

Fig. 10-26. Materials needed for fiberglass lamination.

silicone RTV bathtub sealer. No draft angle is needed and the mold may be taken apart to remove the laminated part.

Thermoformed polyethylene molds also may be used. These do not need mold release because polyethylene is naturally waxy. Large thermoformed molds may need to be backed up with plaster so they will hold their shapes better. A number of these are available from hobby shops. This type of mold is described in more detail in Chapter 11.

Glass fibers applied with a glass chopper should be put on in thin layers. Polyester resin should be applied between each layer. Stipple the resin into the glass fibers with an inexpensive natural bristle brush and apply more chopped fibers. Build up these layers until the laminate is thick enough.

Operation Sequence for Fiberglass Lamination

There are six steps in the fiberglass lamination process:

1. Get the mold(s) ready.
2. Get the fiberglass materials ready.
3. Measure, mix, apply, and cure the gel coat.
4. Measure, mix, and apply the polyester resin and glass materials.
5. Cure the laminate and remove it from the mold.
6. Trim the laminate.

The laminating sequence is shown in Table 10-1. Also, **read the resin manufacturer's instructions very carefully. Check the amount of catalyst recommended.** Make sure to cover the benchtops before you start. Wear protective clothing and disposable plastic gloves that will protect the hands from the plastic resin and glass fibers. Assemble all the materials needed before you start.

Safety Note
Use polyester resin only in a well ventilated room.

Checklist of Materials and Equipment Needed

Assemble the following materials and equipment for fiberglass lamination:
1. Polyester resin.
   a. Gel coat (air-inhibited).
   b. Laminating resin (non-air-inhibited).
2. Catalyst.
3. Paper cups or other disposable measuring containers.
4. Color pigments (optional).
5. Eye droppers for measuring catalyst.
6. Mixing sticks or coffee stirrers.
7. Molds.
8. Mold release or carnauba wax (Trewax is one kind).
9. Acetone or lacquer thinner for cleanup.
10. Cheap natural bristle brushes. (Soak the paint off the handles with acetone first.)
11. Plenty of wiping rags.
12. Several cans to clean brushes in.
13. Bench covers.
14. Protective clothes.
15. Plastic gloves (disposable).
16. Glass mat, cloth or roving.
17. Styrene monomer to thin resin, if needed.

TABLE 10-1
Operation Sequence for Fiberglass Lamination

Sequence/Action	Reason	Troubleshooting
❶ Prepare the mold(s): Apply mold release to the mold, Fig. 10-27. (a) Paste type — rub on and polish out two coats. Use a soft cloth. (b) Film forming (PVA) — brush, dip, spray, or rub on a thin coat. Allow to dry completely.	(a) Gives a fine detailed finish. Use for smooth shiny surfaced molds. (b) Gives a better release but tends to collect fine particles of dirt when drying. Use for more difficult release situations.	Part sticks to the mold: (a) Mold release not properly applied. (b) Not enough mold release. (c) Mold release burned out by overheated lamination. Use less catalyst.
❷ Prepare the fiberglass material, Fig. 10-28. (a) Glass cloth and mat — cut to shape about 1" (25.4 mm) larger than the mold. Cut with a tin snips, large scissors, or sharp knife. Use about two layers of mat for the hot pad or clipboard. (b) Decorative cloth or mat — cut to the size of the glass material.	(a) To allow enough extra material for shifting, etc. (b) For a decorative surface treatment — use decorative mat, colored burlap, or cotton print cloth. Iron the cotton cloth before use.	
❸ Apply the gel coat: (a) Measure the gel coat resin. It will take about one fluid ounce per square foot (0.03 ml per cm^2). (b) Add a color pigment **only if** a solid color is wanted. (c) Measure and mix in 8 to 12 drops of MEK peroxide catalyst **per ounce** (0.27 to 0.40 drops per ml) of resin, Fig. 10-29. Mix well. If cobalt promotor must be added, use a separate eye dropper. CAUTION: **Do not mix cobalt and catalyst directly together. An explosion may result.** (d) Brush on the gel coat to an even .025" to 0.35" (0.63 mm to 0.89 mm) layer (about the thickness of tablet-back cardboard), Fig. 10-30. Allow the gel coat to **cure** about 1 to 1½ hours. (e) **Clean the brush in acetone or lacquer thinner.**	(a) This will cover the mold about .025" to .035" thick. (b) Pigments are normally opaque. **If decorative cloth is used do not use color pigments in the** gel coat. (c) 2% catalyst is normal for gel coats. 8 drops per fluid ounce (0.27 drops/ml) = 2% catalyst. (d) The gel coat keeps the cloth or glass materials away from the surface. It gives a good surface finish to the product. (e) **Cured** resin cannot be cleaned from the brush.	Resin does not cure: (a) Catalyst too old. Use fresh catalyst. Store it in refrigerator. (b) Not enough catalyst used. Use more next time. Thin layers need more catalyst. Part delaminates: Surfacing agent not removed between cured coats.

(continued on page 204)

Lamination 203

Fig. 10-27. Putting on the mold release.

Fig. 10-28. Materials ready to use for FRP lamination project.

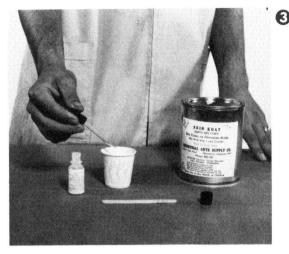

Fig. 10-29. Mixing the catalyst.

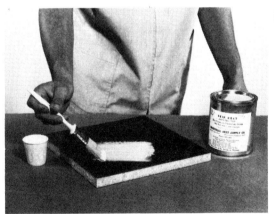

Fig. 10-30. Brushing on the gel coat.

TABLE 10-1 (Cont.)

Sequence/Action	Reason	Troubleshooting
④ Apply the lamination layer: (a) Measure out 8 ounces (0.25 ml per cm²) of laminating resin per square foot of mold area. Mix in thoroughly 32 to 48 drops of MEK catalyst per 8 ounces (236 ml) resin or 4 to 6 drops per ounce (0.13 to 0.20 drops per ml) of resin. (b) Brush a generous coat of resin over the cured gel coat. Add the decoration or glass materials to the fresh resin, Fig. 10-31. (c) Add decoration at this point (optional). (1) Decorative mat. (2) Cotton print cloth **face down.** Brush resin over the print cloth. (3) Metallic stars, metal flakes, or other decorative materials. (4) Color pigment in resin (optional). (d) Add fiberglass mat (or other glass material) at this point. Lay them flat. Brush or stipple resin into the glass material until all air (whiteness) disappears, Fig. 10-32. **Clean the brush.** (e) A second lamination layer may be added to the fresh resin or later after it is cured. **Remove any surfacing agent (wax)** if the first layer is cured. Glass cloth, mat, or chopped fibers may be used for this and/or the other lamination layers. **Clean the brush.** (f) A second mold may be placed over the back of the lamination layers. It should have a cured gel coat on it, also. Weights can be put on it to compress the laminate, Fig. 10-33, or the molds may be put in a press.	(a) 4 drops/ounce (0.13 drops/ml) = 1% catalyst 6 drops/ounce (0.20 drops/ml) = 1½% catalyst 8 drops/ounce (0.27 drops/ml) = 2% catalyst **Check manufacturer's instructions.** (b) Form base for the lamination layer. (c) To give the surface a decoration if desired. Disregard the decoration if an opaque pigment is used in the gel coat. (d) This is the reinforcement layer. Trapped air weakens the product and spoils the looks. (e) The second lamination layer gives added strength. More layers can be added if necessary. (f) The second mold gives a smooth surface to the back.	Resin does not cure: (a) Humidity too high. Use more catalyst. (b) Room too cold. Warm up the room OR use more catalyst. (c) Paintbrush not cleaned well — acetone or lacquer thinner still in the brush. Clean and wipe the brush dry. (d) Add a small amount (about 1 drop per fluid ounce (0.03 drops/ml) of polyester resin) of cobalt promoter to the resin. This will help cure resin that has been stored a long time. Resin cures too fast: (a) Room too warm. (b) Too much catalyst. **Thick layers need less catalyst.**

(continued on page 206)

Lamination 205

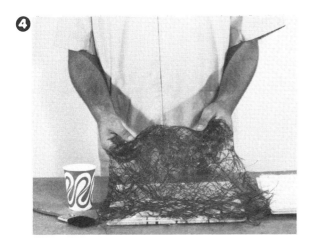

Fig. 10-31. Laying on the decoration.

Fig. 10-33. Double flat mold with cement block weight.

Fig. 10-32. Saturating the mat with resin.

Section III FORMING AND LAMINATION

TABLE 10-1 (Cont.)

Sequence/Action	Reason	Troubleshooting
5 Cure, remove from mold and trim: (a) Curing should take 1 to 3 hours. (b) Pry the part from the mold with a stick, Fig. 10-34, or give a few light taps with a **rubber mallet** to loosen the part. See Fig. 10-35. (c) Trim the part with a jig saw, hacksaw, or metal cutting band saw, Fig. 10-36. Tin snips can be used for thin laminates. (d) File and/or sand the edges, Figs. 10-37 and 10-38. Buff to a fine finish.	(a) Curing too fast will crack the resin. (b) Care should be used to not damage the mold. (c) Fiberglass will ruin the teeth on a wood cutting band saw. (d) To give that "final" finish.	A metal cutting band saw should be run slowly to keep from ruining the blade.

For further information about fiberglass lamination, see **Fiber Glass: Projects and Procedures** by Gerald L. Steele, published by McKnight Publishing Company, 1962.

Fig. 10-34. Opening the flat double mold.

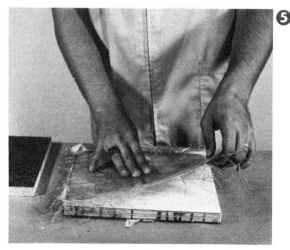

Fig. 10-35. Removing the part from the mold.

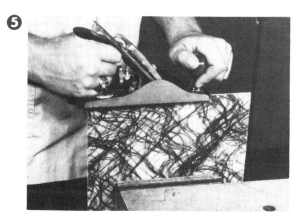

Fig. 10-36. Planing the edges.

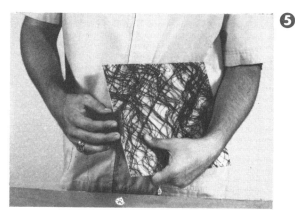

Fig. 10-38. Sanding the edges.

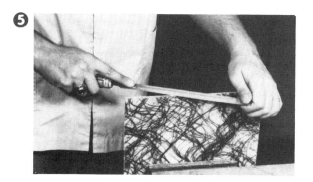

Fig. 10-37. Filing the edges.

```
——  TWO BLOTTER PADS  ——
      ONE POLISH PLATE
      ONE VINYL SHEET
    PHOTOGRAPH OR ARTICLE
      ONE VINYL SHEET
      ONE POLISH PLATE
——  TWO BLOTTER PADS  ——
```

Fig. 10-39. Assembly of a thermoplastic lamination sandwich.

Thermoplastic Lamination — Process Description

Thermoplastic lamination is a process in which a photograph, identification card, leaf specimen, or a similar flat item may be sandwiched between two or more thermoplastic sheets. The object and plastic sheets are combined under heat and pressure into a single product called a **laminate**. The heat causes the plastic to soften and the pressure bonds it tightly together. An invisible **cohesive bond** is formed. See Chapter 14 for an explanation of different types of bonding.

After a short heating time at a low pressure to soften the plastic, the pressure is increased to form a tight bond. The "sandwich" is then cooled **under pressure** to harden the plastic and keep it flat. After removal from the press, the lamination is trimmed on the edges.

Importance and Advantages

Thermoplastic lamination provides a rapid way to permanently protect and seal many valuable articles. It is a process that has some commercial applications but is a **very** small part of the plastics industry. No accurate information on its share of the plastics processing volume is available.

Laboratory Equipment Needed

Either a regular compression molding press or a special thermoplastic laminating press may be used. Each must have heated and cooled platens. Heat is provided by electric heaters in the platens and cold tap water is used to cool them. A drain must also be provided for the cooling water.

Materials for Laboratory Thermoplastic Lamination

Most thermoplastic sheets will work for this process. Clear .010" (0.25 mm) vinyl (polyvinyl chloride) or acetate (cellulose acetate) sheets are used most often. Sometimes dark colored sheets are used for the background with a clear sheet over the top of the article. Any article which is dry and can stand the heat and pressure of lamination may be used in the "sandwich." Thin, flat objects such as leaves, cards, photographs, and instruction sheets are best. There is a limit to the amount of plastic that can be forced around an object in lamination. **The total thickness of the plastic sheets should be about three to four times the thickness of the object enclosed.**[2] If desired, more plastic than this may be used to build up a thicker laminate.

[2]Cherry, Raymond, **General Plastics.** Bloomington, Illinois: McKnight Publishing Company, 1967, p. 255.

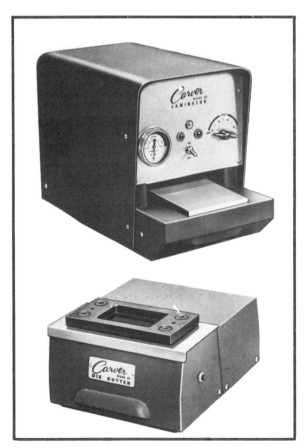

Fig. 10-40. Thermoplastic Laminating Press and Die Cutter

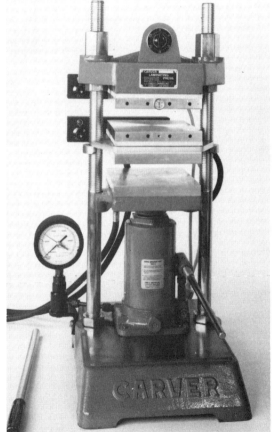

Fig. 10-41. Compression press set up for thermoplastic lamination.

Operation Sequence for Thermoplastic Lamination

There are six steps in the thermoplastic lamination process:

1. Assemble the materials to form the sandwich(es).
2. Place the sandwich(es) between the heated press platens.
3. Apply a low platen pressure for a short time to soften the laminating plastic.
4. Apply a higher pressure for a long time to compress and bond the laminate(s) together.
5. Cool the laminate(s) under pressure.
6. Remove and trim the laminate(s).

The laminating sequence is shown in Table 10-2. Also, **read the manufacturer's instruction manual for exact specifications and instructions.**

Checklist of Equipment and Materials Needed

Assemble the following materials and equipment for thermoplastic lamination:

1. Compression or laminating press with heated and cooled platens.
2. Four (4) cushion blotter pads — tablet back cardboard will do if none other is available.
3. Two (2) chrome polished plates — photographic ferrotype plates will work. Tin plate can be used but will not give as smooth a finish.
4. Vinyl or acetate sheets cut to size. They must be at least ½" (12.7 mm) wider and ½" (12.7 mm) longer than the piece to be enclosed in the sandwich.
5. Object to be laminated.

TABLE 10-2
Operation Sequence for Thermoplastic Lamination

Sequence/Action	Reason	Troubleshooting
❶ (a) Close the press platens. (b) Turn off the water flow. (c) Turn on the press platen heater switch and heat it up to 300°F. (148.8°C).	(a) To get the press ready. The platens will heat faster when they are closed. (b) To let the platens heat up. (c) To heat up the platens.	**Platens do not heat up:** (a) The electricity is not flowing to the heaters. Check the switch, plug, and fuses. (b) The water flow has not been turned off.
❷ While the press is heating, assemble the sandwich(es) for laminating, Fig. 10-42. The following is an example of a single sandwich. **(Top)** 2 cushion blotters. 1 chrome polish plate, face down. 1 or more plastic sheets. **(Center)** Article(s) to be laminated. **(Bottom)** 1 or more plastic sheets. 1 chrome polish plate, face up. 2 cushion blotters. Be sure the plastic sheets are at least ¼" (6.35 mm) larger **all around** than the article being laminated. Multiple sandwiches may be made up and pressed at the same time. One blotter pad and two polish plates (back to back) must be put between each sandwich. **Follow Table 10-3 for heating and bonding times for multiple sandwich lamination.**	(a) Cushion blotters distribute the force of the press evenly. (b) The polish plates put a glossy surface on the plastic laminate. (c) The plastic bonds to and protects the article. Multiple sandwiches may save time. Two polish plates and a blotter pad are needed between each single sandwich to give each sandwich a smooth surface on each face.	**Laminate presses unevenly:** (a) The blotter pads may need to be replaced. (b) The sandwich was not centered right in the press. **Multiple sandwich does not bond:** Not enough heating or bonding time. Check the time recommended in Table 10-3.
❸ Open the press platens when the press reaches 300°F. (148.8°C). Put the sandwich(es) between the two heated press platens, Fig. 10-43. **Be careful. The press platens are hot. Use gloves.** Close the press platens on the sandwich(es), Fig. 10-44.	The press platens must be opened wide enough to let the sandwich be put in. Safety first: Wear insulated gloves when putting the sandwich in the press.	**Sandwich curls:** It should be put in the press quickly to keep it from curling.

(continued on page 212)

Lamination 211

Fig. 10-42. Thermoplastic lamination ready to put in press.

Fig. 10-43. Putting the thermoplastic lamination sandwich into the press.

Fig. 10-44. Closing the press on the thermoplastic lamination sandwich.

TABLE 10-2 (Cont.)

Sequence/Action	Reason	Troubleshooting
④ Apply a low pressure of about 10 psi (0.703 kg/cm²) to the sandwich for 1½ to 16 minutes depending on the number of sandwiches being laminated at a time. See Table 10-3 for correct times. Figure the **total ram pressure** this way: Length in inches (millimetres) × width in inches (millimetres) × 10 (0.703) = TOTAL RAM PRESSURE. A 9" × 12" (237.6 mm × 304.8 mm) lamination will need about 1080 pounds (489 kg) **total ram pressure** for preheating.	Under low pressure the plastic softens and spreads around the object to be laminated. More time is needed for softening when more sandwiches are laminated at one time. The **total ram pressure** must be figured for presses which have gauges which show total ram pressure. **This is not necessary** for some laminating presses on which the gauges show pounds per square inch (g/cm²).	
⑤ Apply 300 psi (21.09 kg/cm²) pressure for 2½ to 20 minutes depending on the number of sandwiches in the lamination, Fig. 10-45. See Table 10-3 for the exact time. A 9" × 12" (228.6 mm × 304.8 mm) lamination will need about 32,400 pounds (14,696 kg or 14.7 Mt) **total ram pressure,** for bonding. If the pressure drops below this as it bonds, it is not necessary to add more pressure.	Under the high pressure the plastic bonds to the object being laminated. The bonding pressure often drops due to the plastic spreading under the heat and pressure.	**Poor bonding:** (a) The bonding pressure is too low. (b) Not enough edge overlap. (c) The bonding temperature is too low. (d) The bonding time is too short.
⑥ (a) After the bonding time is up, turn off the platen heater switch. (b) Turn on the water valve **very slowly** and run the water through the platens to cool them.	(a) The press will not cool properly if the platen heat is not turned off. (b) The platens must be cooled **slowly** to keep them from being damaged.	**Press does not cool off:** (a) The platens may still be heating. (b) The water may not be flowing properly.
⑦ (a) After about 6 minutes of cooling (longer for several sandwiches), the platens should drop to about 100°F. (37.8°C) or 150°F. (65.5°C). Open the platens and take out the laminate(s). (b) Flex the laminate(s) and it (they) will pop off from the polished plates, Figs. 10-46 and 10-47.	(a) Thicker laminates and multiple sandwiches take longer to cool. (b) Flexing the laminate will release it from the polish plates.	
⑧ Close the platens, turn off the water, and turn on the platen switches to start another cycle.		

Adapted from information supplied by the Dake Corporation, Grand Haven, Michigan.

Lamination 213

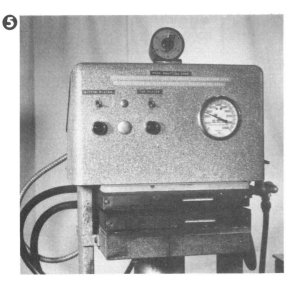

Fig. 10-45. Applying the pressure after preheating.

Fig. 10-47. Removing the laminate from the polished plates.

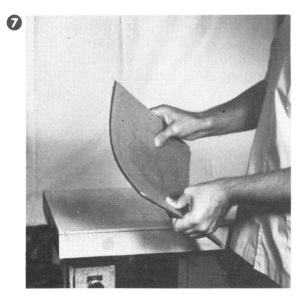

Fig. 10-46. Releasing the thermoplastic sandwich by flexing.

TABLE 10-3
Time, Temperature, and Pressure Needed for Thermoplastic Lamination of Vinyl Plastic Sheets

300°F (148.8°C) Platen Temperature, 10 psi (0.703 kg/cm²) preheat pressure and 300 psi (21.09 kg/cm²) bonding pressure for a 9″ x 12″ (228.6 mm x 304.8 mm) sheet.

Number of Sandwiches	Preheat Time (Minutes)	Bonding Time (Minutes)
1	1½	2½
2	3½	5
3	5	8
4	8	11
5	13	15
6	16	20

Information furnished by the Dake Corporation, Grand Haven, Michigan.

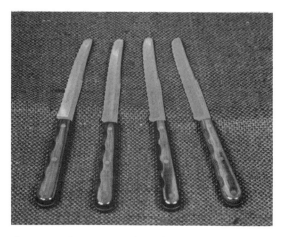

Fig. 10-48. Compreg Laminated Knife Handles

Fig. 10-49. Placing compreg sandwich in the press.

Compreg Lamination — Process Description

Compreg is a laminate much like plywood. It is made from layers of 1/28" (91 mm) thick wood veneer. The wood is soaked in liquid phenolic (phenol formaldehyde) resin for 24 hours and then air-dried for another 24 hours. The treated (impregnated) wood veneer is stacked and compressed at about 800 psi (56 kg/cm² or 5520 kPa) and 300°F (149°C) for 10 to 15 minutes in a standard compression molding press. Compreg gets its name from compressed (com) and impregnated (preg).

Compreg is superior to regular plywood because it is impregnated (saturated) with plastic before it is compressed. The wood fibers are much closer together than in plywood. This makes a solid, dense mass of wood and plastic that is very strong and scratch resistant. Compreg is used for durable knife and kitchen utensil handles, buttons, airplane propellers, and sheet metal forming dies.

Checklist of Equipment and Materials Needed

Assemble the following equipment and materials for compreg lamination:

1. Compression molding press.
2. Metal platen covers, tin plate or ferrotype plates.
3. Metal pan to soak the wood veneer in.
4. Liquid phenol formaldehyde resin.
5. Wood veneer, 1/28" (91 mm) thick.
6. Fluorocarbon mold release.

Read the resin manufacturer's instructions before beginning.

Operation Sequence for Compreg Lamination

There are nine steps in the compreg lamination process:

1. Cut sheets of suitable 1/28" (91 mm) wood veneer to the size of the product to be made. Several small parts may be sawed from a larger finished compreg sheet.
2. Impregnate the wood veneer by soaking in a solution of liquid phenolic (phenol formaldehyde) plastic resin for about 24 hours.
3. Take the veneer out of the phenolic resin and let it air dry for another 24 hours.
4. Heat the press platens to 300°F (149°C).
5. Cover the press platens with metal plates to keep the resin from sticking. Tin plate

or polished ferrotype plates may be used. Spray the plates with fluorocarbon mold release.
6. Stack up the impregnated sheets of veneer like a sandwich. The wood grain normally is alternated 90° like regular plywood is but may also all run the same way. Grain direction depends on the product to be made.
7. Put the veneer sandwich on the bottom plate in the press, Fig. 10-49. **Caution: Wear gloves. The press platens are hot.** Put the top platen protector plate over the sandwich. Be sure the sandwich is in the center of the press platen.
8. Close the press platens and apply a 300 psi (21 kg/cm^2) pressure (length × width of the sandwich × 300 = **total ram pressure**). Cure the laminate for about 12 minutes, or longer if it is quite thick.
9. Open the press and take the compreg laminate out. It may now be cut to any shape.

STUDENT ACTIVITIES
1. Make a clipboard, hot pad, or other flat fiberglass product on a Formica laminated mold.
2. Make products from fiberglass on purchased or laboratory-made molds.
3. Materials experimentation:
   a. Vary the resin types used.
   b. Vary the catalyst percentage on experimental pieces. **Caution: Do not exceed 4% catalyst concentrations.**
   c. Use different fiberglass reinforcing materials. Test the resulting products for strength and durability.
      1) Glass mat
      2) Glass cloth
      3) Chopped roving
      4) Roving
      5) Surfacing mat
      6) Burlap
      7) Cotton cloth
4. Process experimentation (vary **a** and **b**):
   a. Lay-up procedure
      1) Spray-up
      2) Hand lay-up of mat
      3) Hand lay-up of cloth
   b. Molding procedure
      1) Contact molding
      2) Compression molding
5. Design products which can be made by fiberglass lamination.
   a. Flat shapes and profiles
   b. Curved shapes and profiles
   c. Irregular shapes and profiles
6. Construct molds for products which you have designed. Several mold types which may be considered are:
   a. Formica flat forms
   b. Vacuum formed polyethylene
   c. Formica parts screwed together to make a tank or boxlike object.
   d. Turned or built up wooden forms. Seal the surface with polyester resin before use.
   e. Plaster or Ultracal 30 forms. Seal the surface with polyester resin before use.
7. Make knife handles, clipboards, and similar items from compreg lamination.
8. Laminate photos, cards, biology specimens, coins, and similar objects between thermoplastic sheets.
9. Production organization and operation for fiberglass lamination:
   a. Mold making.
   b. Material supply.
   c. Material dispensing.
   d. Cutting and measuring glass materials.
   e. Mold preparation — mold release.
   f. Mixing resin and catalyst.
   g. Spray-up or hand lay-up.
   h. Molding.
   i. Curing.
   j. Part removal.
   k. Trimming.
   l. Cleanup.
   m. Record keeping.
      1) Materials used
      2) Processing variables

3) Parts made
4) Finishing and assembly
5) Labor
6) Rejects
7) Reject repair
10. Look up data on lamination materials in the *Modern Plastics Encyclopedia.*

Questions Relating to Materials and Process

1. Why does the thickness of a fiberglass laminate affect the amount of catalyst used?
2. Why are different types of polyester resins needed in fiberglass lamination? What are they? What do they do?
3. What type of polyester resin is best for gel coats? Why?
4. What type of polyester resin is best for laminating coats? Why?
5. What is the difference between polyester laminating and gel coat resins?
6. What is a surfacing agent and what does it do?
7. Can laminating resin be used for gel coats? If so, how?
8. How does room temperature affect resin cure?
9. How does humidity affect polyester resin cure?
10. Give the advantages of the following:
    a. Fiberglass cloth
    b. Fiberglass roving
    c. Fiberglass mat
    d. Fiberglass chopped roving
11. How is fiberglass reinforced lamination related to glass-filled nylon injection molding?
12. What is compreg?
13. Why is compreg better than plywood?
14. What advantages does compreg offer over other plastic laminates?
15. Why is it necessary to use several thermoplastic sheets when laminating thick objects?
16. Give some of the advantages and disadvantages of thermoplastic lamination.
17. Where is thermoplastic lamination used in industry?
18. Name some new uses for fiberglass lamination in industry.

Casting is a group of processes in which plastic materials are poured into a mold by gravity. They are processed until they harden. This is probably the simplest of all plastics processes. In casting, no force other than gravity and atmospheric pressure is put on the material in the mold. In molding, pressure is used. This pressure factor is the main difference between casting and molding. Casting may be used for both thermoplastic and thermosetting materials. Casting resins include clear casting resins, water-extended polyester (WEP) resins, epoxy, silicone, DAP, urethane, acrylic, and phenolic casting resins.

In casting, the plastic materials are poured into the mold, as in Fig. 11-2. Often, the mold or the material is heated to process the plastic. Some thermoplastic materials are heated before being poured into the mold. They are then cooled in the mold to harden them (a physical change). Thermosetting materials such as polyester resins use internal heating catalysts that cure them (a chemical change) in the mold. In some cases the mold is heated to cure the thermosetting resins.

CASTING plASTIC MATERIAlS

chAPTER 11

Importance of Casting

No figures are available as to the relative position of casting in the plastics industry. It is, however, a method by which materials can be processed with very inexpensive equipment. Some plastics may not be processed any other way. Casting is also important as a mold making technique, especially for silicone RTV, urethane, and mass casting epoxy resins.

Fig. 11-1. Silicone rubber mold used to produce fine detail castings.

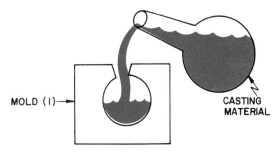

Fig. 11-2. In casting, material is poured into a mold by gravity.

217

Advantages

Casting provides an economical way of forming plastics. Equipment costs are very low. Often equipment can be made or adapted at a low cost. Sometimes many different types of plastic materials may be processed in the same molds. Molds do not have to stand high pressures. They may be made of cast aluminum, machined aluminum, plastic, gypsum cement, silicone RTV, and urethane rubber compounds. Even thermoformed polyethylene and high impact polystyrene molds have been successfully used.

Disadvantages

The main disadvantage of casting is that it is often slower than other molding methods. Molds frequently are filled by hand. Some automatic mixing and mold filling equipment has been introduced. Manual casting methods cost more for labor but less for equipment as compared to many processes.

Industrial Equipment

Several types of industrial casting equipment are available, one of them shown in Fig. 11-3. Resin dispensing equipment mea-

Fig. 11-3. Resin Dispensing Equipment Used for Polyester, Epoxy, Polyurethane, Polysulphide and Silicone Casting Materials.

Fig. 11-4. Hyatt's Billiard Ball—Cellulose Nitrate Cast.

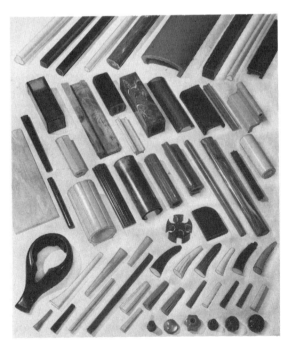

Fig. 11-5. Cast Phenolic Products

sures and dispenses plastic resins to be poured by hand.

Casting Resins — Process Description

Thermoplastic and thermosetting casting resins, used for a number of products in industry, are poured directly into a mold and cured. Probably one of the earliest uses of plastic casting was the first Celluloid (thermoplastic) billiard ball made by John Wesley Hyatt in about 1868, Fig. 11-4. Dr. Leo H. Baekeland's phenolic (thermosetting) plastic was also cast into numerous items, such as phenolic desk pen bases which were made for many years. Phenolic parts are still being cast, Fig. 11-5.

Clear polyester casting resins have been in use for some time. Many hobby craft items such as coin embedments, decorative grape clusters, jewelry, and imitation marble are made from cast polyester resin. See Fig. 11-6. A new polyester resin called **water-extended polyester** (WEP) is used for casting furniture parts, picture frames, thermoforming molds, and lawn and garden ornaments. Another type of polyester casting is called **synthetic wood.** It is made by mixing polyester resin, wood flour (sanding dust or ground pecan shells), and **Microballons**®. The Microballons are tiny, hollow glass balls which reduce the weight of the casting. Cast synthetic wood is used for furniture parts and decorative household objects.

Silicone and urethane casting compounds are often used for mold making both in industry and the laboratory. They are easy to work with. Many metal prototype parts for industry are cast in silicone RTV rubber which will withstand temperatures up to

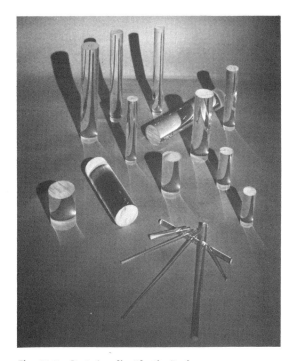

Fig. 11-7. Cast Acrylic Plastic Rods

Fig. 11-6. Cast Polyester Chess Set

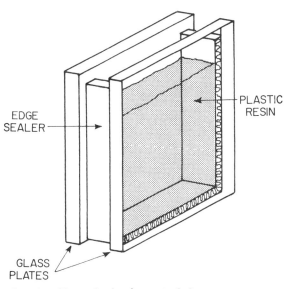

Fig. 11-8. How plastic sheet stock is cast.

900°F (484°C). Silicone and urethane molds for polyester, urethane, and epoxy castings are easily made in the school laboratory. They are discussed later in this chapter.

Epoxy molds for injection molding and other processes are cast from mass casting and metal-filled epoxy resins. This casting method is discussed in Chapter 5.

Acrylic casting resins are commercially cast into acrylic sheet, rod, and tubing. Acrylic sheet plastic is cast between two glass plates. It is seldom cast in the school laboratory, however.

Laboratory Casting of Polyester Resins

Casting of clear polyester resin, water-extended polyester (WEP) resin, and synthetic wood is included in this chapter.

Polyester resins are thermosetting plastics. They polymerize, or cure, by **exothermic** (internal) **heat**. A **promotor** (reactive chemical) is usually added to the polyester resin at the factory. In some cases it must be added at the time of use. It is often a cobalt metal dispersion and causes the resin to be bluish or pink in color. Cobalt is a highly reactive metal in the presence of oxygen. The oxygen from the MEK peroxide catalyst combines with (oxidizes) the cobalt metal. This oxidation causes heat which polymerizes the polyester resin. Oxidation is described in Chapter 10.

A slightly different type of cobalt promotor is used with clear casting polyester resins to avoid the pink or blue color and attain a clearer resin. It is combined with MEK peroxide to produce an exothermic cure. The oxidation works the same way as with other polyester resins.

Clear Polyester Casting Resins

Clear polyester casting resins are normally air-inhibited. Air-inhibited resins do not cure on the surface that is open to the air. This permits one layer of casting resin to stick to another. It also helps to keep the resin clear. If the surface of the **last** layer of clear polyester casting resin does not cure, a small amount of liquid wax may be poured over the surface to keep the air from reaching it. Slight surface tackiness may be cured near a heat source or sanded off.

Polyester resins and catalysts should be stored in a refrigerator when not in use. Refrigeration will greatly lengthen their shelf life. Do not store MEK peroxide at less than 40°F (4.4°C) as it may crystalize.

---- Safety Note ----
Do not store food in the same refrigerator with the resins and catalysts. They may contaminate the food.

Clear polyester resins may be cast in glass, silicone RTV, flexible urethane, latex rubber, thermoformed polyethylene, glazed ceramic, and polyester molds. Glass jars and tumblers may be used for molds. Glass molds may also be bought from handicraft and plastics supply houses. Instructions for casting silicone RTV and flexible urethane molds are included in this chapter. Polyethylene molds may be thermoformed over wood, plaster, or metal or may be purchased from several sources. See Chapter 8 for thermoforming methods. Thermoformed molds may be bought from handicraft stores and plastic supply houses. Glazed ceramic

Fig. 11-9. Several types of molds used for casting polyester resins.

molds are also available from the same places. The casting methods in this chapter may be used to make polyester resin molds. See Chapter 10 for polyester laminating methods.

Operation Sequence for Polyester Casting Resins

There are five steps in the process of casting with clear polyester resins:

1. Select and prepare the mold.
2. Measure the amount of resin needed.
3. Add the right amount of catalyst and mix well.
4. Pour the material into the mold.
5. Cure the resin and remove it from the mold.

Study the sequence for using polyester casting resins, Table 11-1. Also, **read the resin manufacturer's instructions.** Observe any precautions given.

TABLE 11-1
Operation Sequence for Casting Polyester Resins

Sequence/Action	Reason	Troubleshooting
❶ (a) Prepare the molds. Apply a thin coat of mold release to polyester, glazed ceramic, and re-usable glass molds. (b) Silicone RTV, flexible urethane, and vacuum-formed polyethylene molds usually do not need mold release. (c) One-use glass molds do not need mold release.	(a) Mold release insures reuse of the molds after removal. (b) These molds are naturally waxy and self-releasing. (c) One shot glass molds are broken to release the part.	**Polyester sticks to mold:** (a) Mold release not properly applied. (b) Thin spots in mold release. (c) Wrong mold release. **Silicone** and **urethane** molds absorb styrene from polyester. They need to be baked out at 150°F (65.5°C) for 1-2 hours after 10 moldings. This will restore their ability to release the casting.
❷ Read the manufacturer's instructions for the polyester resin. **Check the catalyst and promotor concentration.**	Instructions vary for different resins. Catalyst and promotor concentrations are often different.	**Resin does not cure properly:** (a) Check the catalyst and promotor concentrations. (b) Make sure the resin catalyst and promotor are fresh.
❸ Pour the right amount of casting resin into a paper measuring cup. Pour only enough resin for immediate use. Mold volume can be measured with water. Pour the water out of the mold into a measuring cup to get the volume. Blow the mold dry with an air gun.	Resin has a short "pot life" after it is catalyzed. Resin is wasted if it sets up (polymerizes) in the cup.	**Resin sets up in the mixing cup before use:** (a) Too much catalyst was used. (b) Resin was catalyzed too soon.
❹ Add the right amount of promotor to the resin (if not factory added). Mix the promotor and resin thoroughly. For ½% add about 1 drop per oz. (0.03 drops per ml). 1% = 2 drops per oz. (0.06 drops per ml), 2% = 4 drops per oz. (0.13 drops per ml) of resin. Do not stir air into the mixture. **Caution:** Use **separate** eye droppers for promotor and catalyst.	Promotor must be stirred in first. Air bubbles will remain in the resin and stick to the mold. If the catalyst and promotor are mixed directly together, they may explode.	**Resin does not cure:** (a) Promotor may have been left out. (b) Wrong promotor concentration added. (c) Room temperature too cool. (d) High moisture content in the air.

(continued on page 222)

TABLE 11-1 (Cont.)

Sequence/Action	Reason	Troubleshooting
❺ Add the right amount of catalyst to resin/promotor mix, Figs. 11-10 and 11-11. Use same measuring method as above. **Caution:** Thicker castings require less catalyst. Consult the manufacturer's recommendations. One manufacturer suggests the following: **Casting Thickness** — **Catalyst Concentration** 1/8" or less (3.1 mm or less) — 16 drops per fluid oz. (0.54 drops/ml) 1/8" to 1/4" (3.1 to 6.3 mm) — 12 drops per oz. (0.41 drops/ml) 1/4" to 1/2" (6.3 to 12.7 mm) — 8 drops per oz. (0.27 drops/ml) 1/2" (12.7 mm) and over — 4 drops per oz. (0.13 drops/ml) 1 fluid oz. (29.5 ml) = 2 tablespoons (29.5 ml) Add color concentrate if desired. Stir in without mixing in air. **Caution:** Do not mix promotor and catalyst directly together. Violent explosion may result.	Thicker castings do not lose their exothermic heat as fast as thinner ones. Heat builds up too high and cracks the plastic as it cures. Color concentrates will add color to the resin. Both opaque and clear color concentrates are available.	**Resin cures too fast:** (a) Too much catalyst. (b) Room temperature too hot. (c) Humidity very low.
❻ (a) Pour the resin into the mold, Fig. 11-12. Fill the mold to the top if it is a one-layer casting. Allow to cure. (b) Fill the mold to the desired height for a multilayered casting. Cure first layer. Add a colored layer or the article to be embedded, Fig. 11-13. "Wet" the coin or object to be embedded in the resin first. (c) If wire inserts are used, place them in after plastic resin is poured. To prevent cracking around wires or other embedments, dip the wire or embedment in an anti-cracking solution.	(a) Cure entire part in the one layer if possible. (b) In order to "float" a coin or other object, the lower layer must cure to hold it up. "Wetting" eliminates trapped air. (c) Cracking around wires and embedments is caused by resin shrinkage. It shrinks more than the embedded article, causing cracks.	**Lamination line shows:** This is hard to eliminate if more than one layer is cast. **Air bubbles trapped:** (a) "Wet" the embedment. (b) Vibrate the mold. (c) Pour resin more slowly. **Cracks around embedments:** Shrinkage of resin. Use anticrack solution.
❼ Remove the product from the mold after curing. Curing may take 2 to 6 hours. Rigid glass or ceramic molds may be placed in warm water to help removal if necessary.	Thick casting resins need long cures. Quick cures cause cracks.	**Product sticks to the mold:** See Sequence #1.

Casting Plastic Materials 223

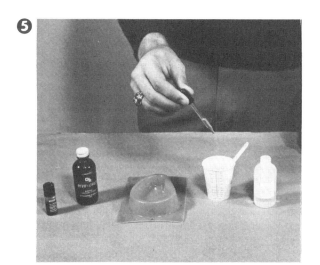

Fig. 11-10. Using an eyedropper to add catalyst to clear polyester casting resin.

Fig. 11-12. Filling a vacuum formed polyethylene mold with clear polyester casting resin.

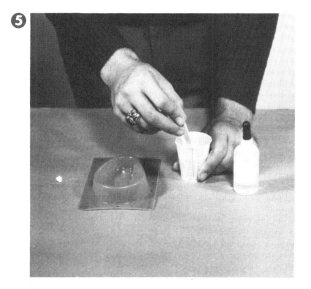

Fig. 11-11. Mixing the catalyst with the resin.

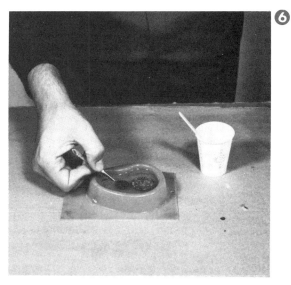

Fig. 11-13. Placing a coin in clear polyester casting resin.

> **Safety Note**
> Do not directly mix catalyst and promotor together. An explosion may result.
>
> Use the materials **only** in a well ventilated room.

Cover the bench tops to keep the resin from sticking. Wear protective clothing, including plastic gloves. Assemble all the materials needed before starting the process.

Checklist of Materials and Equipment Needed

Assemble the following materials and equipment for casting polyester resins:

1. Clear polyester casting resin (air-inhibited).
2. Promotor **if not added at the factory.**
3. Catalyst.
4. Paper cups for mixing and measuring.
5. Color additives as desired.
6. Eye droppers.
7. Mixing sticks.
8. Molds and mold supports (if needed).
9. Acetone or lacquer thinner for cleanup.
10. Mold release.
11. Bench covers and wiping rags.

When using polyester resins, throw away used paper cups and mixing sticks. Destroy sticky bench covers when finished. Clean hands with acetone or lacquer thinner.

> **Safety Note**
> Acetone and lacquer thinner are flammable. Methylene chloride or ethylene dichloride, which are fireproof, may be substituted.

Cast polyester parts may be sanded and buffed, if necessary, to get a final finish. The acrylic plastic buffing procedure may be used.

1. Rough sand to remove deep marks.
2. Fine sand with progressively finer paper. Wet sand to 400 or 600 grit. Always sand at right angles to the last sanding when making each change to finer paper.
3. Buff with a 1725 rpm buffer with a 6" to 8" (152 to 203 mm) loose sewn buffing wheel and acrylic buffing compound.

Water-Extended Polyester Casting Resins

Water-extended polyester (WEP) castings are similar to clear polyester castings. However, water is added to the special polyester resin to cut the cost and make the part lighter in weight. Another advantage is that water-extended polyester castings can be made in large masses without cracking. WEP castings used for lawn and garden ornaments are competitive with concrete products and may be used for a number of useful and decorative indoor articles as well. They may be used in the laboratory for sturdy, long-lasting thermoforming molds.

Water-extended polyester resins are similar to regular polyester resins except they are made so that water may be mixed in. The usual mixture is 50% water and 50% WEP resin. This mixture gives the maximum strength at minimum cost. As much as 75% water may be added, however, without special mixing equipment. As the water concentration increases, the viscosity (thickness) of the mixture also increases. Mixing and pouring high water content (over 50/50) WEP emulsions, or mixtures, presents a problem. Class size quantities of WEP emulsions may be mixed in advance and stored. The catalyst is added to small quantities just before they are poured into the molds.

A cobalt-MEK peroxide promotor/catalyst system is used. The cobalt promotor is usually mixed in at the factory. If the cobalt promotor is added by the user, it is usually a 1.25% mixture by weight. It must be mixed in **before** the water is added. The MEK peroxide is furnished in a separate container and is added to the mixture **last**. The chemical reaction of the catalyst/promotor system is the same as for other polyester resins. (See Chapter 10.) When not in use, polyester resins and catalysts should be stored in a refrigerator in order to extend their life. Do not store MEK peroxide below 40°F (4.4°C) as it may crystalize.

Casting Plastic Materials 225

Fig. 11-14. Cast Water-Extended Polyester Resin Product

Fig. 11-15. Thermoformed polyethylene molds for WEP resin casting.

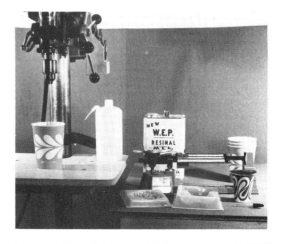

Fig. 11-16. Materials assembled ready to mix WEP resin.

Safety Note
Do not store food in the same refrigerator with polyester resins and catalysts. They may contaminate the food.

The same types of molds used for clear polyester casting, thermoformed polyethylene, silicone RTV, and flexible urethane may be used for WEP emulsions. Thermoformed polyethylene molds are the most economical. All are self-releasing.

Operation Sequence for Water-Extended Polyester Casting

There are six steps in the process of water-extended polyester casting:

1. Calculate the amounts of materials needed.
2. Weigh out the resin and water in **separate disposable containers.** Figure out the amount of catalyst needed.
3. Start up the high speed mixer and add the water **very slowly** into the vortex. Let the mixture stand to permit air to escape.
4. Add the catalyst and stir gently.
5. Pour the mixture slowly into the mold.
6. Allow the part to cure before removing it.

Study the casting sequence in Table 11-2. Read the resin manufacturer's instructions carefully. Check the amount of catalyst recommended. Wear protective clothing, including plastic gloves. Assemble all materials needed before starting the process.

Safety Note
Use polyester resin only in a **well ventilated** room.

TABLE 11-2
Operation Sequence for Casting Water-Extended Polyester

Sequence/Action	Reason	Troubleshooting
1 Figure out the amount of materials needed. (a) Decide on the polyester/water ratio. Try a 50/50 ratio to start. (b) Weigh out, in a disposable paper container, a batch of water equal to about 75% of the volume of the mold to be used. Record this weight. Be sure to subtract the weight of the container.	(a) The ratio of resin to water will help figure the amount of materials needed. (b) Accurate measurement of polyester resin, water and catalyst is needed for good results.	Emulsion sets up before mold is filled: (a) Use cool water for large molds. (b) Pour as soon as catalyst is added. (c) Keep the emulsion cool when storing it.
2 Weigh out a batch of WEP resin **equal** in weight to the batch of water just weighed, Fig. 11-17. Be sure to subtract the weight of the container. Use a widemouth container about 4 times the volume of the resin to be used.	Throw away containers should always be used for this work. Partly cured resin in a cup will affect a new batch of resin. The container needs to be large enough to avoid spilling.	
3 (a) Put the mixer blade in the drill press or drill and set the speed at 2000 rpm. (b) Put the weighed out water in a plastic "wash" bottle. (c) Turn the mixer on and add the water **very slowly** to the vortex of the resin, Fig. 11-18. Continue to mix for about a minute. A polyester color pigment may be added at this point. (d) Clean the mixing blade with acetone when work is completed.	(a) 2000 rpm and the right blade are needed for proper mixing. (b) A "wash" bottle makes an excellent water dispenser. (c) The secret is adding the water **slowly.** It should take several minutes for a 1 quart (0.946 l) mixture. Use only polyester pigments for best results.	Emulsion breaks down: (a) The water was added too fast. (b) Mixing speed too slow. (c) Wrong type of mixer used. (d) Resin added to the water instead of water to the resin.
4 (a) Let the mixture stand for several minutes to 24 hours to get rid of the air in it. (b) Figure out the right amount of catalyst to be used. From ½% to ¾% (0.5% to 0.75%) based on weight is usually recommended. **Caution: Do not mix catalyst and promotor directly together.** (c) Add the catalyst to the resin/water emulsion (mixture) with an eye dropper, Fig. 11-19. Stir the catalyst in with the mixer blade without mixing air into the emulsion. One-half of one percent catalyst (0.5%) is about one (1) drop catalyst per fluid ounce (0.03 drops/ml) of resin.	Exact catalyst concentrations vary with manufacturer. **Read the instructions carefully.** An exact way to determine concentration is to figure 0.5% of the weight of the **resin** used. Weigh out this much catalyst. Draw it up into an eye dropper and add it to the resin/water emulsion.	Resin does not cure: (a) Wrong amount of catalyst. (b) Promotor not in the resin. The blue color in the resin indicates that the promotor has been added at the factory.

(continued on page 228)

Casting Plastic Materials 227

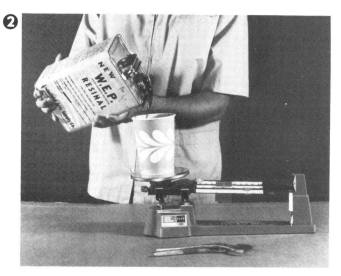

Fig. 11-17. Weighing resin.

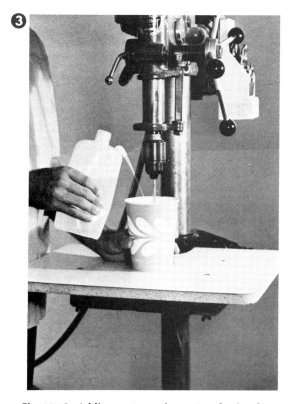

Fig. 11-18. Adding water, using a "wash" bottle.

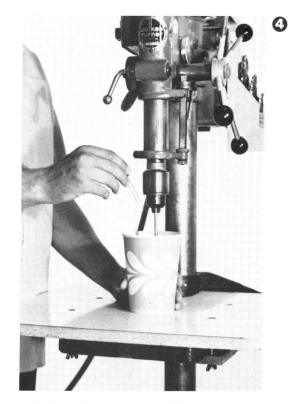

Fig. 11-19. Adding Catalyst to WEP Resin.

228   Section IV   CASTING

TABLE 11-2 (Cont.)

Sequence/Action	Reason	Troubleshooting
❺ (a) Slowly pour the emulsion into the mold, Fig. 11-20. Pour it so as to fill the difficult areas first. Brush some emulsion on fine detailed areas. (b) Vibrate or flex the mold to get out the trapped air.	Pouring slowly avoids trapping air in the mold. Brushing gets resin emulsion into fine details and avoids trapping air. The resin emulsion must be poured as soon as the catalyst is added. Gel time is about 2 minutes.	Air trapped in part: (a) Resin mixture did not stand long enough before use. (b) Wrong kind of mixer was used. (c) Air was trapped when the emulsion was poured.
❻ (a) Let the emulsion cure. A 50/50 mixture should solidify in about 5 minutes. The casting should be about 90% cured in 1 hour. b) The casting may be taken out of the mold in 10 to 20 minutes after pouring, **if it is hard**, Figs. 11-21 and 11-22.	(a) Quicker cures can be made with warm water in the emulsion. Cold water in the emulsion will slow down the cure. (b) The part must be hard enough to hold its shape before being removed from the mold.	**Part warps when taken from the mold:** Part not cured long enough. Leave it in longer.

Fig. 11-20. Filling the mold with WEP resin.

Fig. 11-21. Removing cured WEP resin from the mold.

Fig. 11-22. Completed WEP casting taken from the mold.

Checklist of Materials and Equipment Needed

Assemble the following materials and equipment for casting water-extended polyester:

1. An accurate scale that measures in grams.
2. Several large paper cups or paint pails, 1 pint to 1 gallon in size.
3. A high shear mixer blade with a low pitch to minimize air trapping.
4. A drill press, 2000 rpm electric drill, a malted milk mixer or an electric blender.
5. Water.
6. Water-extended polyester resin with cobalt promotor added.
7. MEK peroxide catalyst.
8. Acetone or lacquer thinner for cleanup. **Caution:** These thinners are flammable. Fireproof methylene chloride or ethylene dichloride may be substituted.
9. Plastic "wash" bottle for adding water slowly.
10. Eye dropper for the catalyst.
11. Mold(s).

Cast Synthetic Wood

Cast synthetic wood with **Microballons®** is usually cast in silicone RTV, polyethylene, or urethane rubber molds. Information on RTV and urethane mold making is given later in this chapter. Polyethylene mold making is described in Chapter 8.

The basic ingredients for cast synthetic wood are thinned polyester resin, wood flour extender (sanding dust), and Microballons. The polyester resin acts as a binder to hold the mixture together. The wood flour extends the resin and makes it look and feel like wood. The Microballons, being very small, hollow glass balls, make the mixture lighter in weight and control the shrinkage.

The amount of styrene monomer used will vary with the polyester resin. Using styrene monomer in the mixture will thin out the polyester resin, making it a pourable mixture. High concentrations of styrene monomer, however, will weaken the resin.

A typical mixture of these materials is as follows, **by weight:**

	Specific Gravity
100 parts catalyzed nonthixotropic[1] or polyester casting resin	1.15
10-25 parts styrene monomer (depending on resin thickness and pourability desired)	
40 parts wood flour (sanding dust) extender or ground pecan shells	1.00
8 parts Microballons	0.35

Operation Sequence for Cast Synthetic Wood

The sequence of operation is nearly the same as for polyester casting except for the addition of the other materials. Follow the same material checklist, adding the styrene monomer, wood flour, and Microballons. Wood flour may be obtained from the school's woodshop. The Microballons may be bought from the manufacturer.[2] Use the following mixing instructions. **Read the manufacturer's instructions before starting.**

1. Weigh an empty disposable mixing container. Record this weight.
2. Measure out a quantity of polyester resin in the container. Weigh the resin. Deduct the weight of the container.
3. Weigh out a quantity of styrene monomer equal to .10 to .25 the resin weight. Add it to the resin and stir in well.

---

[1]Thixotropic and nonthixotropic materials are defined in Chapter 10, page 198.

[2]Microballons is the registered trade mark of Emerson & Cuming, Inc., Canton, MA, 02021 where this material may be purchased.

230 Section IV CASTING

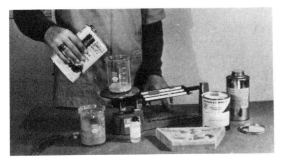

Fig. 11-23. Mixing synthetic wood.

Fig. 11-24. Pouring synthetic wood.

4. Measure out wood flour equal to .40 the resin weight. Slowly add this to the resin while mixing at a slow speed with a low shear mixing blade in an electric drill.
5. Measure out the Microballons equal to .08 times the resin weight. Slowly add these while mixing with the same mixer at the same speed.
6. Measure out and add the catalyst (1%-2% of the resin weight) and mix again.
7. Pour the mixture into a flexible rubber mold. Allow it to cure. Cure time may be as short as three to five minutes.
8. Remove the part by twisting the mold.

The molded synthetic wood piece can be cut, sanded, nailed, glued, or fastened like natural wood. It may also be finished like wood. Formulations for this material vary greatly depending on the product desired. Some experimentation with small batches is suggested before large parts are made.

Tooling Up for Production

Molds for casting may be obtained in a number of ways. Sources for ready-made molds are listed in Appendix B. Methods for making thermoforming molds are described in Chapter 8. Cast aluminum molds may be purchased or made in the laboratory. Aluminum molds may also be made from solid aluminum stock, as described in Chapter 4. Silicone RTV and urethane rubber molds must be made in the laboratory. These molds may be used for several of the processes outlined in this chapter.

Casting Silicone and Urethane Rubber Molds

1. **The Master Pattern or Model.** Select or make a master pattern for your mold. It may be made of wood, metal, plaster, soap, plastic, glass, or wax. Clean the surface of the pattern carefully. Remove grease, oil, chips, dust, or shavings. Blow wood patterns with compressed air. To prevent sticking, plaster and wood patterns must be sealed with a spray lacquer or similar material. Coat the master pattern with the proper mold release. **Check the manufacturer's recommendations.**

2. **Mounting the Pattern or Model.** The master pattern should be mounted on a flat plate or glass surface. Contact cement, silicone bathtub seal, or rubber cement may be used to hold it in place. A metal or Formica®-surfaced frame is placed around the pattern. A ½" (12.7 mm) clearance should be left all around. The frame should be about ½" (12.7 mm) higher than the top of the pattern. Silicone bathtub seal or mask-

ing tape can be used to seal the frame in place.

3. Preparing the Mold-Making Material. Flexible casting molds may be made from a variety of silicone and urethane rubber compounds. General directions for mixing and pouring are given here. Study the **exact mixtures and procedures stated in the manufacturer's specification sheets.** Most of these materials cure at room temperatures. A few need to be cured in an oven at higher temperatures.

a. Select a clean paper cup large enough in which to mix the mold rubber.
b. Weigh the paper cup. Record this weight.
c. Pour a volume of rubber compound into the cup.
d. Weigh the cup and rubber. Subtract the weight of the cup to determine the weight of the rubber.
e. Multiply the weight of the rubber by the catalyst ratio to find out how much catalyst to add. Silicone RTV rubber **often** uses a 10:1 rubber/catalyst ratio.
f. Add the catalyst weight to the rubber and cup weight. This is the weight of the final mixture.
g. Pour in the catalyst.
h. Mix the catalyst/rubber compound well. Use a stirring stick. Do not mix in air bubbles.
i. When completely mixed, the compound should be vacuum de-aired, Fig. 11-26. Put it under a bell jar and turn on the vacuum. If vacuum de-airing is not possible, the compound should be vibrated.

4. Making the Mold. Mold release compounds recommended by the manufacturer, if any, should be applied to the master pattern and frame.

a. Pour the catalyzed rubber compound into one corner of the mold frame. Let it flow slowly over and around the master pattern. This helps avoid trapping air.
b. Stop pouring when the pattern is half covered. Let the material level off.
c. Continue pouring until the pattern is covered about ½" (12.7 mm).
d. Avoid trapped air by brushing the sur-

Fig. 11-25. Pattern and frame mounted and ready to weigh, mix and pour silicone RTV for a mold.

Fig. 11-26. Vacuuming de-airing a silicone molding compound.

face of the mold with rubber before pouring the rest of the compound.
e. Vacuum or vibrate the poured mold compound to remove trapped air.
f. Let the compound stand overnight to cure.
g. Remove the pattern and frame.

STUDENT ACTIVITIES

1. Materials experimentation.
   a. Use various times, colors, and densities of materials.
   b. Use different proportions of water in WEP resins. The base 50/50 ratio may be varied to 75% water/25% resin and vice versa. Bake the water out of cured WEP products (150°F or 65°C) for a lighter product.
2. Process experimentation.
   a. Vary catalyst concentrations **slightly** in casting resins. Observe the variation in polymerization times.
   b. Experiment with different mold material for each casting process.
3. Design products and molds for each casting process.
4. Make molds for casting by thermoforming, rubber casting, and metal machining.
5. Production organization and operation.
   a. Material supply.
   b. Preparing molds — mold release.
   c. Measuring materials.
   d. Mixing materials.
   e. Filling molds.
   f. Curing parts.
   g. Removing parts.
   h. Record keeping.
      1) Materials.
      2) Processing variables.
      3) Assembly and finishing.
      4) Rejects.
      5) Labor.

Questions Relating to Materials and Process

1. List reasons why each of these mold materials are used for specific applications. List the application(s) for each along with reasons for use.
   a. Silicone RTV rubber.
   b. Urethane rubber.
   c. Cast aluminum.
   d. Vacuum formed polyethylene.
   e. Paper.
   f. Glass.
   g. Cast polyester.
2. Which plastic casting materials should be used for the following applications?
   a. Embedding electrical components.
   b. Injection mold making.
   c. Molds for polyester casting resins.
   d. Cast synthetic wood.
3. What products would be **best** suited for the following materials?
   a. Polyester casting resin.
   b. WEP polyester resin.
   c. Silicone RTV materials.
   d. Urethane casting rubber.
4. Give the main characteristics of each of the materials in question 3.
5. How do you decide on the best casting materials to use for each application? What properties are considered?
6. What is the relationship between casting plastics and casting metals, glass, and ceramics?

Expansion is a group of processes in which the plastic material is poured or forced into a mold and then expanded. The expanded plastics, usually called **foams**, are cellular (spongelike) in structure. Expanded plastics may be molded by extrusion, injection, and compression; sprayed on; poured; and expanded from beads.

Plastic foams prove useful in many ways. Flotation devices such as life jackets and rafts are made of foam. Plastic foams are used for insulation in ice buckets, storage tanks, and clothing. They have higher insulation values than many other materials. There are also cushioning, impact absorbing, and packaging applications of the foams, replacing more traditional packaging materials. Nearly all plastics, both thermoplastic and thermosetting, may be expanded (foamed).

The most common plastic foams are made from polystyrene (Styrofoam® and expandable polystyrene beads, a thermoplastic) and polyurethane (a thermosetting plastic). Other common plastic foams are made from polyethylene, cellulose acetate, phenolic, silicone, and epoxy.

Both rigid and flexible foams are produced, Fig. 12-1. Boat flotation cells, refrigerator insulation, building insulation panels, and furniture frames can be made of rigid foams. Flexible foams are used for sponges, furniture padding, padded dash panels for cars, and mattresses.

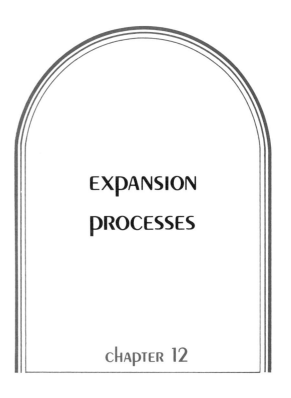

EXPANSION PROCESSES

CHAPTER 12

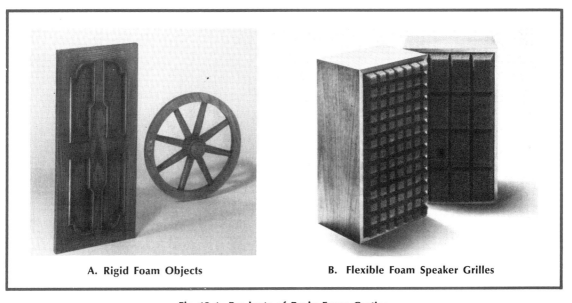

A. Rigid Foam Objects **B. Flexible Foam Speaker Grilles**

Fig. 12-1. Products of Resin Foam Casting

234 Section IV CASTING

Fig. 12-2. Flexible, open-cell urethane foam in gas tank to reduce a possibility of explosion.

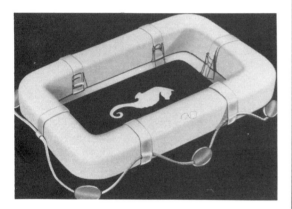

Fig. 12-3. Swim toy made from bead foam.

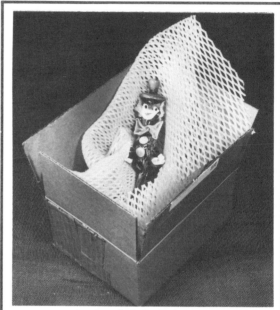

A. Extruded Mesh Foam

B. Spaghetti Foam

Fig. 12-4. Plastic Packaging Materials

Characteristics

Foams may be either an open- or a closed-cell structure. An **open-cell structure** has openings between the cells. The cell walls interconnect. Gases and liquids can move from cell to cell. A sponge is a good example of an open-celled structure. An open-cell foam has been developed which is used in racing car and airplane gas tanks, Fig. 12-2. It reduces the possibility of explosions upon impact or puncture. It takes up only 3% of the volume of the tanks. Foams of a **closed-cell structure** have separate cell walls. Each cell is a separate unit. Gases and liquids cannot move from cell to cell easily. Rigid foams usually have closed cell structures, and flexible foams usually have open cell structures.

Foam products are made in many densities. **Density** is the weight per unit of volume. It is expressed in pounds per cubic foot (g/cm³) for foams. Most plastic foams are low in density (lightweight). Plastic foams range in density from 1/10 to 70 pounds per cubic foot (0.0016 to 1.12 g/cm³). Water weighs 62 pounds per cubic foot (1.0 g/cm³) at room temperature. Most plastic insulation and packaging foams weigh from one to three pounds per cubic foot (0.016 to 0.048 g/cm³). Low densities provide good insulation and flotation characteristics. High densities make stronger, more rigid products. Plastic foams have been developed that have exactly the same density as wood. They also have the feel and strength of wood. These woodlike foams may be nailed, screwed into, and finished like natural wood. Woodlike plastic foams range in density from about 10 to 20 pounds per cubic foot (0.16 to 0.32 g/cm³).

Importance of Foams

Foamed products are common to everyday life. They can be found in cars, the home, and industry. No market figures are available for foamed products as they are reported as a part of other plastics statistics.

Advantages

Plastic foams have many applications. They are used to make lightweight, high-insulation value, cushioning, and flotation products. Foams conserve plastic materials since larger parts can be made with the same amount of raw plastic. There is a choice of a wide variety of techniques for processing foam. Equipment costs vary greatly depending on the process used.

Disadvantages

Many foamed plastics cure or solidify slowly in the mold, due in part to their high insulation value. Slow curing tends to lengthen the cycle time. If molding presses are used for processing, the longer cure cycle increases the molding cost. However, the increased molding cost is often offset by the lower material cost.

Industrial Equipment

Foams may be processed by extrusion, compression molding, injection molding, spraying, pouring, and bead expansion. In some cases, special foam molding equipment is built. In others, conventional extruders or injection molders are adapted for foam molding.

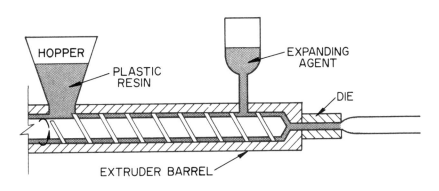

Fig. 12-5. In the screw extrusion of plastic foam, an expanding agent is added in the metering section of the barrel.

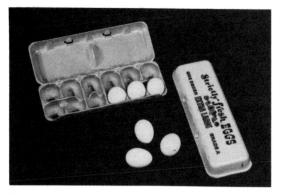

Fig. 12-6. Thermoformed Foam Sheet Egg Cartons

In extrusion foam molding, an expanding agent is added to the plastic melt in the extruder barrel, Fig. 12-5. The material expands after it leaves the extruder die. An example of this is extruded foam sheet that is thermoformed into egg cartons, Fig. 12-6.

In a similar process, a plunger type injection molder is used for expansion, Fig. 12-7. In this method the expanding or blowing agent is added to the plastic in the heating cylinder or barrel.

Another process called **structural foam molding** usually uses an accumulator be-

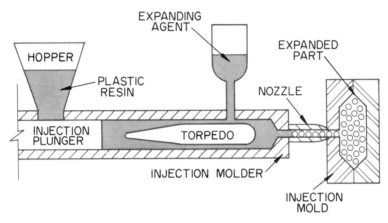

Fig. 12-7. Plunger Injection of Foamed Plastics
The expanding agent is added to the heating cylinder to produce a foam molded part.

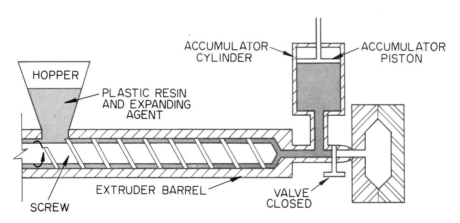

Fig. 12-8. Structural foam molding material is forced into the accumulator.

tween the extruder screw and the mold, Fig. 12-8. The molten plastic and blowing agent are extruded into the accumulator. When the proper size charge is stored in the accumulator, it is quickly forced into the mold where it expands rapidly, Fig. 12-9. A dense, solid plastic skin usually forms outside the foam core. Structural foam products are strong and rigid. This process can be used for furniture frames, shipping containers, pallets, and decorative moldings for the fronts of furniture.

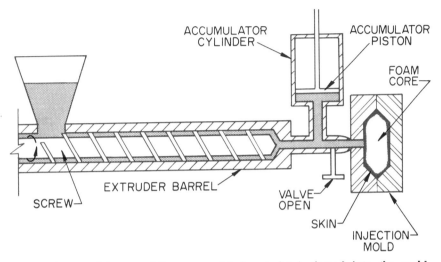

Fig. 12-9. Structural foam molding accumulated material is forced into the mold where it expands.

Fig. 12-10. Structural Foam Molding Press

238 Section IV CASTING

Fig. 12-11. Furniture parts made by the structural foam molding process.

Liquid plastic resin foams may be cast into open or closed molds. The resin components are usually mixed and dispensed with special foam dispensing equipment, Fig. 12-12. Liquid plastic resin foams can also be mixed by hand in small quantities.

Expandable polystyrene bead foam products are commercially molded in specially

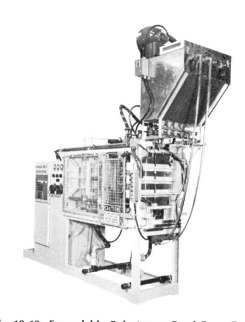

Fig. 12-13. Expandable Polystyrene Bead Foam Press

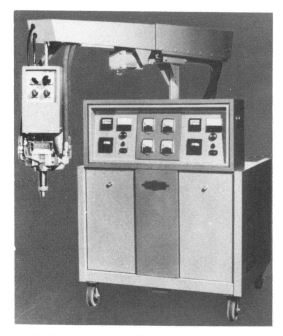

Fig. 12-12. Resin Foam Dispensing Unit

Fig. 12-14. Expandable Bead Foam Mold

designed foam molding presses, Fig. 12-13. Bead foam can also be cast in small molds in a pressure cooker or hot water.

Resin Foam Casting

This chapter includes the casting of two- and three-part rigid and flexible urethane foams. Operation sequences for both are the same, except for the difference in the proportions of each component (part) of the resin system. These also vary among manufacturers.

Urethane resin foams are thermosetting plastics and polymerize by exothermic heat. They work by the chemical foaming method. That is, they foam by the generation or release of a gas in the resin as they are mixed. The gas in the resin is expanded by the exothermic heat. As the gas expands, it makes larger cells. The heat then cures the resin in the expanded (foamed) form.

In some resins, the gas and the promotor are in one component of the resin mix. The gas is absorbed by the resin component much like carbon dioxide in soda pop. It is held under pressure in the container. A catalyst is contained in the other part of the mix. The two components of the resin mix are contained in separate cans. When the part "A" resin component and the part "B" resin component are mixed together, heat is caused by the chemical reaction of the catalyst in part "A" and the promotor in part "B". The chemical reaction causes heat, expands the gas, and cures the foam.

In other resin mixtures, a catalyst is added to the resin. The catalyst reacts with the resin causing (1) a gas to be generated by the chemical reaction between the catalyst and resin and (2) the heat to be formed. The heat then expands the gas which has been generated and cures the resin.

Resin foam densities vary from about one pound per cubic foot to 70 pounds per cubic foot (0.016 to 1.12 g/cm^3). Densities are controlled by the type of foam used and the mixing proportions. Lighter (less dense) rigid closed cell foams usually hold up more weight in water. They also have better insulation values than more dense foams. Because dense foams are more rigid, they are used to form such items as furniture parts and door panels.

Fig. 12-15. Molds for resin foam casting.

Molds for resin foam casting may be of several types. Flexible silicone RTV and urethane rubber work well, but they are the most expensive moldmaking materials. Thermoformed polyethylene and high impact polystyrene are also satisfactory materials. The resin foams release from the silicone RTV, urethane, and polyethylene but stick to the polystyrene. Plaster, gypsum cement, fiberglass, and cast aluminum molds have also been used successfully. These molds require several coats of a mold release. Consult the resin manufacturer's instructions on the proper choice of mold release.

Another way to cast resin foam is called the **foam-in-place method.** The object into which the foam is poured becomes a permanent mold. Examples of this are foamed-in-place insulation, boat flotation chambers, and foam core panels. Duck decoys as well as other products can be made this way in the school laboratory. The outer shell can be thermoformed in two halves. The halves are glued together and filled with foam from the bottom. The outside is painted and weights are added. See Fig. 12-16.

Checklist of Materials and Equipment Needed

Assemble the following list of materials and equipment for resin foam casting:

1. Resin foam components.
2. Disposable cups for mixing.
3. An accurate scale for weighing, graduated in grams.
4. Color additives — dry colorants or polyester pigments.
5. Eye droppers for catalysts (if needed).
6. Mixing sticks (coffee stirrers work well).
7. Power blade for paint mixing and electric drill.
8. Molds.
9. Mold release (if needed).
10. Acetone or lacquer thinner for cleanup.
11. Wiping rags or paper towels.

Fig. 12-16. Thermoformed duck decoy filled with urethane foam.

Operation Sequence for Resin Foam Casting

There are six steps in the process of resin foam casting:

1. Prepare the mold.
2. Figure the volume of the mold.
3. Measure the resin components.
4. Add the resin components together and mix well.
5. Pour the mixture into the mold.
6. Cure the part and remove it from the mold.

Study the operation sequence in Table 12-1. Also, **read the resin manufacturer's directions before starting. Observe any precautions given.** Keep all foam components in a cool, dry place. Many need to be stored and used at less than 75°F (24°C). Flexible foams absorb moisture so they should not be refrigerated. All foam materials should be **capped tightly** after each use. Clean the lid and top of the can well before closing. Clean the mixer blade with acetone or lacquer thinner. **CAUTION: Lacquer thinner and acetone are flammable.** Wear protective clothing.

Safety Note

Use resin foams only in a well ventilated room. They should be placed directly under a ventilating hood when foaming.

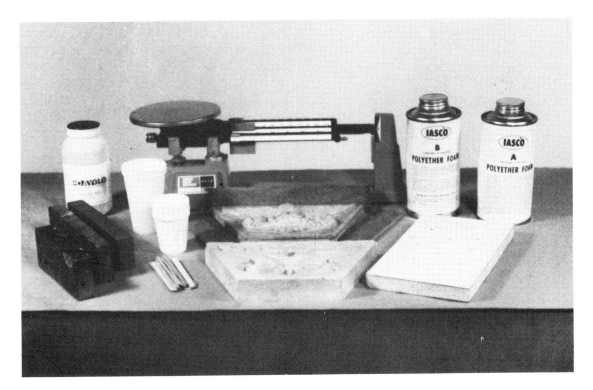

Fig. 12-17. Materials assembled for resin foam casting.

TABLE 12-1
Operation Sequence for Casting Urethane Foams

Sequence/Action	Reason	Troubleshooting
❶ Prepare the mold. (a) Silicone RTV and urethane rubber molds: Mold release is often not needed. A barrier coat may be applied as a release agent, Fig. 12-18. Spray, brush, or wipe it on. (b) Vacuum-formed polyethylene: No release needed. (c) Vacuum-formed high impact polystyrene: No release needed for flexible foams. (d) Fiberglass and cast aluminum molds: Use fluorocarbon mold release. (e) Plaster molds: Coat well with vaseline.	(a) Barrier coats for urethane foams help release them and provide a base for finishes. They may be bought from many major paint companies. See Appendix B. (b) Polyethylene is naturally waxy.	Part sticks to the mold: (a) Wrong mold release used. (b) Mold release not put on right. (c) Part taken from the mold too soon. **Caution:** Let all mold releases dry completely before pouring the foam. Solvents in the mold release can collapse the foam.
❷ Measure the volume of the mold. Fill the mold with water, Fig. 12-19. Pour the water into a square or rectangular container. Figure the volume of the water in cubic feet (cubic centimeters or millilitres — 1 cm³ = 1 ml) by measuring the size of the space it occupies. Add 10% to make sure the mold fills.	The volume should be measured to be sure that the mold is filled in one shot without waste.	Mold does not fill: (a) Not enough resin mixed. (b) Wrong mixture. Material does not foam up enough. (c) Old materials. Gas may be gone from the resin. (d) Solvents in the mold release not evaporated before the resin is poured into the mold.
❸ Calculate and measure the foam components. (a) Multiply the foam density by the mold volume. Foam density is given in pounds per cubic foot (g/cm³). (b) Weigh the mixing cup. (c) Multiply the total resin weight by the percentage of Part "A" (resin component "A"). **Check the manufacturer's instructions for percentages.** (d) Multiply the total weight of the resin by the percentage (%) of Part "B". (e) Repeat "d" above if a third component is used. (f) Add the mixing cup weight to the weight of Part "A". Pour Part "A" resin into the cup until this amount is weighed out, Fig. 12-20. (g) Add the weight of the cup and Part "A" to the weight of Part "B". Set the scale and add "B" to "A" until the total weight is reached. (h) Repeat this if a third part is used. (i) Color pigment may be added at this point.	(a) To figure the weight of the **total resin mix** needed. Each part will be a percentage (%) of this weight. (b) To deduct the cup weight from the mix weight. (c) To get the right weight for Part "A". (d) To figure the right weight for Part "B". (e) To figure the right weight for Part "C" if used. (f) Total weight of the first measurement is the cup weight plus Part "A" weight. (g) To get the right amount of Part "B" in the mixture. Second measure = "A" + "B" + the cup weight. (h) Third measure = "A" + "B" + "C" + cup. (i) To color the resin if desired.	(e) Resin components not figured right. Recheck volume and component calculations.

(continued on page 244)

Fig. 12-18. Applying barrier coat to silicone mold.

Fig. 12-19. Measuring mold volume with water.

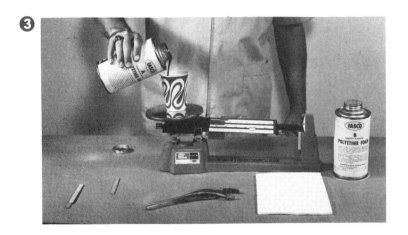

Fig. 12-20. Weighing urethane resin foam.

TABLE 12-1 (Cont.)

Sequence/Action	Reason	Troubleshooting
Example of resin calculation: 90% Part "A" and 10% Part "B"*		

Mold volume   12" x 12" x 4" (30.48 cm x 30.48 cm x 10.16 cm)   =   ⅓ cubic foot (9439 cm³)
 Foam weight   ⅓ cubic foot (9439 cm³) x 3 lb./ft.³
                   (0.048 g/cm³)   =   1 pound or 454 grams (g)
 10% extra foam   454 grams × .10   =   45.4 grams**
 Total needed   454 grams + 45.4 grams   =   499.4 or 500 grams
 Cup weight   20 grams   =   20 grams
 Part "A" weight 500 grams × .90   =   450 grams
 Part "B" weight 500 grams × .10   =   50 grams
 	
 CALCULATION: 20 grams cup weight
                 + <u>450</u> grams PART "A" weight
                    470 grams measurement #1
                 + <u> 50</u> grams PART "B" weight
                    520 grams measurement #2
 	
 *This is **only an example** of component calculation. Each resin has a different proportion of A and B components. Read the manufacturer's instructions for the right ratio.
 **Calculations are given in grams because more accurate weight measurement can be made on a gram scale.		

④ Mix the components. (a) Mix the two parts of the resin very well for about 20 to 30 seconds with a high shear paint mixing blade on an electric drill. The resin may also be mixed by hand with a stick, Fig. 12-21. (b) As soon as the mixing cup feels warm, stop mixing. The foam mix should be poured at once, Fig. 12-22. (c) Clean the mixing blade with acetone or lacquer thinner. (d) Close the mold on the mix, Fig. 12-23. A relief hole may need to be left open to let air and extra resin out. Use weights, clamps, or bolts to hold the mold closed.	(a) To mix the two or three parts of the resin well and start the foaming action. (b) The resin mix will begin to thicken as soon as heat is generated. It will not fill the mold properly if it gets too thick. (c) Trapped air may cause the mold to not fill right. (d) The foam sometimes collapses if a vent hole is not used.	**Air void in the part:** (a) Air trapped in the mold. (b) Resin overmixed. (c) Waiting period too long before resin was poured. **Collapsed foam:** (a) Mold not vented. (b) Wrong kind or too much pigment. (c) Wrong mixture of resin components. (d) Resin overmixed. (e) Mold release solvents not dried out in the mold. (f) Waxed paper cups used.
⑤ (a) Cure the product. Cure time of the foam depends on the resin used, mold temperatures, room temperature, and humidity. **Read the manufacturer's instructions for approximate times.** Some resins need oven cures. Follow the manufacturer's instructions carefully for this. (b) Remove the part from the mold when the surface of the excess resin is tack-free. Open the mold, Figs. 12-24 and 12-25. Trim off the flash with a scissors. Paint the surface if desired.	Curing time and temperature are important. Each resin is designed for a specific one.	

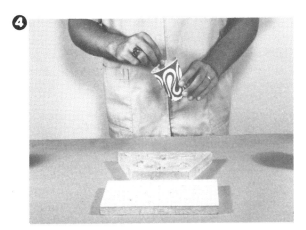

Fig. 12-21. Mixing urethane resin foam.

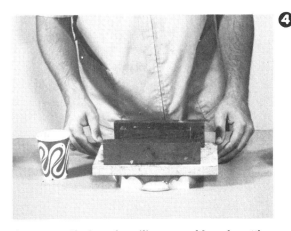

Fig. 12-23. Closing the silicone mold and putting weights on to hold it shut.

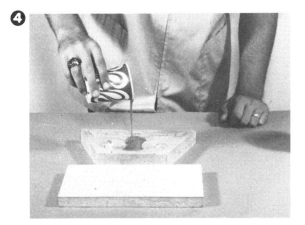

Fig. 12-22. Pouring resin foam.

Fig. 12-24. Opening the mold after curing.

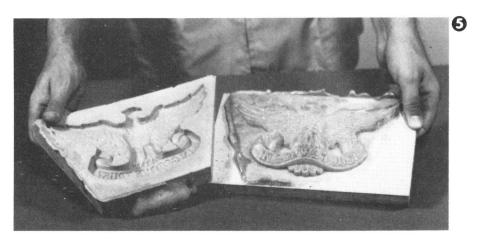

Fig. 12-25. Cured Product and Mold

Expandable Polystyrene Bead Foam Molding

Expandable polystyrene bead foams are thermoplastics. They are small spheres with a porous center. They look like a sponge inside. When the beads are heated, the cell walls soften and a gas inside expands. The softened beads expand and fuse together in a mold. They are cooled and a solid closed cell mass results.

Virgin (unexpanded) polystyrene beads may be expanded up to about 50 times their original size. See Fig. 12-26. Virgin beads are about 0.020" (0.50 mm) to 0.057" (1.44 mm) in diameter. Controlled densities of 0.8 to 20 pounds per cubic foot (0.012 to 0.32 g/cm^3) are possible. The best insulation value is at a density of 2 pounds per cubic foot (0.032 g/cm^3). At 2 pounds per cubic foot (0.032 g/cm^3) density, 1 cubic foot (28 317 cm^3), of expanded foam will support 60 pounds (27.2 kg) of weight in water.

Bead foams are used for familiar things like picnic coolers, ice buckets, packaging cushions, and flotation devices. Typical examples are shown in Figs. 12-27 and 12-28.

Polystyrene beads are usually expanded in machined or cast aluminum molds. In industry these molds are fitted into large presses which look somewhat like big injection molding machines. One is shown in Fig. 12-13. The beads are usually expanded by high pressure steam (20 to 25 psi) (138

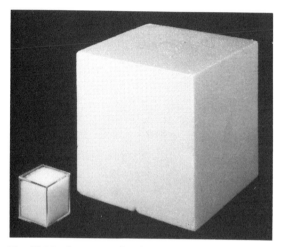

Fig. 12-26. One pound of polystyrene bead foam before and after expansion.

Fig. 12-27. Boat made of expandable polystyrene bead foam.

Fig. 12-28. Flotation ring made from bead foam.

to 241 kPa) and cooled with cold water. The steam and water enter a steam chest or cavity behind the mold. The heating and cooling is transferred through the mold from the steam chest. This is the industrial method most used for processing foam beads. Steam autoclaves or pressure cookers may also be used. In these, the molds are placed inside the steam autoclave for heating. After the heating cycle is completed, the molds are removed and cooled outside the autoclave.

Polystyrene beads are usually pre-expanded before being put into the mold. This brings them up to a usable size. The density of the final product depends upon the pre-expanded density. Pre-expanded beads should be used within three to six days. After pre-expansion, the beads loose their internal gas rapidly. Virgin beads should be stored in a cool place. They may be refrigerated or frozen for longer shelf life.

Laboratory Bead Foam Molding

In the laboratory, polystyrene beads are usually molded in aluminum hand molds. These molds may be purchased from several plastics suppliers or made in the laboratory. The molds are filled with pre-expanded beads and put in either a pressure cooker (autoclave) or a pail of boiling water. After processing for the required time, the molds are taken out and cooled in a pail of cold water.

Fig. 12-29. Commercial pre-expansion setup for polystyrene bead foam.

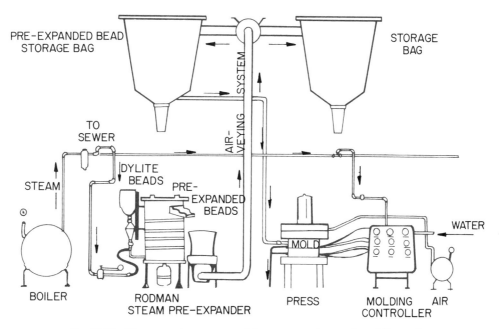

Fig. 12-30. Diagram of a commercial pre-expansion and molding setup.

248 Section IV CASTING

Fig. 12-31. Rodman Pre-Expander for Expandable Polystyrene Beads shown with Sides Removed.

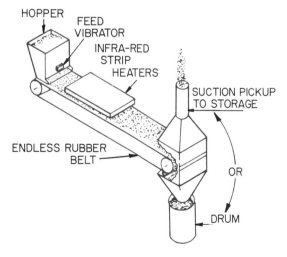

Fig. 12-32. Diagram of a Radiant Heat Pre-Expander

Pre-expansion of the beads in the laboratory may be done either in boiling water or by radiant (dry heat) pre-expanders that are available for the school laboratory.

Operation Sequence for Bead Foam Molding

There are six steps in the process of bead foam molding:

1. Pre-expand the beads.
2. Prepare the mold(s).
3. Fill the mold(s) with pre-expanded beads and close it.
4. Fuse the beads in the mold under heat.
5. Cool the mold and part inside.
6. Open the mold and remove the part.

The operation sequence is shown in Table 12-2. Also, **read the bead manufacturer's instructions.**

Safety Note

When handling the virgin polystyrene beads in large quantities, observe the following precautions:
  a. A volatile pentane gas gathers above large quantities of the beads. **A potential fire or explosion hazard exists.** Open the bead barrel and allow the gas to escape before removing the beads.
  b. Use a paper or plastic container to scoop out the beads. A metal cup might cause a spark of static electricity that will ignite the gas. No apparent danger exists when the beads are packed in small plastic bags.
  c. Do not use virgin or pre-expanded beads near flame of any kind.
  d. Protect the bench tops with cement-asbestos (Transite®) board.

Assemble all the materials required, Fig. 12-33.

Expansion Processes   249

Fig. 12-33. Materials Assembled Ready for Use — Expandable Polystyrene Bead Foam Molding

Checklist of Materials and Equipment Needed

Assemble the following materials and equipment for bead foam molding.

1. Electric stove or hot plate(s) with 2000 to 2500 watt burners, 230 volt, single phase.
2. One or two five-gallon metal pails for pre-expanding beads.
3. A large pressure cooker (21-quart or larger).
4. Two or three five-gallon plastic pails for cooling water.
5. Assorted aluminum molds.
6. Silicone or fluorocarbon mold release.
7. Expandable polystyrene beads, white and/or colored.
8. Dry color pigments for polystyrene beads (optional).
9. One or two large kitchen sieves to remove the pre-expanded beads from the water.
10. Two pairs of insulated gloves.
11. Drying frame or power drier for pre-expanded beads (optional).
12. Plastic storage containers for foam beads.

Rejuvenating Expandable Polystyrene Foam Beads

Old **virgin** polystyrene beads may be rejuvenated (revived) by saturating them with methylene dichloride gas. A box similar to the fluidized bed coater, for which plans are provided at the end of Chapter 15, may be used. The bottle should be filled about ⅔ full of methylene dichloride. A pressure of about 1 to 5 psi (6.9 to 34.5 kPa) should be applied to the inlet tube of the bottle. This will generate bubbles of gas which then enter the chamber under the porous plate and penetrate up through the beads reviving them. These beads should be used soon after treatment. Pre-expanded beads do not seem to respond to this treatment. See Figs. 12-42 and 12-43.

TABLE 12-2
Operation Sequence for Polystyrene Bead Foam Molding

Sequence/Action	Reason	Troubleshooting
❶ Pre-expansion — hot water method: (a) Bring about 1 gallon (4 l) of water to boil in a 5 gallon (20 l) pail. Pour a few ounces of virgin beads into the boiling water, Fig. 12-34. Stir until the proper bead density is obtained, Fig. 12-35. (b) Remove the pail from the burner. Dip the beads out with a sieve. Place them in the drying rack. **Heat gun and traveling belt method:** (a) Heat gun and traveling belt pre-expanders are available or may be built. This is a dry type of pre-expansion. Follow the manufacturer's instructions for their use.	(a) To bring the beads to the proper density for molding. (b) To prepare beads for loading into the mold. Allows excess water to drain off. (a) Dry-type pre-expansion produces a bead which can be **poured** in a mold immediately.	**Beads do not pre-expand:** (a) Beads are too old. The gas has escaped. (b) Water not hot enough.
❷ Prepare the mold. Dry the mold out with a blast of compressed air. Spray with mold release about once each 10 times mold is used or when parts stick. Use a fluorocarbon or silicone mold release. Paste wax may also be used.	To prevent sticking. The mold release should dry several minutes before use.	**Parts are stuck to the mold:** (a) Needs more mold release. Clean the mold with acetone before applying more mold release. (b) Change to a different type mold release. (c) Increase cooling time.
❸ Fill and close the mold. (a) **Dry Fill Method:** Close the mold and bolt or clamp shut. Open the filler hole and pour in dry pre-expanded beads, Fig. 12-36. Shake down until full. Close filler hole. (b) **Wet Fill Method 1.** Wet beads may be put in one open mold and built up to fill the other half of the cavity. Beads must be kept off the parting line. (c) **Wet Fill Method 2.** Fill each half of the mold with wet beads, Fig. 12-37. Place a piece of sheet metal over one. Put the two mold halves together and slide out the sheet metal, Fig. 12-38. See also Fig. 12-39.	(a) Dried beads may easily be poured into molds with large openings. (b) Allows quick use of wet expanded beads. (c) Allows a uniform fill of spherical objects. Note: Be sure molds are completely filled in all methods.	Mold not filled out, dry fill: Beads not shaken enough. Mold not filled, wet fill: Not enough beads packed in.

(continued on page 252)

Expansion Processes 251

Fig. 12-34. Pouring in virgin beads, using the laboratory pre-expansion, hot water method.

Fig. 12-37. Wet fill of polystyrene beads in a mold.

Fig. 12-35. Stirring to get uniform expansion.

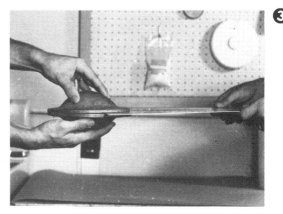

Fig. 12-38. Pulling the slip sheet, two-person method.

Fig. 12-36. Dry filling a mold with expanded polystyrene beads.

Fig. 12-39. Pulling the slip sheet, one-person method.

TABLE 12-2 (Cont.)

Sequence/Action	Reason	Troubleshooting
❹ (a) **Fusing in the Mold, Autoclave Method,** Fig. 12-40: Bring 1 to 2 inches (25 mm - 50 mm) of water to a boil in a pressure cooker. Load several molds in the cooker. Close the cover and release valve. Bring to about 20 psi (138 kPa). Hold this pressure for 20 to 30 seconds. Remove the cooker from the stove and open the release valve. Let out **all the steam** and **then** open the lid. **Caution: Steam is hot. It burns. Handle carefully.** (b) **Fusing in Boiling Water:** Molds may be placed in pail 3/4 full of boiling water to fuse the beads. A longer heating cycle is needed. Twenty to 30 minutes is often needed for this method.	To fuse or weld the beads together in the mold to produce a solid part. Note: Thicker parts require longer fusing time. For thick parts (footballs, duck decoys, etc.) hold 20 psi (138 kPa) pressure for about 1 minute or longer. May be used where a pressure cooker or autoclave is not available.	Beads do not fuse: (a) Not under pressure long enough. (b) Steam not entering mold. Drill several small (#60) holes in the mold. (c) Dry beads more before filling mold. (d) Cool mold more slowly.
❺ **Cool the mold.** Put on gloves. Take the molds from the pressure cooker or water pail. Use a pair of pliers to pick up the molds, Fig. 12-41. Drop the molds in a bucket of cool water and let cool, the longer the better.	(a) Prevent burns by using pliers and gloves. (b) Long cooling periods are needed for thick objects. The thicker the part the longer it must cool.	Part expands after removal from mold: Cool it longer. Part collapses after removal from mold: (a) Do not heat it as long. (b) Pre-expansion density too low. (c) Cool more gradually.
❻ **Remove the part from the mold.** When you can hold the mold in your bare hands, open it. Remove the part with a blast of compressed air. Trim off any flash with a scissors.	Water tension in molds often holds the part in. An air blast removes it.	

Expansion Processes   253

4

Fig. 12-40. Fusing bead foam in an autoclave (pressure cooker).

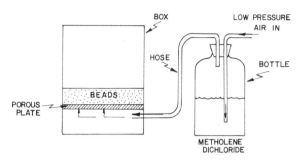

Fig. 12-42. The device used to rejuvenate polystyrene beads.

5

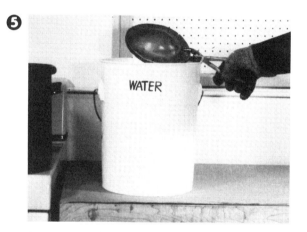

Fig. 12-41. Cooling the mold in water.

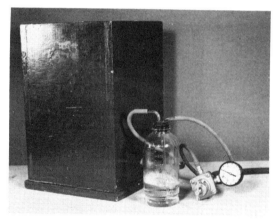

Fig. 12-43. Using methylene dichloride to rejuvenate beads. This unit is the same as that described in Fig. 15-33.

STUDENT ACTIVITIES

1. Process available resin and bead foams in school-made or purchased molds.
2. Materials experimentation:
   a. Pre-expand foam beads to different densities. Mold products from the different density beads and compare them.
   b. Mix different colors of foam beads to obtain color variations.
3. Process experimentation: Process foam beads in the mold for different time periods. Compare the results for over- and under-processed parts.
4. Design products and make molds for resin and bead foam processes. Cast or machine aluminum molds for bead foam. Cast resin foam molds from urethane or silicone rubber. Resin foam molds may also be cast from plaster and aluminum.
5. Production organization and operation:
   a. Obtaining material supply.
   b. Preparing the molds.
   c. Measuring the materials.
   d. Mixing materials or pre-expanding beads.
   e. Filling molds.
   f. Curing parts.
   g. Removing parts.
   h. Record keeping.
      1) Materials.
      2) Processing variables.
      3) Assembly and finishing.
      4) Rejects.
      5) Labor.

Questions Relating to Materials and Process

1. Which of the foam materials should be used for the following applications?
   a. Egg cartons.
   b. Refrigerator insulation.
   c. Flotation chambers in a canoe.
   d. Floating swim toys.
   e. Furniture frames.
   f. Package cushioning.
   g. Dashboard padding.
   h. Mattresses.
2. Name some products that would be best suited for the following materials.
   a. Structural foams.
   b. Extruded foam sheet.
   c. Injection molded foam.
   d. Bead foam.
   e. Rigid urethane foam.
   f. Flexible urethane foam.
3. Give the main characteristics for the various types of foams.

Process Description

Thermofusion, meaning **heat fusion,** is a name given to a group of similar plastics processes. It is the melting and sticking together of plastics particles by means of heat. In these processes, plastics are fused (melted) together in or on a mold surface. No force, other than gravity and atmospheric pressure, is added to the plastic material during fusion. Thermofusion molding is similar in some ways to casting except that it is often used to make hollow parts. Casting methods are usually used to make solid parts, although they may also be made by thermofusion.

There are several processes in the thermofusion group. They are divided into two types:
1. Static (stationary) and
2. Dynamic (in motion).

In the **static** processes, the mold is stationary (static) during the fusion of the plastic. In the **dynamic** processes, the mold rotates (turns) during the fusion. The **static** type of processes includes dip casting and coating, full mold casting, slush casting, and static powder molding (Engle® process). The **dynamic** type of thermofusion processes includes single-axis rotational molding (Heiser® process), two-axis rotational molding, and centrifugal casting. Either powdered or liquid plastics may be used in most of these processes.

THERMOFUSION

CHAPTER 13

Static powder molding, single-axis and two-axis rotational molding are technically **casting** processes because the molds are filled by gravity. In addition, no force other than gravity and atmospheric pressure is applied to the materials during processing. They are all, however, called molding pro-

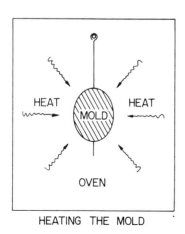

HEATING THE MOLD

DIPPING

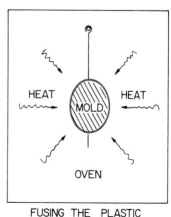

FUSING THE PLASTIC

Fig. 13-1. Dip Casting

255

```
                    THERMOFUSION
        ┌───────────────┴───────────────┐
Static (Stationary)         Dynamic (In Motion)
1. Dip casting and          1. Single-axis rota-
   coating                     tional molding
2. Full mold casting           (Heiser® process)
3. Slush casting            2. Two-axis rota-
4. Static powder               tional molding
   molding (Engle®          3. Centrifugal
   process)                    casting
```

cesses more often than casting by the plastics industry. For this reason they will be called **molding** processes in this chapter. Since centrifugal casting is seldom used for plastics now, it will not be discussed.

Thermofusion Materials

The plastic material used in thermofusion may be either dry powder or liquid. Almost any plastic that may be finely powdered is used. Powdered plastics are usually available in coating and molding grades. **Coating grades** stick to the product, while **molding grades** are designed to release from the mold. Finely ground polyethylene is a dry powdered plastic. It is probably the most popular thermofusion plastic at present.

Vinyl plastisol is used as a liquid thermofusion plastic. Finely ground polyvinyl chloride plastic powder is mixed with a liquid called a **plasticizer**. The plasticizer makes the vinyl plastic softer and more flexible. When heat is added to the plastisol, the suspended vinyl powder melts (fuses) and the plasticizer cures (polymerizes) with the vinyl.

Vinyl plastisol may be prepared in various hardnesses, depending on the amount of plasticizer used. The more plasticizer used, the softer the vinyl product. Hardness of vinyl plastisols can be measured by a durometer, a hardness rating instrument used for rubber and rubberlike materials. The higher the durometer number, the harder the plastic or rubber.

Vinyl plastisols usually do not stick to the surface being coated. This helps to remove parts from a mold. If adhesion is desired, a plastisol primer is usually used. Some special formulations of self-bonding plastisols are available.

Dip Casting and Coating

Dip casting and coating is a thermofusion process in which a hot metal mold or product — about 400°F (205°C) — is dipped into a cold (room temperature) liquid plastic. This process may also be called **dip molding**. The plastic used is usually a vinyl plastisol. The heat in the mold melts the plastic that is next to the mold. When a layer of plastic is built up (fused), the mold or product is taken out of the liquid plastic. The mold or product with the layer of plastic on it is then put in an oven at about 300°F to 350°F (149°C to 177°C). The plastic is completely fused and smoothed in the oven. The fused part is then cooled and removed from the mold. When the fused plastisol is removed from the mold, the process is called **dip molding**. When the fused plastisol is left on the part permanently, the process is called **dip coating**.

Importance of the Process

Dip casting is a minor process in the plastics industry today. Usually only fairly small parts are made by this method. However, the process can be highly mechanized, or made automatic.

Advantages

Flexible parts with thin walls may be made by dip casting. Examples are plastic gloves and stretch boots. Parts with thicker walls such as bicycle handle grips and coin purses are also made by this process.

Disadvantages

Dip casting materials must be stirred often to keep an even color on the parts that are being dipped. Dip casting is a fairly slow process. Long mold heating and part curing

Thermofusion

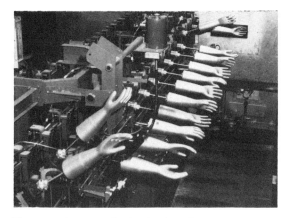

Fig. 13-2. Automated Dip Casting Line

Fig. 13-3. Forms enter tank of vinyl.

times are needed. The uniformity of the part thickness depends on the mold heating and dipping time. The mold only controls the shape of the inside of the part. Part thicknesses vary more by this method than processes such as injection molding.

Materials Used in Industry

Dip casting is usually done with vinyl plastisols. Various hardnesses may be used, depending on the product to be made. Hardness is controlled by the amount of plasticizer added to the vinyl.

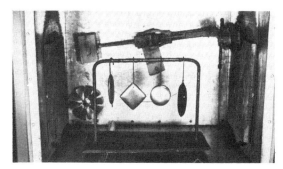

Fig. 13-4. Rotomolding machine set up for dip casting.

Industrial Equipment

Production lines for automatic dip casting have been developed. They include a preheat oven, dip tank, fusing oven, cooling station, and a stripping station. The molds or parts to be dip cast are carried through this line on a conveyor belt. They are stripped (removed from the mold) by compressed air. Similar lines have been set up for dip coating except that the coating is not removed from the product.

Fig. 13-5. Materials assembled for dip casting.

Laboratory Equipment

A good heating oven is probably the most important piece of laboratory equipment needed for the dip casting process. A rotational molder or an electric kitchen stove oven may be used. Other equipment includes the molds and a device for suspending them in the oven.

TABLE 13-1
Operation Sequence for Plastisol Dip Casting and Coating

Sequence/Action	Reason	Troubleshooting
❶ Mix the plastisol well. Use a paint paddle or power mixer, Fig. 13-6. Add color pigments and mix again.	To mix the plastisol and colors and make a uniform material.	Dipping makes a thin coating: (a) Plastisol not mixed well. (b) Wrong type of plastisol used.
❷ Remove the air bubbles from the plastisol by letting it stand 24 hours, jiggling, or vacuum de-airing it.	Air bubbles will show up in the surface of the part if they are not removed from the liquid plastisol.	**Bubbles in the surface of the finished part:** Remove the air bubbles from the liquid plastisol before dipping.
❸ Preheat the mold(s) or part(s) to be dipped for about 20-30 minutes at 400° F (204° C), Fig. 13-7. Check the manufacturer's instructions for exact time and temperature.	The mold must soak up enough heat to fuse a layer of plastisol.	Dipping makes a thin coating: The molds are not preheated long enough or hot enough.
❹ Put on insulated gloves. Take a mold from the oven with a pair of pliers, Fig. 13-8.	Gloves protect the hands from burns. Pliers are needed for protection too.	**Safety first:** Always use gloves and pliers when handling hot molds.
❺ Dip the mold or object in the plastisol for about 1½ minutes, Fig. 13-9. **Do not touch the sides of the plastisol can with the mold.** Let the plastisol stop dripping before returning the mold to the oven, Fig. 13-10.	The longer the mold is dipped in the plastisol, the thicker the layer it will get. **Dipping time may vary with different brands of plastisol.**	Dipping makes a thin coating: The mold may not be dipped long enough.
❻ Put the dipped mold back in the oven to cure, Fig. 13-11. Cure (fuse) the plastisol about 15 minutes at about 350°F (177°C). Check manufacturer's instructions for exact time and temperature.	To fuse the plastisol. **Temperatures and time for curing may vary with different brands of plastisol.**	**Plastisol tears or crumbles when being removed from the mold:** Plastisol was not cured long enough.
❼ Put on insulated gloves. Take the mold from the oven with a pair of pliers.	**Safety first:** Always use gloves and pliers to handle hot molds.	**Plastisol turns black or burns in the oven:** The oven is too hot for the plastic being used.

(continued on page 260)

Fig. 13-6. Mix plastisol with an electric drill and paint mixing paddle blade.

Fig. 13-9. Dip the hot mold in the plastisol.

Fig. 13-7. Preheat dip molds in the oven.

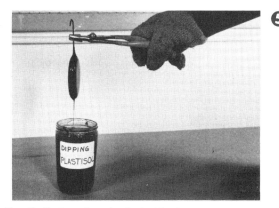

Fig. 13-10. Let the plastisol drip free.

Fig. 13-8. Remove a preheated dip mold from the oven with pliers and gloves.

Fig. 13-11. Put the dipped mold back in the oven.

TABLE 13-1 (Cont.)

Sequence/Action	Reason	Troubleshooting
⑧ Put the cured part and mold in cool water, Fig. 13-12. Cool it for at least 10 minutes.	Safety first: Always cool the plastisol and mold long enough to keep from getting burned when removing the part from the mold.	
⑨ (a) Cut around the wire ends with a sharp knife. (b) Cut a slit from end to end with a straight edge and a sharp knife, Fig. 13-13.	The end holes which are formed by the wire keep the slit from tearing further.	
⑩ Carefully slip the part off the mold, Fig. 13-14.		The part tears when it is removed from the mold: The part was not cured long or hot enough.

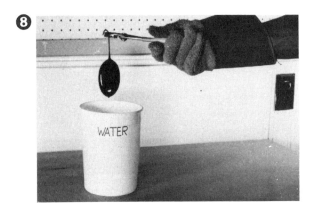

Fig. 13-12. Put the cured part in water to cool.

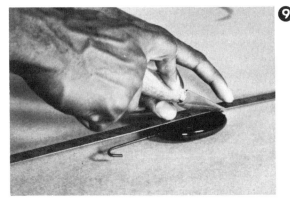

Fig. 13-13. Slit the coin purse from end to end.

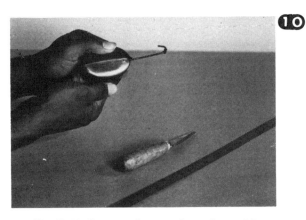

Fig. 13-14. Remove the part from the mold.

Operation Sequence for Dip Casting and Coating

A number of interesting objects can be dip cast or coated in the school laboratory. Molds may be bought or are easily made for coin purses, key cases, comb cases, fishing fly cases, funnels, and similar objects. Tool handles and other metal parts may be dip coated to give a smooth plastic surface.

There are ten steps in process of dip casting. Study the operation sequence in Table 13-1. Also, **read the labels on the cans of plastisol.** Read any other instructions by the manufacturer. The mold preheat temperatures, preheat times, cure times, and cure temperatures are not the same for each plastisol used.

colored products may be made by dipping the mold in one color part of the time and the rest in another color. Use a shorter time for the first color and longer time for the second. More plastisol is picked up by the mold at first. A second layer may also be **added after** the first layer is cured in the oven. Three layers may also be done this way. Remember that the layer thickness is controlled by the time the mold is held in the plastisol and the amount of heat in the mold.

When dip coating tool handles, use a tool dip plastisol and tool dip primer. Some primers must be air-dried for 24 hours before dipping. Hang the tool with a wire. Preheat the tool, dip it, cure it, and cool it as above. Check the directions furnished with the material for exact times and temperatures. Several new air-dry tool dip plastisols are now available. They do not need oven cures.

Checklist for Materials and Equipment Needed

Assemble the following materials and equipment for dip casting and coating:

1. Insulated gloves.
2. Pliers, long nose type preferably.
3. Several aluminum molds or other objects to dip.
4. Heating oven (electric stove or rotational molder).
5. Plastisol material for dip casting or coating.
6. Paint mixer blade.
7. Electric drill, about 1000 rpm.
8. Pail of water.
9. Knife and straight edge.
10. Can of lacquer thinner or acetone.

General Operating Suggestions

Vinyl plastisols usually are sold uncolored. The desired color pigment is added to the plastisol before use. It should be thoroughly mixed with a power mixer and let stand to remove the air bubbles before use. Multi-

Full Mold Plastisol Casting

Vinyl plastisols may also be cast in open molds to form solid objects. This process has many of the same limitations, advantages, and disadvantages as vinyl dip casting. It should not be used for extremely thick castings, however. The process is usually not automated. The main difference between dip casting and full mold casting is the type of mold used. The plastisol is cast **over** a dip mold and **into** a full mold. A common example of full mold plastisol casting is bait casting.

Plastic fishing bait may be cast in an aluminum mold with a special soft vinyl plastisol. The bait casting plastisol has more plasticizer than most dip casting plastisols. Bait casting plastisol makes a soft, flexible, lifelike part. Bait casting plastisols may be colored in the same way as other plastisols. Filling the molds is easier if the colored plastisol is put into squeeze bottles.

The flexible bait casting plastisol is poured in a cool (room temperature) mold, cured in the oven, cooled, and removed from the mold.

> **Safety Note**
>
> Do not mix different kinds of plastisols together. They are usually blended at the factory for specific types of casting. Read the label on the plastisol container for specific manufacturer's instructions before starting.
>
> Wear insulated gloves when handling hot molds.

Checklist for Materials and Equipment Needed

Assemble the following materials and equipment for bait casting:

1. Insulated gloves.
2. Pair of pliers.
3. Bait casting molds.
4. Bait casting plastisol (colored).
5. Heating oven.
6. Pail of water.
7. Knife.

Operation Sequence for Full Mold Bait Casting

There are five steps in the process of full mold bait casting. Study the operation sequence in Table 13-2.

Fig. 13-15. Materials assembled ready for bait casting.

TABLE 13-2
Operation Sequence for Full Mold Bait Casting

Sequence/Action	Reason	Troubleshooting
❶ Pour the bait casting plastisol into each mold cavity, Fig. 13-16. The mold should be at room temperature. Do not overfill the cavities. Plastisol expands when it heats. Fish hooks may be cast into the plastisol if they are wanted.	Mold must be at room temperature to start. As the plastisol heats, it expands and may run out of the mold, causing flash.	Flash over the edge of the mold: Mold filled too full.
❷ Put on insulated gloves. Put the filled molds in the oven, Fig. 13-17. Cure them at 325-350°F (163° to 177°C) until the plastisol **changes color completely**.	When the plastisol cures it turns a darker color. The larger the part, the longer the cure time.	Part tears when it is removed from the mold: It is not cured enough.
❸ Put on insulated gloves. Take the mold(s) from the oven. Put it in water to cool, Fig. 13-18.	Safety first: Always use insulated gloves to put molds in and take them out of the oven.	
❹ Pull the cooled plastic from the mold. Use a sharp knife to get under the edge of the plastic to get it loose. Pull from one end, Fig. 13-19.	The plastic must be pulled with an easy steady motion to keep from tearing it when it is removed from the mold.	Part is brittle: It is overcured.
❺ Trim off any flash.		

Thermofusion 263

Fig. 13-16. Pour the bait plastisol into a cold mold cavity.

Fig. 13-18. Put mold in water to cool.

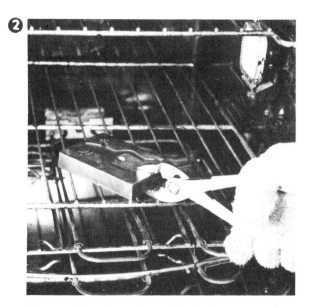

Fig. 13-17. Put the filled mold in the oven, using pliers and gloves.

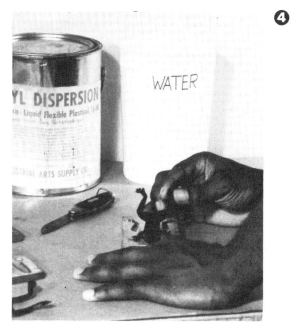

Fig. 13-19. Remove the part from the mold.

Slush Casting

Slush casting of plastisols is similar to full mold plastisol casting except that a heated, hollow, open mold is used. A variety of hardnesses of vinyl plastisols may be used, depending on the product produced. The plastisols may be colored in the usual manner. Hollow parts such as drinking tumblers, dice boxes, football tees and door stops may be made this way.

The vinyl plastisol is poured into the heated mold and allowed to solidify on the mold walls. When the desired wall thickness is reached, the extra plastisol is poured back in the container to be reused. The mold is then put in the oven where the plastisol fuses. The mold is taken from the oven, cooled, and the part removed.

Operation Sequence for Slush Casting

There are six steps in the process of casting. Study the operation sequence in Table 13-3.

TABLE 13-3
Operation Sequence for Slush Casting

Sequence/Action	Reason	Troubleshooting
❶ Preheat the slush casting mold to 350° to 400°F (177° to 205°C) for 20 minutes. Check the material manufacturer's instructions for exact time and temperature.	To get the mold hot enough to fuse a layer on the inside of the mold.	Plastisol does not build up enough on the mold: Preheat the mold longer or hotter.
❷ Put on insulated gloves. Take the mold from the oven and fill it with vinyl plastisol, Fig. 13-20. Let the plastisol fuse to the desired wall thickness. About 2 to 5 minutes is necessary, depending on the mold.	Safety first: Gloves must be worn to handle hot molds.	
❸ Using the insulated gloves to handle the hot mold, pour the extra plastisol back in the container, Fig. 13-21.	After the desired wall thickness is reached the extra plastisol may be returned to the container for reuse.	
❹ With gloves on, put the mold back in the oven. Allow the plastisol to cure at about 400°F (205°C) for 15 to 20 minutes, depending on the wall thickness. Check the plastisol manufacturer's instructions. After the plastisol is cured, a second layer may be added to the hot plastisol. Extra plastic can again be poured out.	Cure time is dependent upon wall thickness. The thicker the wall, the longer the cure time.	
❺ With insulated gloves on, remove the mold from the oven and cool it in water, Fig. 13-22.	Cooling helps shrink the part for easy removal.	
❻ Remove the part from the mold, Figs. 13-23 and 13-24.		Plastisol crumbles when removed from the mold: The plastisol was not cured long enough.

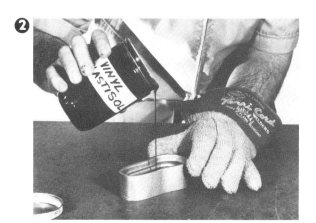

Fig. 13-20. Fill the slush casting mold with vinyl plastisol.

Fig. 13-22. Cool the mold.

Fig. 13-21. Pour the excess vinyl plastisol back.

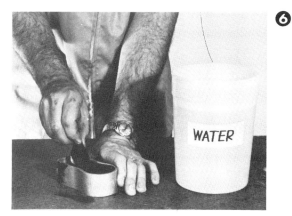

Fig. 13-23. Remove the slush cast part.

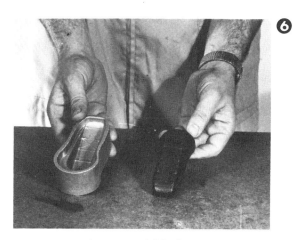

Fig. 13-24. Finished Part

> **Safety Note**
> Do not mix different kinds of plastisols. Read the manufacturer's instructions before use.
> Wear insulated gloves when handling hot molds.

Checklist of Materials and Equipment

Assemble the following materials and equipment for slush casting:

1. Insulated gloves.
2. Vinyl plastisol and color pigments.
3. Slush casting mold(s).
4. Heating oven.
5. Pail of water.

Manual Rotocasting of Plastisols

If a rotational molder is not available, closed hollow parts may be made by manual Rotocasting. This process is similar to rotational molding except that the mold is rotated in two directions by hand. An ordinary kitchen oven or similar heating device may be used to heat the mold and cure the part.

In manual rotocasting, a plastisol which is more rigid than dip or bait plastisol is often used. However, dip casting plastisol also may be used. It is colored in the same manner as the other plastisols. The vinyl plastisol is poured into a heated mold, and the mold is closed and bolted shut. The mold is rotated in two directions by hand until the plastisol is fused in an even layer on the inside of the

TABLE 13-4
Operation Sequence for Manual Rotocasting

Sequence/Action	Reason	Troubleshooting
❶ Preheat the aluminum mold in the oven at about 350° to 400°F (177° to 205°C) for 15 to 20 minutes, Fig. 13-25. Check the material manufacturer's instructions for exact time and temperature.	To get the mold hot enough to fuse the pastisol when it is poured into the mold.	Plastisol does not build up enough on the mold: Mold not preheated long enough or hot enough.
❷ Put on the insulated gloves. Take the mold from the oven and partially fill it with vinyl plastisol. Bolt the mold together. If a fill hole is provided, pour the plastisol into the hole and plug it.	Safety first: Use gloves when handling hot molds.	
❸ Let the plastisol build up a layer (fuse) on the inside of the mold for 1½ to 2 minutes. **Rotate the mold in two directions during the fusing**, Fig. 13-26.	Rotation will let the plastisol fuse in an even layer on the inside of the mold.	Plastisol builds up in an uneven layer: (a) The mold was not rotated. (b) One side of the mold was hotter than the other.
❹ Put the mold back in the oven at about 400°F (205°C). Leave it in the oven for 20 to 30 minutes. Check manufacturer's instructions.	To completely fuse the plastic.	Plastisol crumbles when removed from the mold: The plastisol was not cured long enough.
❺ Put on the insulated gloves. Take the mold from the oven and cool it in water.	To cool the part. Safety first: The part and mold must be cool to prevent burns.	
❻ Open the mold and take the part from it, Fig. 13-27. Trim the part as needed.		

mold. The mold is put back in the oven and the plastisol is cured. It is then removed from the oven, cooled, and the part taken out.

Safety Note

Do not mix different kinds of plastisols. Read the label on the container before using the material.

Always wear insulated gloves when handling hot molds.

Checklist of Materials and Equipment Needed

Assemble the following materials for manual rotocasting:

1. Insulated gloves.
2. Pair of pliers.
3. Rotocasting plastisols and color pigments.
4. Rotocasting molds.
5. Heating oven.
6. Pail of water.

Operation Sequence for Manual Rotocasting

There are six steps in the process of manual rotocasting. Study the operation sequence in Table 13-4.

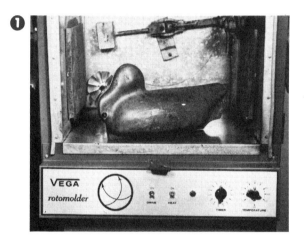

Fig. 13-25. Preheat the mold for manual rotocasting.

Fig. 13-26. Let the plastisol build up on the inside of the mold.

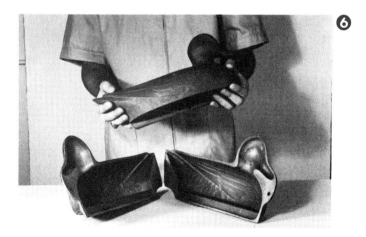

Fig. 13-27. Open the mold and remove the part.

Static Powder Molding (Engle® Process)

The static powder molding (Engle®) process is a stationary (static) molding or casting method. In this process dry plastic powder is put into a cool (room temperature) mold usually made of aluminum or steel. After filling, the top of the mold is covered with a heat insulating cover. The mold is placed into an oven which is preheated to 500°F (260°C) or more. The plastic powder next to the inside wall of the mold melts as the mold heats up, Fig. 13-28. The powder soon fuses into a layer of solid plastic. When the right thickness of plastic is built up on the inside wall of the mold, the mold is removed from the oven. The cover is taken off and the unmelted plastic is poured out. The mold is put back in the oven without a cover. It is left in the oven until the inner surface is smooth. The mold is then taken out and water-cooled.

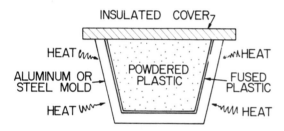

Fig. 13-28. Diagram of static powder molding.

Fig. 13-29. 60-Gallon Static Molded Tank

Importance of the Process

Static powder molding is one of several thermofusion powder molding processes. No breakdown of the use of plastic powders by process is now available. It is rapidly being replaced by rotational molding. The importance of static molding is in its advantages. Tanks, trash containers, and other open-ended containers, both large and small, are made this way. It is a process that can easily be done in the laboratory.

Advantages

Almost any size part can be molded by the static powder molding process. It is limited only by the size of the mold and oven available. See Fig. 13-29. Tanks as large as 500-gallon (1893 l) are possible with this process. Equipment and molds are not very expensive for the size of the part made. Many molds are welded up out of sheet metal. Short production runs can be made where tooling costs are low. No mechanical turning of the mold is needed.

Disadvantages

The fused wall sags some during smoothing. Often the bottom of the part is much thicker than the top. Only low melt index resins — 1.0 to 5.0 — should be used. Using low melt index resins cuts down the sag on the walls. Static powder molding is costly because it requires much hand labor. It is difficult to automate. The plastic powder often gets dirty from being poured in and out of the mold. This process is patented and requires a license for commercial use.

Materials Used in Industry

The Engle® static molding process uses the same type of materials as rotational molding. Polyethylene of high, medium, and low density is the most popular plastic used. Cellulose acetate-butyrate, polystyrene, polycarbonate, vinyl, and nylon are also used. Liquid vinyl plastisols are used in a similar process called **plastisol slush casting.** Slush casting, described earlier, is like slip casting of ceramics.

Industrial Equipment

All that is needed for industrial production by the static powder molding process is a large oven, a mold, and a method of handling the mold. The device for handling the mold must be able to take it in and out of the oven. It must also be able to tip the mold over for dumping the extra plastic out. A specially designed forklift truck is often used for large molds.

Laboratory Equipment

Little equipment other than a mold and a good oven are needed for static powder molding in the school laboratory. A rotational molder oven or an electric stove is adequate. A mold may be cast of aluminum, turned from solid aluminum, or spun from aluminum. A good mold can be made from an aluminum drinking tumbler which can be bought locally. The outside of the mold should be sprayed with high heat-resistant flat-black paint to help the mold absorb heat faster. The paint can be purchased at an automobile supply store. The cover for the mold should be a flat piece of ½" to 1" (12 mm to 25 mm) thick, soft asbestos (Marionite®) board.

Checklist of Materials and Equipment

Assemble the following materials and equipment for static powder molding:
1. Mold(s).
2. Oven-rotational molder or electric stove.
3. Powdered plastic — low melt index polyethylene (1.0 - 5.0).
4. Heavy insulated gloves.
5. Thick piece of soft asbestos sheet — ½" to 1" (12 mm to 25 mm) thick.
6. Dry color pigment (optional).
7. Fluorocarbon spray mold release.

Operation Sequence for Static Powder Molding

There are six steps in the process of static powder molding. Study the operation sequence in Table 13-5.

Static Molding of Thermoplastic Pellets

Thermoplastic injection molding pellets may be used in a simple static thermofusion process. Decorative items may be fused from pellets on a Teflon®-coated cookie sheet or in aluminum jello molds. The plastic pellets are fused in an oven. Color variations within one plastic family can be used to create designs. All types of thermoplastic pellets may be used. Different thermoplastics families should not be mixed in one product, however, as the pellets may not fuse to each other.

Acrylic plastic injection molding pellets are often used for this process. They produce a rigid part. They are sold under various trade names by many craft and hobby shops. One such name is "Cookin' Crystals." More flexible parts may be made from polyethylene or polypropylene pellets. Polystyrene and ABS pellets may also be used. Artificial flowers, candleholders, hot pads, and place mats may be made by this method.

Procedure:
1. Use a Teflon-coated cookie sheet or some other metal plate or object for a mold. Tin plate, aluminum foil or a ferrotype plate may also be used.
2. If the mold is not Teflon-coated, spray it with fluorocarbon mold release.
3. Spread a layer of clear plastic pellets on the plate or mold.
4. Create a design by adding colored plastic pellets over the clear pellets.

TABLE 13-5
Operation Sequence for Static Powder Molding (Engle® Process)

Sequence/Action	Reason	Troubleshooting
❶ Spray the inside of the mold with 2 coats of fluorocarbon mold release. Wipe lightly with a soft cloth after each coat dries.	To keep the part from sticking. Fluorocarbon mold release will not break down at high temperatures (up to 500°F or 260°C).	Part sticks to the mold: (a) The wrong mold release was used. (b) Not enough mold release was used.
❷ Fill the mold full of plastic powder, Fig. 13-30. The powder may be colored or natural. A small amount of dry color pigment may be mixed with the powder in a jar first if color is desired. Cover the mold with ½″ to 1″ (12 mm to 25 mm) soft abestos sheet.	The mold must be full to get a complete part. A very small amount of color pigment is needed to get good coloring. The asbestos cover keeps the plastic from fusing over the top.	Plastic fuses over top: (a) Plastic was not covered. (b) The cover board was a poor insulator.
❸ **Put on the insulated gloves.** Put the filled, covered mold in the preheated oven, Fig. 13-31. Cure for about 2 to 3 minutes at 400°F to 450°F (205°C to 232°C).	(a) **Safety first:** Protect hands and arms with insulated gloves. (b) The heat penetrates the mold and fuses a layer of powdered plastic on the inside of the mold.	Fused layer is too thin: (a) Leave the mold in the oven longer. (b) Use a higher oven temperature.
❹ **Put on the insulated gloves.** Take the mold from the oven and remove the asbestos cover. Pour out the unfused plastic, Fig. 13-32.	To remove the extra plastic from the mold.	Fused layer is too thick: (a) Take the mold from the oven sooner. (b) Use a lower oven temperature.
❺ Put the mold back in the oven **without the cover** for a few minutes until the plastic smoothes on the inside, Fig. 13-33. Remove as soon as smoothness occurs.	This lets the plastic completely fuse and smooth on the inside.	Fused layer sags to the bottom of the mold: Use a lower melt index plastic — 1.0 to 3.0.
❻ **Put the gloves on again.** Take the mold out of the oven. Cool it in water. Trim the top edge as needed.	The part will shrink from the mold as it cools.	Part sticks to the mold: (a) Add more mold release. (b) Cool longer. (c) Use compressed air to help remove it.

Thermofusion 271

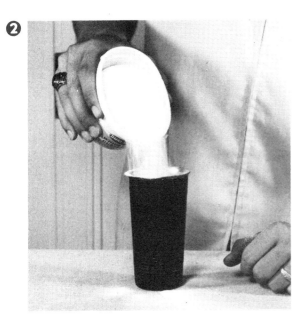

Fig. 13-30. Fill the mold with powdered plastic.

Fig. 13-32. Pour out excess plastic.

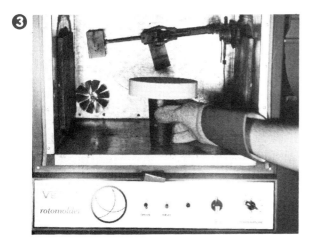

Fig. 13-31. Put in oven with an insulated cover over the mold.

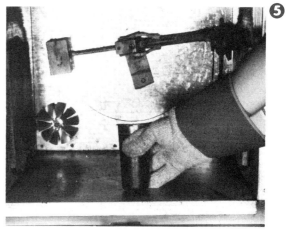

Fig. 13-33. Return mold to oven to smooth out.

Fig. 13-34. Static molding of thermoplastic injection pellets on a Teflon-coated cookie sheet or in aluminum molds. Several completed products are shown in the foreground.

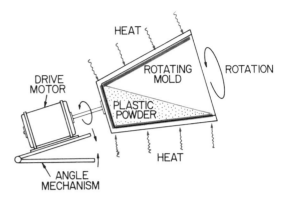

Fig. 13-35. Single-Axis Rotational Molding

5. Put the mold in an oven at about 375°F (190°C) until the pellets are fused together. The length of fusion time and temperature will vary depending on the thickness of the layers, the plastic used, and the surface texture desired. Some experimentation may be needed when materials are varied.
6. Remove from the oven and cool. **Caution: Wear gloves to prevent burns.**
7. Remove the product from the mold after cooling.

Single-Axis Rotational Molding (Heiser® Process)

Single-axis rotational molding or casting (Heiser® Process) uses a heated metal mold (325°F to 475°F or 163°C to 190°C). The hot mold is open on one end and rotated (turned). The rotating axis (shaft) of the mold is at an angle (tipped up) from the horizontal. The heated mold is turned on its axis until the plastic fuses on the inside walls, Fig. 13-35. Any extra powder is then dumped out and the mold is heated again to smooth the inside surface. The part is water-cooled and removed from the mold. The whole process may be done either inside or outside an oven.

Importance of the Process

Single-axis rotational molding is another of the thermofusion powder molding processes. It can produce tanks and other open-ended containers, both small and large. No definite figures on the production by this process are available. Commercially, it is being replaced by two-axis rotational molding.

Advantages

Equipment for single-axis rotational molding is less costly than for two-axis rotational molding. Single-axis rotational molding gives more uniform wall thickness than static powder molding and generally produces the same types of products. Also, single-axis rotational molding requires less labor than the static powder molding process.

Closed containers also can be made by single-axis rotational molding. Two molds are processed at the same time. The open ends are joined together while the plastic is hot. This **double** Heiser process may be practical for short production runs. It requires smaller equipment than two-axis rotational molding does for the same size part.

Disadvantages

Single-axis rotational molding requires more labor and is less mechanized than two-axis rotational molding. The Heiser process is patented and requires a license for commercial use.

Thermofusion 273

Fig. 13-36. Shop-Built Single-Axis Rotational Molding Machine

Materials Used in Industry

The single-axis rotational molding process uses the same materials as static and two-axis rotational molding. Polyethylene is probably the most commonly used material.

Industrial Equipment

Very little information is available on industrial single-axis rotational molding. Much of the equipment used appears to be custom fabricated.

Laboratory Equipment

No commercial laboratory equipment for single-axis rotational molding is now available. Several single-axis rotational molders, similar to the one shown in Fig. 13-36, have been built in school laboratories. The mold is turned by a barbecue grill motor. Heat is provided by an electric stove burner under the mold. The axis is tilted by moving the entire upper unit up and down.

Checklist for Materials and Equipment Needed

Assemble the following materials and equipment needed for the single-axis rotational molding process:

1. Mold(s).
2. Single-axis rotational molder.
3. Powdered plastics — low melt index polyethylene (1.0 - 15.0).
4. Heavy insulated gloves.
5. Dry color pigment (optional).
6. Fluorocarbon mold release spray.

Operation Sequence for Single-Axis Rotational Molding

There are five steps in the process of molding by the single-axis rotational method. Study the operation sequence in Table 13-6.

TABLE 13-6
Operation Sequence for Single-Axis Rotational Molding
(Heiser® Process)

Sequence/Action	Reason	Troubleshooting
❶ Spray the inside of the mold with fluorocarbon mold release. Polish it lightly with a soft cloth.	To keep the plastic from sticking. Fluorocarbon mold release is needed for high heat applications.	**Part sticks to the mold:** (a) Not enough mold release used. (b) The wrong kind of mold release used.
❷ Put the mold on the machine. Heat the mold to 350°F to 475°F (177°C to 246°C). The mold should be turning as it heats.	The mold must be preheated so the plastic will stick to it as the mold turns.	**Uneven mold coating:** Mold not turned as it is preheated.
❸ Pour a small amount of low melt index (1.0 - 15.0) plastic powder into the turning mold, Fig. 13-37. Let the wall thickness build up slowly. Color may be mixed with the plastic before it is put in the mold.	The plastic layer builds up slowly. Extra plastic powder in the mold will run out. The plastic should roll or slide as the mold turns.	**Part builds up unevenly:** (a) Add plastic a little faster. Let the plastic build up in the mold. (b) Change the tilt angle of the mold. (c) Use different melt index plastic.
❹ Tilt the mold up to pour out the extra plastic powder. Keep the mold turning with heat until the part gets smooth, Fig. 13-38.	The extra powder must be taken out to keep it from building up unevenly.	**Plastic turns dark when smoothing out:** (a) Use a lower heat for smoothing. (b) Remove the mold sooner.
❺ Put on the insulated gloves. Remove the mold from the heater of the machine. Cool it in water. Remove the part from the mold and trim if necessary.	Cooling solidifies the plastic and shrinks it away from the mold.	**Plastic sags as it smooths out:** Use a lower melt index plastic. **Plastic gets ripples as it smooths out:** (a) Use a higher melt index plastic. (b) Use a higher heat as it smooths out.

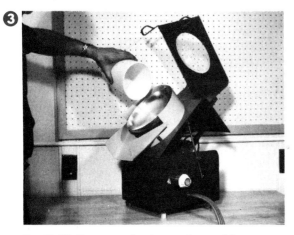

Fig. 13-37. Pour powder in rotating mold.

Fig. 13-38. Completed part ready for removal from machine.

Two-Axis Rotational Molding

Two-axis rotational molding or casting is a process for making completely closed hollow products such as balls, hobby horses, and tanks. The parts may be either rigid or flexible. An opening in the part may be made after it is produced.

A hollow, two-part mold is used to make the product. The cool (room temperature) mold is filled about ⅓ to ½ full of plastic powder or plastisol. The mold is closed, clamped together, and put in a heated oven (400°F to 700°F or 204°C to 371°C). The exact oven temperature depends on the plastic material used. The mold is rotated in two directions at once — on a primary axis and a secondary axis. See Fig. 13-39. The two directions, or axes, are at right angles to each other. As the mold turns, the heat melts and fuses the plastic material in an even layer inside the mold. When all the plastic is fused, it is cooled by air and water spray as the mold continues to turn. The plastic shrinks away from the mold during cooling. The mold is then opened and the part is removed. The mold must be dried before another part is formed.

Importance of the Process

Two-axis rotational molding is probably the most important thermofusion process in use today. Production of parts by this process accounts for about 1.5% of all plastics. It is expected to increase to about 2.5% or 3% by 1980 as more new markets continue to be found.

Advantages

Two-axis rotational molding produces little or no scrap. Wall thicknesses are usually uniform. The two halves of the hollow part have no weak weld lines and parts are nearly strain-free. Totally enclosed parts from very tiny to 1200″ x 3600″ (30 m to 92 m) are now possible. Even larger parts will be made as larger equipment becomes available. Only one heating step is required in the two-axis

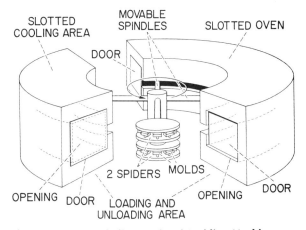

Fig. 13-40. Four-Spindle Rotational Molding Machine

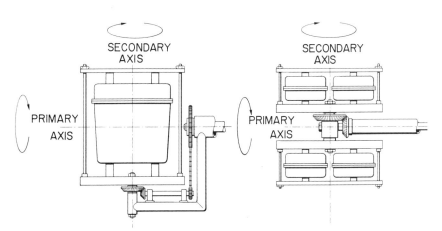

Fig. 13-39. Two-Axis Rotational Molding Machine
A model with an offset arm is shown at left and with a straight arm at the right.

rotational molding process. It is often a less expensive process than single-axis or static molding because several parts may be made at one time, cutting labor cost. Two-axis rotational molding may be used for both liquid and dry plastic materials.

Disadvantages

Commercial equipment for the two-axis rotational molding process is usually large and expensive. Also, it requires a very large amount of floor space. Even though the molder is somewhat automated, hand loading and unloading are still needed.

Materials Used in Industry

Powdered thermoplastic materials such as high, medium, and low density polyethylene, ethylene-vinyl acetate, polystyrene, polypropylene, polyvinyl chloride, cellulosics, and nylon may be used. Special grades of vinyl plastisols are available for rotational molding flexible parts.

Industrial Equipment

Industrial two-axis rotational molding equipment varies in size from 12" x 12" (300 mm x 300 mm) to 1200" x 3600" (30 m x 92 m). Sizes are listed by the overall dimensions of the maximum part size the machine can produce. The machines are also rated by the maximum weight the arm can hold. Machines may have one to several arms, called **spindles,** on each of which one or more molds are placed. Two-axis rotational molders

Fig. 13-41. Three-Spindle Rotational Molding Machine

A. Large Tank

B. Table and Chairs

C. Birdhouse

Fig. 13-42. Rotomolded Products

usually have an oven, a cooling chamber, and a loading area. Note the multi spindles in the two-axis rotational molders, Figs. 13-40 and 13-41.

Laboratory Equipment

Several two-axis rotational molding machines for the laboratory are available. See Fig. 13-44. One, which provides the largest mold size and has a geared secondary axis, is the most versatile (useful in a variety of ways). Larger and more uniform parts are possible with such a machine. Ready-made molds are available for most machines, either as standard equipment or extra accessories. Molds for rotational molding may be spun, cast, or machined from aluminum. They should be painted on the outside with high heat-resistant flat-black paint to help absorb heat faster.

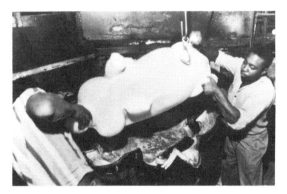

Fig. 13-43. Removing a large, rotationally molded part from the mold.

Fig. 13-45. Materials assembled for rotational molding.

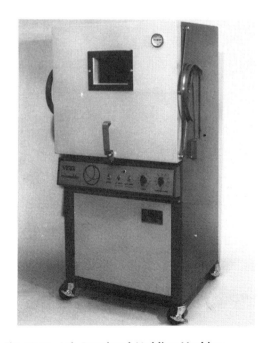

Fig. 13-44. Lab Rotational Molding Machine

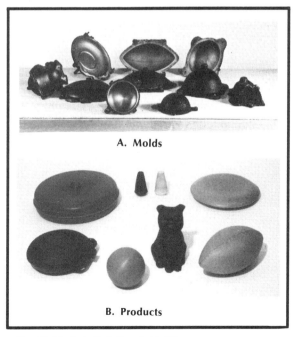

A. Molds

B. Products

Fig. 13-46. Lab Rotational Molding

Checklist of Materials and Equipment Needed

Assemble the following materials and equipment for molding by the two-axis rotational process.

1. Rotational molder.
2. Assortment of molds.
3. Fluorocarbon mold release.
4. Powdered or plastisol plastic. (Polyethylene with melt index 2.0 to 50.0 and density .912 to .965, depending on the product. Other powders may also be used. Vinyl plastisol, rotational molding type.)
5. Heavy insulated gloves.
6. Dry color pigment for powdered plastic or liquid color for plastisol.
7. Soft face mallet.

Operation Sequence for Two-Axis Rotational Molding

There are 10 steps in the process of two-axis rotational molding. Study the operation sequence in Table 13-7.

TABLE 13-7
Operation Sequence for Two-Axis Rotational Molding

Sequence/Action	Reason	Troubleshooting
❶ Turn on the rotational molder oven. Warm it up to 450°F to 550°F (232°C to 288°C). Blow the mold dry with compressed air, Fig. 13-47.	So the molds will take on heat as soon as they are put in the oven. Wet molds will produce poor parts.	Molding takes too long: (a) Preheat the oven longer. (b) Use a higher temperature.
❷ Spray the inside of the mold halves with fluorocarbon mold release, Fig. 13-48. Polish lightly with a soft cloth.	To keep the plastic from sticking. Several parts may be made before adding more mold release.	Part sticks in the mold: (a) Add mold release. (b) Cool the part longer. (c) Release the part with compressed air.
❸ Fill one half of the mold about 2/3 to nearly full of plastic, Fig. 13-49. The exact amount can be determined by trial and error.	The more plastic in the mold, the thicker the part.	Part too thin: Put in more plastic. Part too thick: Put in less plastic.
❹ Add a **small amount** of dry color pigment to the powdered plastic, Fig. 13-50. The amount you can get on the end of a small screw driver blade should be enough for most molds.	Too much color could weaken the plastic. It also wastes an expensive ingredient.	
❺ Close the mold, Fig. 13-51. Fasten it together, Fig. 13-52. If it has bolts, do not overtighten. Finger-tight is enough. Shake the mold to mix the color.	The mold could be bent by overtightening the bolts. Color should be mixed.	Mold gets bent: Bolts may have been overtightened.

(continued on page 280)

Fig. 13-47. Blow the mold dry.

Fig. 13-50. Add color pigment.

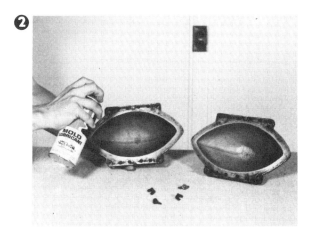

Fig. 13-48. Spray mold release.

Fig. 13-51. Close the mold.

Fig. 13-49. Fill the mold.

Fig. 13-52. Fasten the closed mold.

TABLE 13-7 (Cont.)

Sequence/Action	Reason	Troubleshooting
6 Put on the insulated gloves. Open the rotational molder oven and put the mold in the mold holder, Fig. 13-53. Make sure the mold is held tight. Close the door right away. **Note:** Vinyl plastisols often leak at the parting line when rotationally molded. This can often be avoided by heating the filled mold in the oven for about 2 minutes before starting to rotate the mold. This allows the plastisol to thicken enough to keep from leaking during rotation.	**Safety first:** The oven is very hot. **Gloves must be worn.** The molds must be held tight in the mold holder to keep from falling off.	**Mold comes off:** Mold not clamped in mold holder right.
7 Turn on the rotation molder motor. Set it for a slow speed for large molds and faster for smaller molds. Set the timer for about 10 to 15 minutes. Use longer times for thicker or larger parts.	Slow speeds let the plastic fall to each end easily. Large molds take longer for the plastic to fall.	**Parts have voids in ends or corners:** (a) Slow speed lets the plastic fall into corners. (b) Use a higher melt index material.
8 Put on insulated gloves. Turn off motor when the cure time is over. Take the mold out and air-cool it for 2 to 4 minutes, turning it in 2 directions by hand. Be careful not to get burned. Use a fan or low-pressure air jet (30 psi or 207 kPa) to help cool it.	The mold must be cooled slowly to prevent cooling shock to the plastic. Water will be sucked into the plastic if it is dipped into water right away.	**Parts have holes or water in them:** Mold put in water too soon. Air-cool first.
9 Put the air-cooled mold in a pail of luke warm water for about 5 minutes, Fig. 13-54. Then put in a pail of cold water for 5 minutes more.	Longer, more gradual cooling makes a better part. Sides do not suck in. The part gradually shrinks away from the mold.	**Sides of the part suck in:** Parts cooled too fast.
10 Tap the parting line of the mold lightly with a soft-face mallet to open the mold. The part should fall out, Fig. 13-55. Use compressed air to blow out parts which stick. **Do not pry the mold open with a screw driver.**	Prying with a screw driver damages molds.	

Thermofusion 281

Fig. 13-53. Put the mold in the oven and lock in the arm.

Fig. 13-54. Put in water to cool.

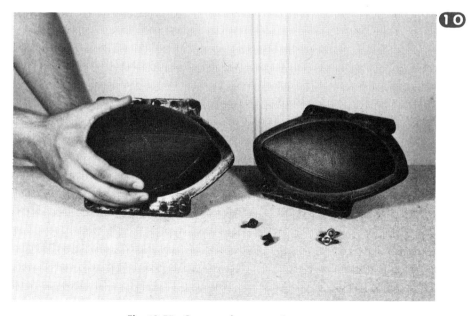

Fig. 13-55. Open and remove the part.

STUDENT ACTIVITIES

1. Dip-cast products on prepared molds in vinyl plastisol.
2. Dip-cast products on prepared molds in more than one color of vinyl plastisol to get a layered effect.
3. Experiment to find the best dipping time for a given plastic at a given mold temperature. Preheat several molds and dip each one for a different length of time. Cure each and cool. Remove from the molds and compare them. Determine which dipping time makes the best part.
4. Dip-coat tool handles with tool-dip plastisol. Coat the handles with the proper tool-dip primer first. Follow the procedure for dip coating. Do not strip the coating after it is cured.
5. Cast fishing bait and other lures with bait casting plastisol. Mix colors to make your product lifelike.
6. Find or make a mold for static powder molding. Mold several parts using different heating and curing times.
7. Build a single-axis rotational molder similar to that described in the text. Mold parts from different shaped spun or cast aluminum molds.
8. Mold hollow parts by the manual rotocasting technique.
9. Mold decorative items by static molding of thermoplastic pellets.
10. Slush-cast door stops, football tees, and similar items.
11. Rotationally mold either powdered polyethylene or vinyl plastisol into products of many shapes. Use color pigments and metallic flakes to get varied decorative effects.

Note: Prepare a log sheet, like the one shown in Appendix A, page 394, for your experimentation with rotational molding processes. Save this record of your experimentation for later reference.

Questions Relating to Thermofusion

1. What is the main difference between dip casting and bait casting plastisol?
2. What kind of products are made in industry by dip casting?
3. To what other thermofusion process is dip casting similar?
4. What is common to all thermofusion processes?
5. Name the thermofusion processes that powdered plastics may be used for.
6. Name the thermofusion processes that vinyl plastisols may be used for.
7. How are thermofusion processes similar to casting processes?
8. What is the advantage of single-axis rotational molding over two-axis rotational molding?
9. What is the advantage of two-axis rotational molding over single-axis rotational molding?
10. What are the main advantages of static powder molding (Engle® process) over rotational molding?
11. Name the processes these materials are most likely to be used for:
    a. Low density, low melt index, molding grade powdered polyethylene.
    b. Low density, high melt index, molding grade powdered polyethylene.
    c. Highly plasticized (low durometer) vinyl plastisol.
    d. Hard (high durometer) vinyl plastisol.
    e. Medium hardness (medium durometer) vinyl plastisol.
    f. Low density, coating grade, low melt index, powdered polyethylene.

The primary processes of the plastics industry were discussed in earlier chapters. Chapters 14 and 15 describe the secondary or finishing processes of the industry.

Plastic products made by the many primary processes (injection, compression, extrusion, and the others) often need additional work. This can include flash cleanup, buffing, sprue cutting, assembly, and decoration. Various methods for doing these operations are explained in these chapters. Some are similar to those used for woods and metals.

Machining Plastics

Machining is a process in which machine-operated tools are used in finishing operations. Many standard sizes of plastic sheets, rods, and tubes are machined before use in the construction of plastic parts and products. Commercially, however, only a limited number of these parts or products are machined. They are made only for short production runs, for one-of-a-kind items, or for those products that can be made no other way. It is usually less expensive to tool up and mold a part for production runs than to machine it.

Machining is a secondary operation that, if used, follows the molding process. Many molded items require parts or flash to be removed by sawing, shearing, die cutting, and heated tool cutting. Holes are sometimes drilled in parts rather than molded in. Parting lines or trim lines may be filed and sanded in some operations. Scraping is used

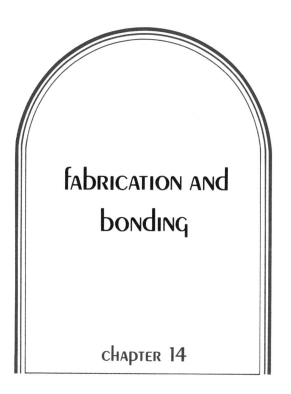

fAbRicATioN ANd bONdiNg

chApTER 14

to remove flash on soft, low-melting temperature thermoplastics. Sanding and buffing are quite often used after other machining operations, especially on harder plastic materials such as acrylics and thermosetting plastics. Tumbling is a popular method of removing flash from compression and transfer molded parts, Fig. 14-1. It is more economical than sanding and polishing large numbers of parts.

Fig. 14-1. Commercial Tumbler

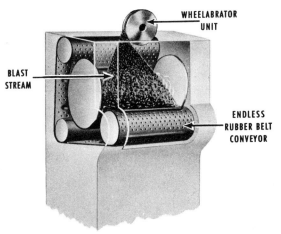

Fig. 14-2. Interior of a commercial tumbler.

Cutting Plastics

Plastic materials are often cut on woodworking and metalworking machinery. Common cutting operations on these machines include sawing, shearing, and routing. Hand woodworking and metalworking tools can also be used for these operations.

Safety Note

Safety glasses should always be worn during all power cutting operations.

Sawing

Circular saws, radial arm saws, band saws, jigsaws, and routers are used for cutting operations on plastic rods, tubes, and sheets. Special overarm saws are used in plants where much plastic sheet is cut, Fig. 14-3.

Straight line cutting of plastic sheet stock is done on circular and radial arm saws, Fig. 14-4. Fine-tooth (12 teeth per inch), hollow-ground or slightly set circular saw blades work best for plastic sheet, rods, and tubes. When the softer thermoplastics are cut, extra tooth clearance on circular saw blades is often necessary.

A variety of straight and curve cutting may be done on the band saw, Fig. 14-5. The work must be held steady to avoid the chipping or breaking of brittle plastics. Shark-tooth wood/metal band saw blades with four to six teeth per inch work best on most plastics. They allow ample room for chips and they cool well.

Carbide-tipped blades should be used to cut glass-reinforced and high pressure laminates, such as Formica®, Micarta®, etc. Cutting with ordinary circular saw or band saw blades on these plastics will dull the teeth rapidly. Metal-cutting band saw blades **at metal-cutting speeds** are also recommended for high pressure laminates and fiberglass.

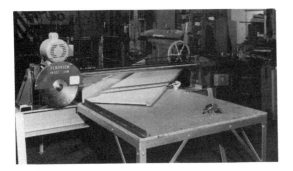

Fig. 14-3. Sawing plastic counter tops with an overarm saw.

Fig. 14-4. Cutting on the circular saw. (Safety guard removed for illustrative purposes.)

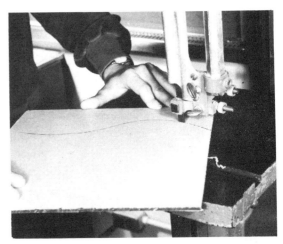

Fig. 14-5. Cutting plastic on the band saw.

Fabrication and Bonding 285

Fig. 14-6. Hand sawing plastic.

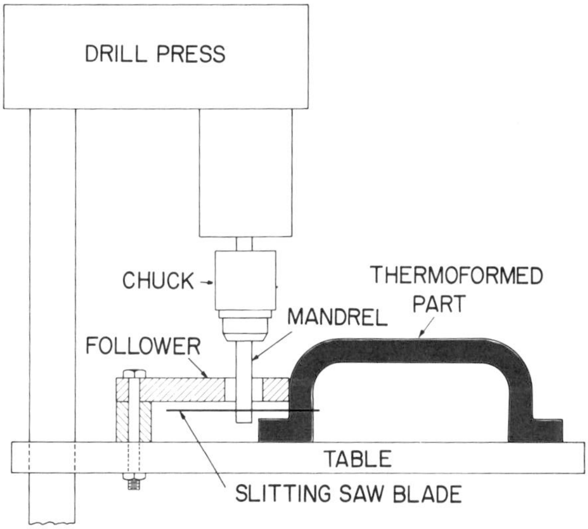

Fig. 14-7. Slitting saw on the drill press.

Fine lines and sharp curves may be cut on the jigsaw. Skip tooth blades or blades with good set are necessary. Thermoplastics with low melting point often melt as they are cut and the blades plug up. This problem can be avoided by slower cutting. The chips must not be allowed to pile up on the blade.

Handsaws, backsaws, and coping saws may be used for hand cutting of thermoplastics, Fig. 14-6. Fine-toothed blades work best. Thermosetting plastics, especially high pressure and fiberglass reinforced laminates, should be cut with a hacksaw if done by hand.

Slitting saws may be used on a drill press for cutting thermoformed parts, Fig. 14-7. A 1½″ to 2″ (38 mm to 51 mm) diameter slitting saw mounted on a mandrel is ideal. A guide around the mandrel helps the operator follow the shape of the part. A fine, clean cut can be made. Cutting should be done against the rotation of the saw blade. As always during any cutting operations, **safety glasses** should be worn by the operator.

When holes with large diameters are to be cut in plastic, a hole saw should be used, Fig. 14-8. A regular metal cutting hole saw can be attached to a drill press or electric

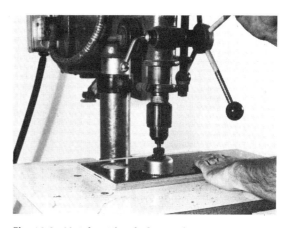

Fig. 14-8. Metal cutting hole saw in use.

power hand drill. The hole saw should be fed into the plastic slowly, using a slow spindle speed on the drill press or other driving unit, especially for large holes.

Routing

Routers are ideal for production cutting of thermoplastic sheet products. Guided routers work especially well on thick thermoformed plastics. The router is usually mounted cutter up in a table. See Fig. 14-9. A **template**, or

286 Section V FINISHING OPERATIONS

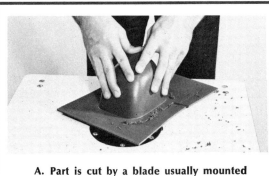

A. Part is cut by a blade usually mounted cutter-up in a table.

B. Part is placed over the template for the cutting procedure.

Fig. 14-9. Using a Router

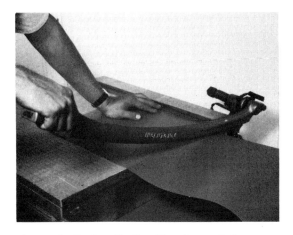

Fig. 14-10. Cutting Plastic with a Paper Cutter

Fig. 14-11. Squaring Shear Cutting Plastic

guide for cutting, is made of ¼" (6.35 mm) double-tempered hardboard (Masonite®). It is made so that the outside edge will be slightly inside the line to be cut around the plastic part. A template guide is screwed to the router base plate and the bit is raised up through the template guide. A jig is fastened to the template to hold the part in place and in alignment with the template. The template and jig with the plastic part over them are then guided around the template guide on the router base until the part is cut out.

Shearing

Most **soft** thermoplastic sheet stock can be cut by shearing, Fig. 14-10. It is a better method of machining than many sawing operations because it does not create heat. Shearing is also safer and makes a smoother cut. Brittle plastics like acrylic are hard to shear.

Heavy duty scissors or tin snips may be used for thicknesses up to about .100" (2.54 mm). An ordinary paper cutter works very well for thermoplastic sheet and film up to about .030" (0.76 mm). A sheet metal squaring shear should be used for sheet stock over .030" (0.76 mm), Fig. 14-11.

Die Cutting

Thermoformed and other sheet plastic parts are often trimmed or cut out by die

Fabrication and Bonding   **287**

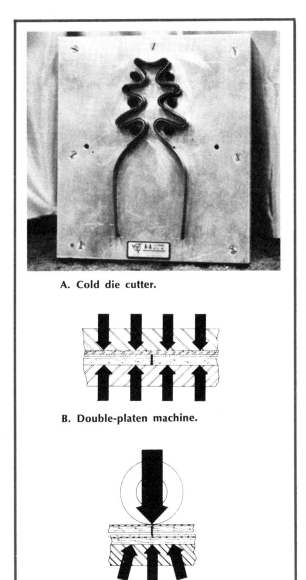

A. Cold die cutter.

B. Double-platen machine.

C. Single-roller platen machine.

Fig. 14-12. Cold Die Cutting

cutting. Two general types of die cutting are used: **cold dies** and **hot dies.** Cold dies are used on plastic up to about .040" (1 mm) thick. Hot die cutting is usually required for plastic over .040" (1 mm) thick. Cold dies are forced through the plastic either with a rapid stroke or by high pressure. Cold die cutters tend to crack or shatter thicker plastic. Hot dies melt their way through the plastic.

Cold die cutting machines are of two general types: (1) dual platen and (2) roller. Steel rule dies, like those used in the graphic arts industry, are mounted on the press platen and forced through the plastic. See Figs. 14-12 and 14-13.

In double-platen cold die cutting, the steel rule die, which is mounted on the movable press platen, is brought down against the plastic material (view B of Fig. 14-12). The die is forced through the plastic by high pressure. Double-platen presses require very heavy equipment to produce the high compressive forces needed. Some of these presses are hydraulic while others are mechanical. The hydraulic machines are like large compression molding presses. The mechanical machines, often called **clicker presses,** are similar to punch presses. They bring the dies down very rapidly. The chief disadvantage of double-platen die cutting is the limited size of or number of parts which can be cut in one stroke due to the great pressures required.

Roller die cutting may be classified into two types: (1) single-roller and (2) double-roller. Each type uses the same type of cold cutting dies as the double platen press.

A single-roller die cutter is built like a graphic arts proof press. A single steel roller is moved across the top of the steel rule die, (view C of Fig. 14-12). It forces the cold die through the plastic sheet over a rigid steel bed plate. Some deflection of the bed plate, however, may cause unequal or varied pressure across the width of a wide die.

Double-roller cold die cutters concentrate the force on the die between two heavy steel rollers. The two rollers are moved simultaneously across the length of a floating bedplate, (view C of Fig. 14-13). They progressively force the die through the plastic. The shearing action is said to compensate for small variations in the material and dies. The double-roller type press may out-produce double-platen machines by as much as 3-to-1.

A. Double-roller cold die cutting machine with the chase up, showing the steel rule die.

B. Same machine with the chase down, showing the upper roller.

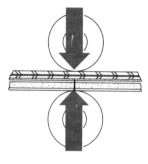

C. Diagram of the double-roller die cutting principle.

Fig. 14-13. Double-Roller Die Cutting

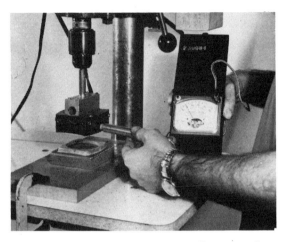

Fig. 14-14. Hot die cutter and cutting jig mounted on a drill press. Temperature is checked with a pyrometer.

Fig. 14-15. The air-operated, double-moving platen, hot die cutting press shown here is laboratory constructed. It will cut two parts at once.

Hot die cutting uses a heated die which melts its way through the plastic. Because of this, much lower pressures are required to force the die through the plastic. Hot die cutting may be done in the school laboratory with ease. Simple hot cutting dies may be made at low cost. Steel rule die stock (obtained from the graphic arts laboratory) can be bolted to an aluminum block. The block

Fabrication and Bonding 289

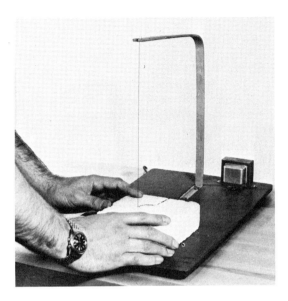

Fig. 14-16. Hot wire cutter used on thermoplastic foam.

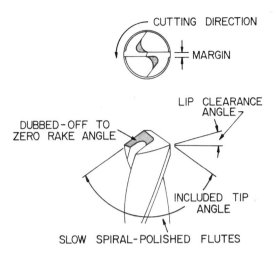

Fig. 14-17. Drill Modification, or Changes for Plastics

should be made in the shape to be cut. Internal or external heaters may be used. The die may be heated by the platen of a compression press or with separate heaters. Ceramic soldering iron heater cartridges, which are inexpensive, can be put into the aluminum mounting block. Better, metal-clad die heating cartridges may be purchased from die supply companies or heater unit manufacturers. See Appendix B for sources. A parallel circuit between the heaters should be used. For good temperature control, a variable power supply or thermostatic switch should be added. Flexible band heaters are also available. They may be stretched, or glued with silicone RTV, along the outside. Asbestos insulating bands may be needed around the outside to prevent burns and contain the heat. Blade temperatures of about 350°F (177°C) are needed to cut high-impact polystyrene. Temperatures for other materials will vary according to their melting points. The temperature may be checked with a portable pyrometer (heat measuring instrument), Fig. 14-14.

Hot die cutters may be mounted on a drill press, the upper platen of a thermoformer, a compression press, or a die cutting press, Figs. 14-14 and 14-15. The die block and hot cutting blades should be well insulated from the press to prevent excess heat loss. With a hot die cutter mounted on the upper platen of a thermoformer, the formed parts may be cut right on it. Many industrial thermoformers have die cutting stations either on them or next to them.

Hot Wire Cutting

Heated tool cutters are often used for thermoplastic foams, Fig. 14-16. Styrofoam and expanded polystyrene bead foam may be cut easily with a hot wire cutter. Such cutters are readily available in hobby craft stores at low cost.

Drilling

Most plastics may be drilled with metalworking drill bits fastened to drill presses, electric hand drills, or manual hand drills. Ordinary metalworking drill bits work best. Brittle plastics often need blunter point angles and smaller relief angles. A special set of metalworking bits in fractional sizes should be kept on hand for this. They should be sharpened with a flat lip and low lip relief angle as shown in Fig. 14-17.

TABLE 14-1
Guide to Speeds for Drilling Plastics

Drill Size	Speed for Thermoplastics (rpm)	Speed for Thermosets (rpm)
No. 33 and smaller	5000	5000
No. 17 through 32	3000	2500
No. 1 through 16	2500	1700
1/16" (1.5 mm)	5000	5000
1/8" (3.1 mm)	3000	3000
3/16" (4.7 mm)	2500	2500
1/4" (6.3 mm)	1700	1700
5/16" (7.8 mm)	1700	1300
3/8" (9.4 mm)	1300	1000
7/16" (10.9 mm)	1000	600
1/2" (12.7 mm)	1000	600

Drill speeds vary in the same proportion as in drilling metal. Soft materials are drilled at higher speeds than hard materials. Smaller drill bits require higher speeds. Large bits need slow speeds. See Table 14-1.

Filing

Large amounts of flash or scrap can be removed from plastic parts by filing. Also, edges can be straightened on thermoformed and other sheet plastic parts. Filing works best on harder plastics because it does not generate much heat. Files need to be cleaned often during use on plastics. Rub a file card over the file in the direction of the cuts for adequate cleaning.

Ordinary metalworking files will work for most plastics. Single- and double-cut mill and half-round files will do a great share of the work. A special shear-tooth file, which does not clog as easily as others, is available for soft plastics.

Files should be chosen for plastics in the same way they are for wood and metal. Coarse files are used to remove lots of material. Fine files should be selected to give a smooth finish.

Safety Note

Files should always have handles to prevent getting hurt. The tang of a file will cut the palm of the hand easily if a handle is not used.

Scraping

Hand scrapers and knives are often useful for removing flash or trimming edges on soft thermoplastics. Many times they eliminate the need for sanding. Scrapers which are pulled across the work, should be sharpened just as they would be for wood scraping.

Sanding

After excess materials are removed by cutting, filing, or scraping, the plastic part is usually sanded. Hard plastics such as acrylics, phenolics, and polyesters should be sanded down to a fine finish before they are buffed. Very soft plastics such as low density polyethylene are hard to sand and buff. They soften under the frictional heat generated in these operations.

Machine Sanding. Many plastics, usually harder or higher density plastics, will withstand machine sanding. Coarse grit (60 to 100) open-coat sandpaper should be used. Further hand sanding is usually necessary after belt or disc sanding. Fine machine sanding on hard plastics may be done with a high-speed (10,000 orbits per minute) portable orbital sander. Such units sand well without loading the sandpaper.

Hand Sanding. Sanding should be done in stages beginning with coarse garnet or production sandpaper. Usually 100 to 150 grit is coarse enough to begin hand sanding. Sanding direction is changed 90° each time a finer grit is used. Final sanding should be done with 400 or 600 grit wet-or-dry sandpaper with water, Fig. 14-18. The plastic will then be ready for buffing.

Buffing and Ashing

Buffing and ashing polishing work is done with a rotating cloth or flannel wheel and fine abrasive. In ashing, a loose-sewn wheel and water-pumice compound are used for rough polishing. High polish is achieved by dry compound buffing. Most plastics may be buffed and ashed. Soft, low-melting-point thermoplastics, such as low density polyethylene, are hard to polish, however. Machine polishing causes frictional heat

Fig. 14-18. Wet Sanding Acrylic Plastic

Fig. 14-19. Buffing Plastics

which softens some plastics, causing them to stick to the wheel and not buff up.

A 6" to 8" (150 mm to 200 mm) loose-sewn cotton or flannel buffing wheel is recommended for general buffing of plastics. It should be driven at 1725 rpm or less with slower speeds used for low melting point plastics. The buffing wheels may be mounted on a double-shaft motor, drill press, or lathe.

Red rouge or tripoli buffing compound is suitable for rough, fast buffing. It is applied by holding the bar against the turning wheel. The wheel should rotate toward the operator, Fig. 14-19, the part being moved around as it is buffed. If it is not moved constantly, the plastic will soften from the frictional heat. Plenty of buffing compound should be kept on the wheel.

A final shine can be put on the part with white diamond or a similar finishing compound with the same procedures being used. A separate buffing wheel is required for each buffing compound.

Hand buffing may be done with a soft clean cloth and liquid buffing compounds such as metal polishes, pumice, tripoli, or liquid and paste auto-body paint cleaners. Compounds are rubbed on the plastic until a shine develops. This is a slow way to polish plastic and the surface must be well prepared by fine sanding. It is recommended that the surface be wet sanded to 600 grit before hand polishing.

Fig. 14-20. Automatic Buffing of Telephone Parts

Solvent Polishes

Many plastics may be polished by solvents. This is usually done only on the edges or in holes, however. Recommended solvents for each plastic material should be obtained from the plastic manufacturer. They vary greatly from plastic to plastic. Solvent polishes are rubbed on, applied as a vapor, or used as a soak.

Solvents may be rubbed onto plastics with a rag or other applicator to polish the surface or edge. Plastics may also be polished by holding them over a container which gives off solvent vapors. The vapor application is often used industrially.

Solvent polishing of acrylic plastics by soaking is common in school laboratories. It is usually used for drilled holes. After the hole is drilled, it is cleaned of chips and then filled with ethylene dichloride for about 30 seconds, Fig. 14-21. The solvent is then poured out, and the part is dried overnight in a well ventilated room.

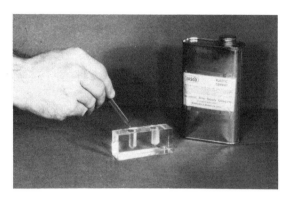

Fig. 14-21. Solvent Polishing Plastics

TABLE 14-2
Smoothing and Polishing Summary

Process	Typical Applications
Filing	To remove flash, shape parts, remove large amounts of plastics
Scraping	To remove flash, semi-finish edges, remove saw marks, smooth edges of tough plastics such as high impact styrene, ABS
Sanding	To remove flash, semi-finish edges, produce matte finish on sheet surface, remove deep scratches, shape large areas as on reinforced-plastic tooling
Buffing	To produce high luster, "reach" into scratches and depressions
Tumbling	To remove flash from large numbers of parts, polish large numbers of parts

Cementing Plastics

Often plastic parts can be joined together by cementing. There are two types of cementing: **adhesive** and **cohesive**.

Adhesive Cementing

Adhesion is the joining of two plastic materials with a gluelike adhesive. Adhesive cementing is similar to gluing wood. A film of the adhesive stays between the two materials and holds them together. Two materials which are alike or unlike can be joined in this way. The adhesive must be able to stick to both materials.

There are three types of adhesive cements: bodied adhesives, elastomeric adhesives, and reactive adhesives.

Bodied adhesives are sometimes called "dope cements." They are solvents with a small amount of the parent plastic dissolved in them. These cements can be spread like glue on a joint. They will not run off the joint. Bodied adhesives are not as neat to use as solvent cements, especially on clear plastics, and therefore should be used where they will not show. Styrene model airplane cement is an example of a bodied adhesive.

Elastomeric adhesives are rubber adhesives. They are made from natural, synthetic, or reclaimed rubber. An elastomeric adhesive is often used where a flexible joint is needed. Two unlike materials may be joined by it. Silicone bathtub seal is an example, Fig. 14-22. Elastomeric adhesives will stick to almost any nonwaxy material.

Reactive adhesives are thermosetting cements. They are usually two-part adhesives which must be mixed together just before use. An example is two-part epoxy cement, Fig. 14-23. Reactive adhesives are quite often used for thermosetting plastics.

Fabrication and Bonding   293

Fig. 14-22. Silicone Bath Tub Seal

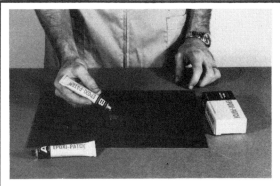

A. An equal portion of Part B of the epoxy is added to Part A.

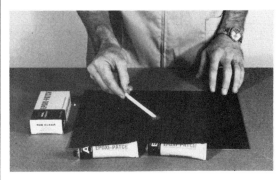

B. The two portions are mixed together to form the adhesive.

Fig. 14-23. Mixing Epoxy

Cohesive Cementing

Cohesive cementing is the softening of the two plastic parts with a solvent. Once softened, the molecules in the two plastic parts intermingle upon contact. When the solvent evaporates, the plastic hardens and the two parts are joined as one. Cohesion is most often used for two similar plastics. In some rare cases, unlike plastics may be cohesive-cemented. The solvent must be able to soften both plastics. The molecules of the two must intermingle in order to bond. Thermal welding and heat sealing are also forms of cohesive bonding that are discussed later in this chapter.

There are two types of cohesive cements: solvent cements and monomeric cements.

Solvent cements are those which will dissolve or soften the surface of the two parts to be joined. Generally, solvent cements will work only on thermoplastics. A number of solvents is usually available for each plastic.

Monomeric cements are made from the same monomer as is the plastic to be joined. Thus, a polystyrene monomeric cement would be made from styrene monomer. The monomer is catalyzed just before use and it cements by polymerization.

Solvent Cementing

Solvent cementing is probably most used in the laboratory for bonding acrylic plastic parts. It is used in the building trades to bond ABS and PVC sewer and water pipe. Parts to be solvent-cemented should have clean joints. They should also have good contact between parts. Plenty of contact surface is necessary for good joints. Solvent cements may be put on by brush, hypodermic needle, an eye dropper, or dipping. Follow the procedures below:

1. **Prepare the joint.** Be sure the parts fit together well.

TABLE 14-3
Adhesives for Plastics

TABLE I. Adhesive types for bonding nonplastics to thermoplastics*

Surfaces	Acetal	Cellulosics[1]	Ethyl cellulose	Nylon	Polycarbonates	Polyethylenes	Polymethylmethacrylate	Polypropylenes[2]	Polystyrene	Polyurethane	Polyvinyl chloride	Tetrafluoroethylene
Ceramics	23	4	14	4, 23	23, 36	3, 41	3, 4	1, 41	41, 42	4	4, 5	23
Fabrics	23, 4	4, 5, 42	14	3, 4	23, 36	3, 41	4	1, 41	5, 3	5, 36	4, 5, 42	22
Leather	23, 4	4, 5, 42	14	3, 4	23, 36	3, 41	3, 4, 42	1, 41	31, 36, 5	4, 5	4, 5, 41, 42	22
Metal	23, 4	3, 4	14	3, 23	23	3, 41, 31	3, 4	1, 2	31	4, 5	4, 3, 36, 15	23, 22
Paper	23, 4	42	14	4, 41	36	41	42	1, 2, 41	31, 36, 5	5, 36	42	23, 22
Rubber	4	1-5	14	3, 2	36, 5	3, 41	1-5	1, 2, 41	2, 6	5, 36	4, 5, 15	23
Wood	23	4	14	23, 36	3, 41		3, 4, 42	1, 2, 41	31, 36	36	4, 42, 36	23

*Adhesive number code is shown at right, below.
[1]Cellulose acetate, cellulose acetate butyrate, cellulose nitrate.
[2]Special surface treatment recommended.

TABLE II. Adhesive types for bonding nonplastics to thermosets*

Surfaces	Diallyl phthalate	Epoxies	Melamine	Phenolics	Polyesters	Polyethylene terephthalate	Urea
Ceramics	5, 24	23, 31	3	3	3	36	4
Fabrics	36	4	4	4	4	5, 36	4, 42
Leather	31, 36	4	3, 4	3, 4	5	5, 36	3, 4
Metal	31	23, 31	4	3	5	36	3, 4
Paper	31, 36	4	41, 42	42	41	5, 36	42
Rubber	3	4	3, 4, 2	3, 4	1-5	36, 13	1-5
Wood	31, 36	23, 31	3	3, 42	3	36	3, 42

*Adhesive number code is shown at right.

TABLE III. Adhesive types for bonding plastics to plastics*

Surfaces	Adhesive type(s)	Surfaces	Adhesive type(s)
Acetal	23, 4	Polyvinyl chloride	4, 11, 5, 42, 36
Cellulosics[1]	14, 4, 36, 5	Tetrafluoroethylene	23, 22
Ethyl cellulose	14	Diallyl phthalate	4, 23, 31, 3, 36
Nylon	23, 22, 3, 36	Epoxies	4, 23, 31, 3, 36
Polycarbonates[2]		Melamine	3, 4, 23, 31, 36
Polyethylenes	23, 41, 31	Phenolics	4, 23, 31, 3, 36, 5
Polymethylmethacrylate	13, 6, 2, 36, 5, 31	Polyester-fiberglass	4, 23, 31, 36
Polypropylenes	23, 31, 41	Polyethylene terephthalate	5, 36
Polystyrene	23, 13, 2, 6, 5, 31, 36	Urea	4, 23, 31, 3
Polyurethane	5, 23, 4, 36		

*Adhesive number code is shown at right.
[1]Cellulose acetate, cellulose acetate butyrate, cellulose nitrate.
[2]No adhesive type recommended, solvent cementing preferred method.

Adhesive number code for Tables I, II and III

Elastomeric
1. Natural rubber
2. Reclaim
3. Neoprene
4. Nitrile
5. Urethane
6. Styrene-butadiene

Thermoplastic
11. Polyvinyl acetate
12. Polyvinyl alcohol
13. Acrylic
14. Cellulose nitrate
15. Polyamide

Thermosetting
21. Phenol formaldehyde (phenolic)
22. Resorcinol, phenol-resorcinol
23. Epoxy
24. Urea formaldehyde

Resin
31. Phenolic-polyvinyl butyral
32. Phenolic-polyvinyl formal
33. Phenolic-nylon
36. Polyester

Miscellaneous
41. Rubber lactices (water based) (natural or synethetic)
42. Resin emulsions (water based)

Reprinted by permission of **Modern Plastics Encyclopedia**, McGraw-Hill, Inc.

2. **Apply the solvent cement.**
   a. **For small areas** or thin joints, the solvent may be put on with a brush, eye dropper, or hypodermic needle, Fig. 14-24. Hold the joint together and run a small amount along the edge of the joint. Do not get it on the sides of the plastic. Capillary attraction (pulling force resulting from surface tension) will draw the solvent into the joint. It may be necessary to put it on from both sides.

Fabrication and Bonding   295

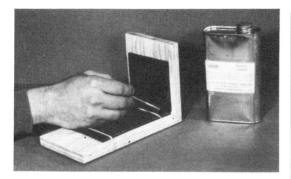

Fig. 14-24. Brush Method of Solvent Cementing

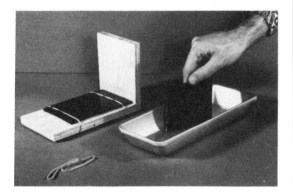

Fig. 14-25. Soak Method of Solvent Cementing

A. Use an eyedropper to flood part of piece.

B. Place two parts together.

Fig. 14-26. Flooding Method of Solvent Cementing

b. **For large areas,** soaking, dipping, or flooding may be needed.
   1) **Soaking.** Put the end of the part to be joined in a flat pan with a shallow layer (1/16" or 1.587 mm) of solvent in it, Fig. 14-25. Felt in the bottom of the pan will help keep the solvent from oversoaking the plastic.
   2) **Dipping.** Dip the joint into a pan of solvent for a few seconds, take out, and put the joint together.
   3) **Flooding.** For large areas such as for acrylic solvent laminating, lay one piece down flat. With an eye dropper, flood part of it, Fig. 14-26A. Put the second piece over the first and gradually close the joint pushing the air out ahead, Fig. 14-26B.
   4) **Hold the joint together** until the solvent evaporates. Use just enough pressure to keep air out of the joint. Do not clamp too tightly. A fixture for holding the joint straight may be needed. It can be made from scrap wood as shown in Fig. 14-24.

Mechanical Fastening

A variety of mechanical fastening devices may be used to hold plastic parts together securely. They are divided into two groups: extrinsic mechanical fasteners and intrinsic mechanical means.

296 Section V FINISHING OPERATIONS

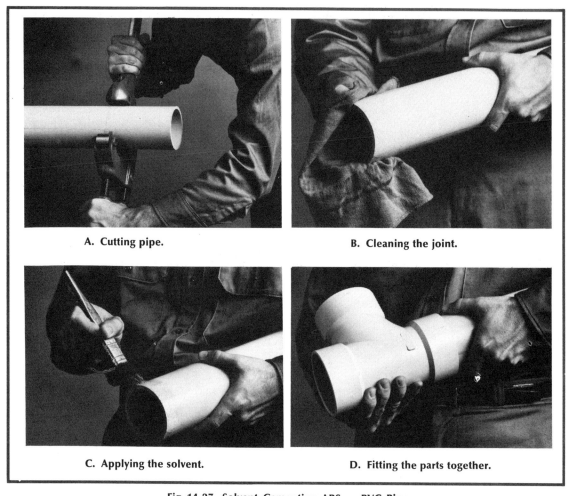

A. Cutting pipe.

B. Cleaning the joint.

C. Applying the solvent.

D. Fitting the parts together.

Fig. 14-27. Solvent Cementing ABS or PVC Pipe

Extrinsic Mechanical Fasteners

Extrinsic fasteners are those devices that are used on the outside of a product to hold the parts together. These include screws, nuts and clips, hinges, rivets, and staples. Most of these mechanical fasteners can be purchased in hardware stores. Few require special knowledge or procedures for use.

Screws. Self-tapping screws (those that form their own threads) can be used to fasten plastics. There are two types: thread-forming (Fig. 14-28) and thread-cutting (Fig. 14-29). **Threading-forming screws** squeeze and dis-place a material as the thread is formed. They are generally used for thermoplastics. **Thread-cutting screws** form threads by actually cutting away the material. They are used for thermosetting plastics. Threads may be tapped into such thermosetting plastics as phenolic, melamine, and urea. Standard threaded bolts and screws may then be used.

Speed Nuts and Spring Clips. Either speed nuts or spring clips are used with screws, bolts, or studs to hold two flat pieces of material together, Fig. 14-30. A **speed nut** is a one-piece self-locking spring steel fast-

Fabrication and Bonding   297

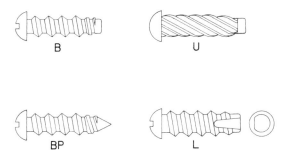

Fig. 14-28. Four U.S.A. standard thread-forming screws most often recommended for use in thermoplastic materials.

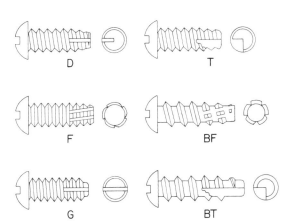

Fig. 14-29. Six types of U.S.A. standard thread-cutting screws are recommended for use in thermosetting plastic materials.

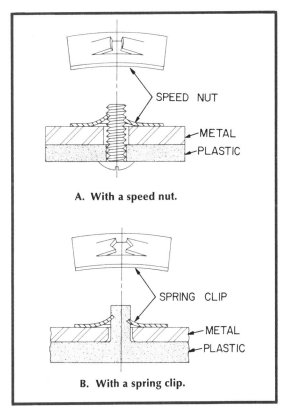

A. With a speed nut.

B. With a spring clip.

Fig. 14-30. Holding Two Flat Pieces Together

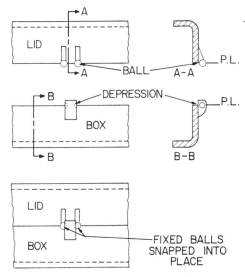

Fig. 14-31. Ball Grip Hinge Design
This type of hinge is used on small boxes and containers.

ener that is used with a threaded bolt to hold the parts together. A **spring clip**, sometimes called a speed clip, can be snapped over a bolt or plastic stud to lock the parts together securely.

Hinges. Two parts can also be held together by a hinge. There are different types of hinges. Some are molded right into the parts as shown in Fig. 14-31. Other hinges can be fastened onto the parts to hold them together, such as piano hinges and other metal pin hinges. Also used are integral hinges which are either molded right into the part or formed and fastened on, Figs. 14-32 and 14-33. They may be extruded, injection molded, or coined (shaped by pressure). The most popular choice of plastic for integral hinges is polypropylene. Other materials used are polyethylene, nylon, and acetal. See Figs. 14-34, 14-39, 14-40, and 14-41.

Rivets. Using rivets is one of the oldest methods of fastening two parts together. One of the newer types of rivets is the pop rivet, Fig. 14-35. Pop rivets fasten plastics to plastics or to other materials.

Staples. Modern staple guns make stapling easy, Fig. 14-36. Some of the newest plastics can be stapled with ease. Soft, thin thermoplastic sheet or film can be fastened to other materials by stapling.

Intrinsic Mechanical Fasteners

Plastic parts may be put together without external (outside) fasteners. Products may be molded with the fastening device built into the parts to be held together. These devices

Fig. 14-32. Integral hinge injection molded glove box interior.

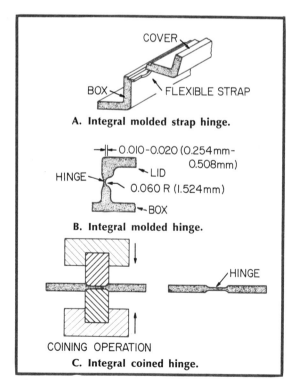

Fig. 14-33. Three types of integral plastic hinges.

Fig. 14-34. Integral Hot Coined Hinge Formed in the Laboratory

Fig. 14-35. Pop Riveting Plastics

Fabrication and Bonding 299

are called intrinsic mechanical fasteners. Three types of fasteners are used: snap fit, press fit, and heat staking.

Snap Fit. Two types of snap fit fasteners may be molded into the parts: snap in and snap on. See Figs. 14-37 and 14-38.

Press Fit. Press fit fastening is forcing of two parts together. The internal part is molded slightly oversize. Such a fastener should be used only for stiff, tough plastics.

Heat Staking. Heat staking is similar to fastening with spring clips or rivets. A plastic stud is molded into the part as shown in Fig. 14-39. It is pushed up through the second

Fig. 14-36. Stapling Plastic

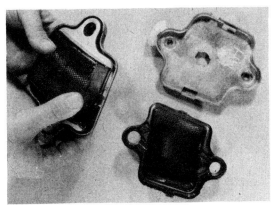

Fig. 14-38. Snap Fit Assembly

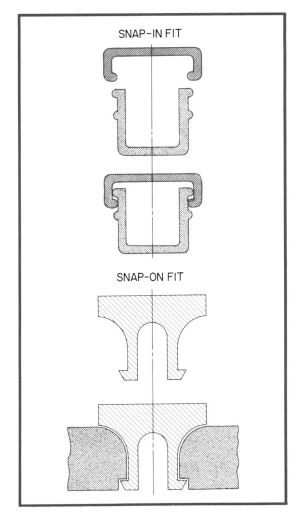

Fig. 14-37. Two types of snap fit fasteners.

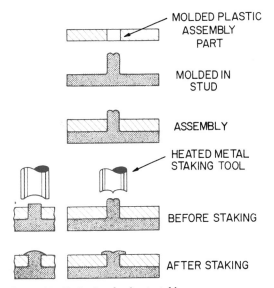

Fig. 14-39. Fastening by heat staking.

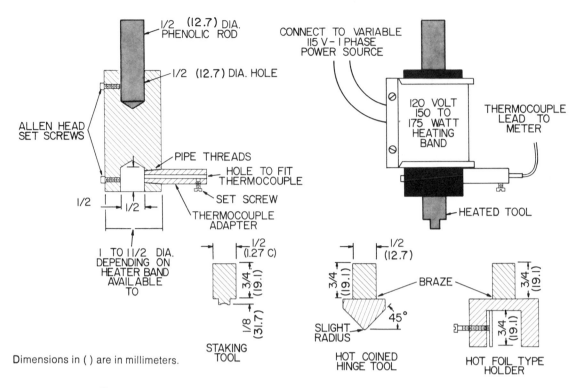

Fig. 14-40. Construction details for a heated tool drill press attachment.

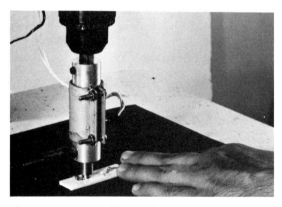

Fig. 14-41. Heat staking a finding (pin back) to a name tag.

part. A heated tool is pressed down over the stud, forming a head. When the plastic cools, the head holds the parts together. Shrinkage often results in loose parts when they are heat-staked. Parts cannot be taken apart without breaking the joint. If they are taken apart, they must be reassembled another way.

A heat staking tool can be made from an electric soldering iron or to fit a drill press. The construction of a simple one is shown in Fig. 14-40. It may also be used for making a plastic hinge and for hot foil stamping which is described in Chapter 15.

Ultrasonic Staking. A faster, more solid method of staking thermoplastics is ultrasonic staking. A vibrating tool, rather than a heated tool, forms the head. The vibrating tool, or ultrasonic horn, moves up and down at 20,000 times per second. The ultrasonic vibration generates frictional heat in the plastic stake under the horn. The heat forms the head on the stake. See Fig. 14-42, page 303.

The ultrasonic horn is driven up and down a very short distance (.003″ to .005″) (0.076 mm to 0.127 mm). A power supply changes 60 cycle per second alternating electric current to 20,000 cycle per second pulsating direct current. An electrical converter changes

TABLE 14-4
Ultrasonic Welding Characteristics*

Table should be used as a guide only since variations in resins may produce slightly different results.

Material	% Weld strength*	Spot weld	Staking and inserting	Swaging	Welding Near field[†]	Welding Far field[†]
General-purpose plastics						
ABS	95-100+	E	E	G	E	G
Polystyrene unfilled	95-100+	E	E	F	E	E
Structural foam (Styrene)	90-100[a]	E	E	F	G	P
Rubber modified	95-100	E	E	G	E	G-P
Glass filled (up to 30%)	95-100+	E	E	F	E	E
SAN	95-100+	E	E	F	E	E
Engineering plastics						
ABS	95-100+	E	E	G	E	G
ABS/polycarbonate alloy (Cycoloy 800)	95-100+[b]	E	E	G	E	G
ABS/PVC alloy (Cycovin)	95-100+	E	E	G	G	F
Acetal	65-70[c]	G	E	P	G	G
Acrylics	95-100+[d]	G	E	P	E	G
Acrylic multipolymer (XT-polymer)	95-100	E	E	G	E	G
Acrylic/PVC alloy (Kydex)	95-100+	E	E	G	G	F
ASA	95-100+	E	E	G	E	G
Methylpentene	90-100+	E	E	G	G	F
Modified phenylene oxide (Noryl)	95-100+	E	E	F-P	G	E-G
Nylon	90-100+[b]	E	E	F-P	G	F
Polyesters (Thermoplastic)	90-100+	G	G	F	G	F
Phenoxy	90-100	G	E	G	G	G-F
Polyarylsulfone	95-100+	G	E	G	E	G
Polycarbonate	95-100+[b]	E	E	G-F	E	E
Polyimide	80-90	F	G	P	G	F
Polyphenylene oxide	95-100+	E	G	F-P	G	G-F
Polysulfone	95-100+[b]	E	E	F	G	G-F
High-volume, low-cost applications						
Butyrates	90-100	G	G-F	G	P	P
Cellulosics	90-100	G	G-F	G	P	P
Polyethylene	90-100	E	E	G	G-P	F-P
Polypropylene	90-100	E	E	G	G-P	F-P
Structural foam (Polyolefin)	85-100	E	E	F	G	F-P
Vinyls	40-100	G	G-F	G	F-P	F-P

Code: E = Excellent, G = Good, F = Fair, P = Poor.

*Weld strengths are based on destructive testing. 100+ % results indicate that parent material of plastic part gave way while weld remained intact.

[†]Near field welding refers to joint ¼" (6.3 mm) or less from area of horn contact; far field welding to joint more than ¼" (6.3 mm) from contact area.

[a]High-density foams weld best.

[b]Moisture will inhibit welds.

[c]Requires high energy and long ultrasonic exposure because of low coefficient of friction.

[d]Cast grades are more difficult to weld due to high molecular weight.

TABLE 14-5
Compatibility of Plastics for Ultrasonic Fabrication

	ABS	ABS/polycarbonate alloy (Cycoloy 800)	ABS/PVC alloy (Cycovin)	Acetal	Acrylics	Acrylic multipolymer (XT-polymer)	Acrylic/PVC alloy (Kydex)	ASA	Butyrates	Cellulosics	Modified phenylene oxide (Noryl)	Nylon	Polycarbonate	Polyethylene	Polyamide	Polypropylene	Polystyrene	Polysulfone	PPO	PVC	SAN-NAS
ABS	■	○			■	○	○	○									○				○
ABS/polycarbonate alloy (Cycoloy 800)		■	○		○	○	○	○					■								○
ABS/PVC alloy (Cycovin)	○	○	■		○	○	○	○												○	
Acetal				■																	
Acrylics	■	○	○		■	○	○	○					○								○
Acrylic multipolymer (XT-polymer)	○	○	○		○	■	○	○									○				○
Acrylic/PVC alloy (Kydex)	○	○	○			○	■	○											○		
ASA	○	○	○		○	○	○	■									○				○
Butyrates	○								■												
Cellulosics										■											
Modified phenylene oxide (Noryl)											■						■	■	■		○
Nylon												■									
Polycarbonate		■			○								■					■			
Polyethylene														■							
Polyamide															■						
Polypropylene																■					
Polystyrene	○				○		○				■						■				○
Polysulfone											■							■			
PPO											■								■		
PVC			○				○													■	
SAN-NAS	○	○			○	○		○			○						○				■

■ —Denotes compatibility.

○ —Denotes some, but not all, grades and compositions compatible.

Fabrication and Bonding 303

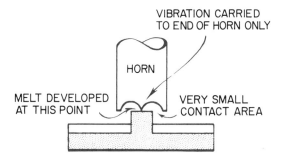

Fig. 14-42. Ultrasonic Staking
As with ultrasonic welding and inserting, ultrasonic staking uses the same principle of creating localized heat by applying high-frequency vibrations. Usually, staking involves the assembly of metal and plastic.

Fig. 14-43. Ultrasonic staking of medallion to taillight lens.

20,000 cycle per second pulsating direct current into 20,000 cycle per second mechanical vibrations. It moves the horn up and down at this speed. The horn is pressed against the plastic and causes the plastic to heat up.

Ultrasonics are used for heat sealing and welding, also. Not all thermoplastics may be ultrasonically processed. See Tables 14-4 and 14-5. Ultrasonic heat sealing and welding are described later in this chapter.

Heat Sealing Plastics

Heat sealing is used to fasten two thin pieces of thermoplastics film or sheet together. The two layers are melted together by heat and pressure, causing a cohesive bond. Most heat sealing is done in the textile industry where raincoats, baby pants, billfolds, notebook covers, beach toys, blister packs, and litter bags are sealed. Thin plastic film such as dry cleaner bags and sandwich bags are also heat sealed. Heat sealing as a means for packaging of all types of goods is on the increase. It is a fast process.

There are four methods of heat sealing: thermal sealing (heated tool), impulse sealing, dielectrical sealing, and ultrasonic sealing.

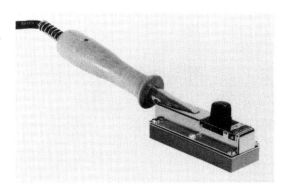

Fig. 14-44. Hand heat-sealing iron for wide seals.

Thermal Sealing

Thermal (heated tool) sealing is the simplest, least expensive, and most popular method of sealing plastic parts. However, the success of this seal depends on the skill of the operator. It is called thermal because its source of heat is continuous.

A heated tool much like a small clothes iron or soldering iron is used, Fig. 14-44. Another tool is the heated wheel. Also, hot plates are used in meat and produce markets to package many food products by heat sealing. Some two-sided heaters are used for thicker materials, Fig. 14-45. They usually clamp the work together between two heated jaws. These heaters are sometimes called bar sealers. Thermal sealing may also be done by shaped dies, Fig. 14-46. These seal a package all around to the shape desired.

Two of the most popular heat sealing plastics are polyethylene and cellophane. Many other flexible plastic films can also be heat sealed with success.

Hand-held heat sealing irons, such as those described, are usually quite inexpensive. They are easily used in the school laboratory.

Thermal Sealing Procedure. Select a piece of clean plastic film. Visqueen® polyethylene sheeting, bread wrappers, dry cleaners' bags, sandwich bags, and plastic garbage bags may be used. Polyethylene lay-flat tubing for heat sealed packages may be purchased in large rolls. Grocery store or meat market packaging film also may be used.

Use the following procedure for hand-held heat sealers.

1. Plug the sealer in and warm it up.
2. Lay two pieces of plastic film together or use lay-flat plastic tubing.
3. Press the iron or roller over the plastic film. Move the iron (or roller) at a uniform speed over the length of the seal.
4. Take the iron off and let the joint cool.

Moving the iron too slowly or pressing too hard will often cut or melt through the film. Moving too fast will not make a solid seal. A felt pad is used under the rotary wheel heat sealer to cushion the impact of the wheel, Fig. 14-49. It helps prevent cutting through.

Impulse Sealing

Impulse sealing does not use continuously heated sealing bars. In this tool, the bar temperature suddenly rises and falls while the plastic film is clamped together. The plastic is heated and cooled under pressure. Less heat reserve is available in this method. It is usually used for films .010" (0.254 mm) or less in thickness.

The most common type of impulse sealer uses a flat Nichrome ribbon on one of the

Fig. 14-45. Clamp type thermal heat sealer or bar sealer.

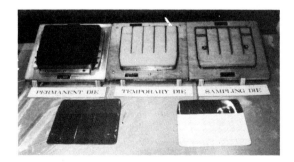

Fig. 14-46. Sealing with heat dies. Parts in the foreground.

Fig. 14-47. Hand-Held Rotary Heat Sealer

bars. It is then stretched over one jaw and covered with a Teflon® fabric which keeps the plastic from sticking to it. The other bar has a soft pad. The two jaws or bars are brought together over the two pieces of plastic film. A short burst (impulse) of electricity is run through the Nichrome wire. It is controlled by a timer. The heat rises and falls rapidly causing a seal. When the joint is cool, the bars or jaws are opened.

Other methods of impulse heat sealing include a Nichrome ribbon on edge and a continuous band type. The edge ribbon type will make a very narrow seal, Fig. 14-49. The continuous band type is similar to a double band saw, Fig. 14-50. Two bands move along, stretched between two wheels. The work is fed between the bands. Heating and cooling areas are provided. Long continuous seals can be made by the continuous band process.

Impulse sealers produce quick, neat, strong seals on thin films. As with thermal

A. Pulling out the film.

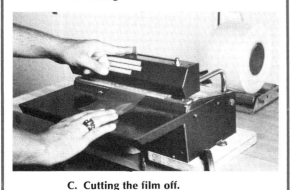

B. Sealing.

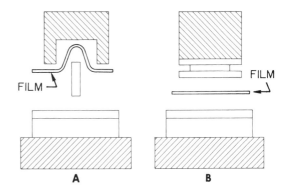

Fig. 14-49. Heat Sealing
      A. Edge sealer.
      B. Flat sealer.

C. Cutting the film off.

Fig. 14-48. Impulse Heat Sealer

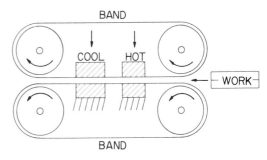

Fig. 14-50. Continuous Band Sealing
      The work is fed between the bands.

Fig. 14-51. Heat Sealed Product

Fig. 14-53. Ultrasonic Sewer (Sealer)

Fig. 14-52. Vinyl plastic pocket-saver made by heat sealing. This product can easily be made in the school laboratory.

heat sealing, most plastic films can be sealed by this method. Impulse sealers are more expensive than hand-held heated tool sealers. They are used for packaging all types of products: food, hardware, candy, and many others.

Impulse Sealing Procedure. Clean plastic materials like those mentioned for thermal heat sealing must be used.

1. Plug the machine in. No warmup is needed.
2. Adjust the impulse time. Thicker materials require longer impulse times. Try a sample or two to get the right setting. Longer sealing times than necessary tend to burn up the bar covers.
3. Hold the two plastic films together over the lower jaw.
4. Close the jaws over the plastic. The heat impulse is controlled by a timer. If it does not seal well, increase the impulse time. If the seal burns through, decrease the time.
5. Hold the pressure on the plastic a few seconds, while it cools. Release the pressure.

Ultrasonic Sealing

Ultrasonic sealing works the same way as ultrasonic staking which was explained earlier. The ultrasonic vibrations of the horn hammer the two plastic sheets or films together. The pressure and vibration cause heat. The heat fuses the two parts into one as the mole-

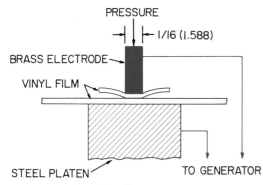

Fig. 14-54. Arrangement for sealing film with a dielectric heat sealer.

cules intermingle. Ultrasonic heat sealers are quite expensive.

Dielectric Sealing

In dielectric sealing, a high-frequency (radio frequency) current is passed through the plastic pieces held between two pressure jaws. The plastic resists the passing of the high-frequency energy through it. The resistance causes heat which seals the plastic as it is held under pressure. See Figs. 14-54 and 14-55. Dielectric heat sealing is used almost exclusively on polyvinyl chloride products. Special frequencies can be used to process some other plastics such as cellulosics and nylon.

Many dielectric sealers radiate so much radio frequency energy that they must be shielded in a special room. If not shielded, they make static which will bother radios, television sets, and aircraft radio equipment.

Welding Plastics

Many plastic parts are assembled by plastic welding rather than by cementing. Very thin as well as thick plastic parts can be held together by the various welding methods. Only **thermoplastics** may be welded, however.

There are four plastic welding methods: hot gas welding, hot plate welding (fusion), spin welding (friction), and ultrasonic welding.

Fig. 14-55. Dielectric Heat Sealer

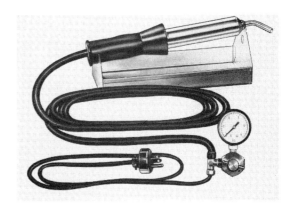

Fig. 14-56. Hot Gas Welding Gun

Hot Gas Welding

In hot gas welding, a stream of compressed air or water-pumped nitrogen is heated by an electric heat cartridge as it passes through a welding gun. The heated

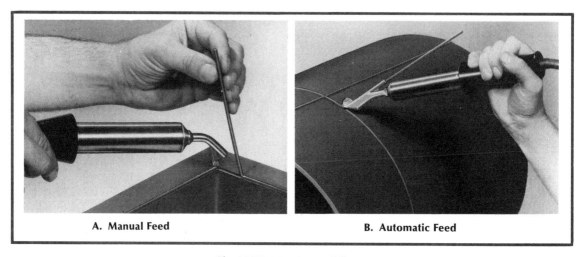

Fig. 14-57. Hot Gas Welding

gas is directed onto the pieces to be welded and the welding rod. See Fig. 14-57. All three are heated to a point where the rod flows into the two parts being welded. Heat fusion results as the molecules of the parent plastic intermingle with those in the rod. The rod used is exactly the same material as the parent material. Both materials should be the same. Dissimilar materials cannot be welded together. This process is similar to gas welding of metals. A hot gas instead of a flame is used, however.

Hot gas welding provides rapid and economical construction of one of a kind or short production run parts. Large tanks, ducts, and electronic component cabinets are made this way. Large acid tanks and duct work made from PVC are examples. They resist the acids for long periods of time.

Polyvinyl chloride sheet is probably the simplest to weld. Compressed air is used as the heated gas and a PVC rod forms the filler rod. Other materials can also be hot gas welded. Many of them require water-pumped nitrogen as the hot gas. Temperatures, materials, and welding gases for various plastics are listed in Table 14-6.

Hot gas welders are relatively inexpensive. Industrial units are used in the school laboratory.

TABLE 14-6
Thermoplastic Welding Chart

	PVC	H.D. Poly-ethylene	Poly-pro-pylene	Penton	ABS	Plexi-glass
Welding Temperature	525°F	550°F	575°F	600°F	500°F	575°F
	273°C	288°C	302°C	315°C	260°C	302°C
Forming Temperature	300°F	300°F	350°F	350°F	300°F	350°F
	149°C	149°C	177°C	177°C	149°C	177°C
Welding Gas *W.P. — water pumped nitrogen	Air	WP* Nitrogen	WP* Nitrogen	Air	WP* Nitrogen	Air

Fabrication and Bonding 309

Hot Gas Welding Procedure. The plastic sheets to be welded are first cut to size. This can usually be done on a circular saw. The edges to be welded are then smoothed and beveled. This may be done on the jointer or power sander. All the parts should be put together to make sure they fit. Figure 14-58 shows how joints may be welded. Use the following procedure for welding:

1. Attach the gas hose and **turn on the gas first**. Regulate the gas flow to about 5 psi (34.5 kPa). **If the gas is not flowing first, the heating element will burn out.**
2. Plug in and turn on the heating element. Let the welder warm up. Adjust the heat by adjusting the gas flow. The more the gas pressure the lower the temperature. The less gas pressure the higher the temperature.
3. Clip the filler rod at about a 60° angle on one end with a diagonal cutter, Fig. 14-59.
4. Put the end of the rod at the starting end of the joint (bevel toward the direction of the weld), Fig. 14-60.
5. Direct about 60% of the heat from the welder tip onto the plastic sheets and 40% on the rod. Move the tip of the welder up and down slightly as you weld. This will keep hot spots from developing.
6. Push the rod down into the beveled joint. A slight flow or wave of parent material should go ahead of the rod.
7. Cut the rod off at the end of the joint.
8. Repeat this procedure until the bevel is filled with rod. Several passes are usually needed to make a good joint.
9. When finished, unplug the electrical current **first. Let the torch cool down with gas flowing through it.**
10. Turn the gas off **when the torch is cool.**

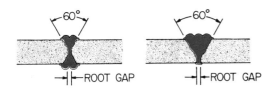

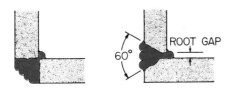

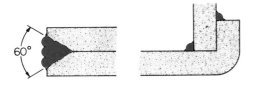

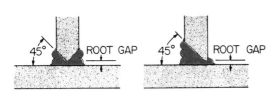

Fig. 14-58. Recommended joints for welding.

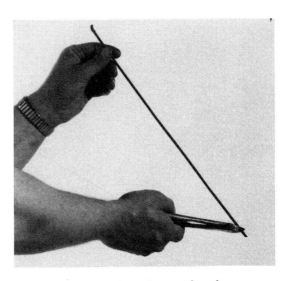

Fig. 14-59. Clipping the rod at a 60° angle.

Spin Welding

Spin welding is the welding of thermoplastics by **friction**. Usually a round object is spin-welded, Fig. 14-62. One half of the object is mounted in a stationary chuck. The other half is mounted in a rotating (spinning) chuck. The spinning chuck is lowered down and rotated. When the two halves of the plastic rub against each other, the friction causes heat. The heat fuses the thermoplastic parts together. When the weld is complete, the rotation is stopped. A slip clutch is used on the rotating chuck to stop it in time. Similar plastics are used in the two halves of the part.

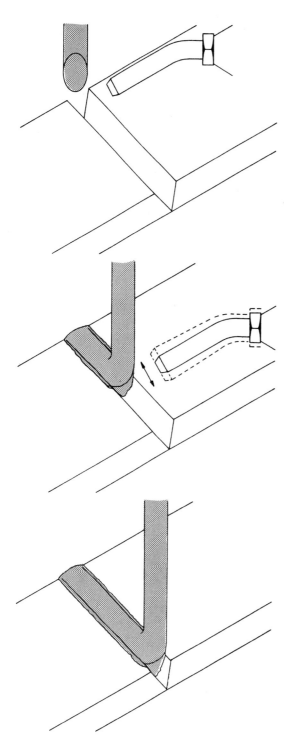

Fig. 14-60. How the weld is made.

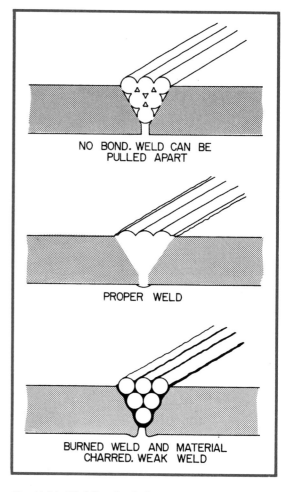

Fig. 14-61. Welding Analysis

Fabrication and Bonding   311

Spin Welding Procedure. A lathe or drill press may be used for spin welding in the laboratory. Simple upper and lower chucks can be made of wood or metal. Each should be coated with silicone bathtub seal or a similar substance for good grip.

1. Mount the two chucks on the drill press or lathe.
2. Put the two halves to be welded in the correct chucks.
3. Turn the machine on at its slowest speed. Bring the two chucks together, Fig. 14-64.
4. Hold until the parts fuse together.
5. Release the pressure and remove the parts.

Fig. 14-62. Spin Welded Parts

A. Parts being brought together.

B. Parts in contact.

C. Finished products.

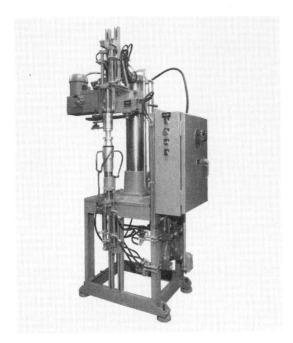

Fig. 14-63. Single-Spindle Spinwelder

Fig. 14-64. Laboratory Spin Welding

312   Section V   FINISHING OPERATIONS

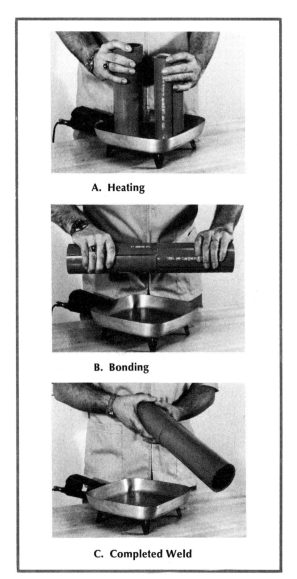

Fig. 14-65. Hot Plate Welding in a Frying Pan

Fig. 14-66. Industrial Fusion-Welded Parts

Fig. 14-67. Hot Blade Welder

In some cases, a heated blade is held between the two parts, Fig. 14-67. When the parts are hot enough, the blade is slipped out and the parts fuse together. This process requires fairly thick sections of plastic to work well. A fluorocarbon mold release should be applied to the blade or hot plate to keep the plastic from sticking.

Ultrasonic Welding

Ultrasonic welding is similar to ultrasonic staking and sealing. The two parts to be welded together are vibrated at 20,000 cycles per second by the ultrasonic horn. The heat generated by the vibrating of the two parts together under slight pressure fuses them.

Ultrasonic welding is one of the fastest ways of joining plastics. Not all the thermo-

Hot Plate Welding

Hot plate welding is often called **fusion welding.** Two pieces of similar thermoplastic can be set on a hot plate until they begin to melt. They are quickly taken from the hot plate and put together. They are held in place until they are cool. An electric fry pan is also suitable for fusion welding. See Fig. 14-65.

TABLE 14-7
Methods of Fastening Plastic Materials

PLASTIC MATERIALS	Mechanical Fasteners	Adhesives	Spin Welding	Solvent Welding	Thermal Welding	Ultrasonic Welding	Remarks
THERMOPLASTICS							
ABS	G	G	G	G	G	E-G	Surface treatment a must for adhesive bonding
Acetals	E	F	G	NR	G	G	
Acrylics	G	G	F-G	F-G	G	G	Molded acrylics difficult to solvent weld
Cellulosics	G	E	E	G	G	G-P	CA should not be solvent welded to CAB or CAP
Chlorinated Polyether	G	G	NR	G	G	P	
Fluorocarbons	G	G	NR	NR	NR	NR	PFEP, PVF$_2$, can be spin or thermally welded; PVF$_2$, can be ultrasonically welded
Nylons	G	F	G	NR	G	G	
Phenoxies	F	G	G	G	G	E	
Polycarbonates	G	G	G	G	G	E	
Polyethylenes	P	P-F	G	NR	G	G-P	Surface treatment a must for adhesive bonding
Polyamides	G	F-G	NR	NR	NR	NR	High temperature adhesives required
Polyphenylene Oxides	G	G	E	E-G	G	G	
Polypropylenes	P-F	P-F	E	NR	G	G-P	Surface treatment desirable for adhesive bonding
Polystyrene	F-G	G	E	G	G	E-P	Impact grades difficult to solvent weld
Polysulfones	G	G	VG	G	E	E	
Polyurethanes	NR	G	NR	G	NR	NR	
Vinyls	F	F-G	F	G	G	F-P	
THERMOSETS							
Alkyds	G	G	NR	NR	NR	NR	
DAP	G	G	NR	NR	NR	NR	
Epoxies	G	E	NR	NR	NR	NR	
Melamines	F	E	NR	NR	NR	NR	Material very notch sensitive
Phenolics	G	E	NR	NR	NR	NR	
Polyesters	G	E	NR	NR	NR	NR	
Silicones	F-G	G	NR	NR	NR	NR	
Ureas	F	G	NR	NR	NR	NR	Material very notch sensitive

E: Excellent; G: Good; F: Fair; P: Poor; NR: Not Recommended

From **Plastic Product Design** by Ronald Beck, copyright 1970 by Litton Educational Publishing, Inc. Reprinted by permission of Van Nostrand Reinhold Company.

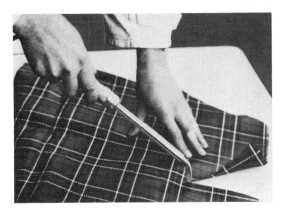

Fig. 14-68. Hot Blade Cutter

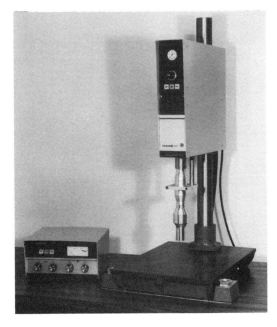

Fig. 14-69. Ultrasonic Welder

plastics may be welded this way, however. Tables 14-4 and 14-5 show those plastics that can be ultrasonically welded and staked.

Joints to be ultrasonically welded need to be specially designed. An **energy director**, Fig. 14-70, is molded into one half of the joint. The energy director contacts the other half of the part, creating frictional heat in the center of the joint first. The heat spreads out from this point completing the welded joint. Note the different types of joint designs in Fig. 14-70. Ultrasonic welders are comparable in price to ultrasonic heat sealers.

STUDENT ACTIVITIES

1. Join or laminate several pieces of acrylic plastic together with solvent cement.
2. Join several pieces of ABS or PVC sewer or water pipe together with solvent cement.
3. Build plastic airplane, car, or naval models with bodied cement.
4. Make an integral coined hinge from polyallomer, polypropylene, or polyethylene sheet plastics. Use a heated wedge-shaped tool in a drill press similar to the one in Fig. 14-34. Flex the hinge several times while it is cooling to align the molecules in the hinge.
5. Use heat staking to fasten molded products together. This may be used to stake the clip on the back of a name tag.
6. Use a heat sealer (heated tool, impulse sealer, or ultrasonic sealer) to package molded parts, mass-produced products, or gifts. Make pocket savers from vinyl plastic sheet.
7. Heat-seal a quantity of water into a bag.
8. Use the hot gas welder to weld sample joints.
    a. Butt joint.
    b. "T" joint.
    c. Top joint.
    d. Corner joint.
9. Collect plastic products that have been heat sealed, welded, cemented, or fastened. (a) Group them as to type of fastening used. (b) Make a display board of the different fastening methods.
10. Weld up a product, such as a radio cabinet, made from plastic sheet.
11. Design a product, such as a practice golf ball or fishing float, which can be spin-welded. Injection mold or thermoform the halves, trim, and spin-weld them together.

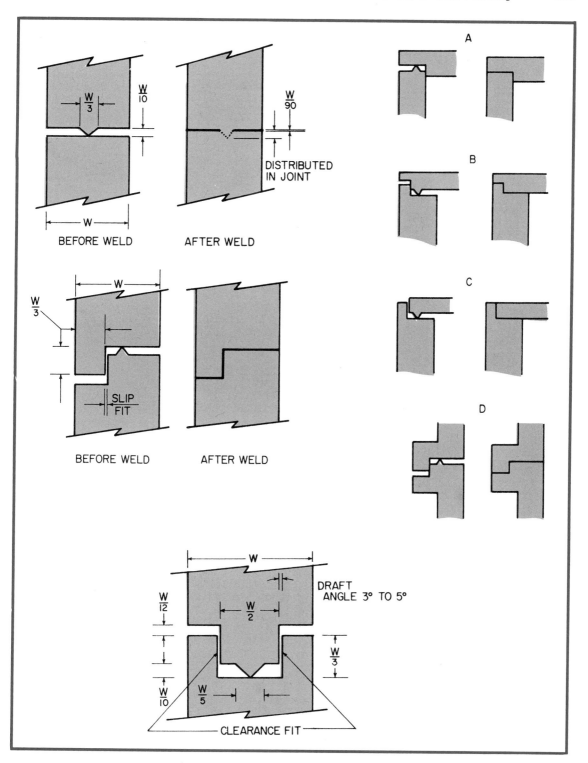

Fig. 14-70. Joint designs for ultrasonic welding.

> 12. Weld two pieces of ABS or PVC pipe together by hot plate welding. An electric frying pan or commercial hot plate, coated with fluorocarbon mold release, may be used for the hot plate.

Questions Relating to Materials and Process

1. Name the process best suited for joining in the following plastics applications even if it is not available in the laboratory.
   a. Lay-flat polyethylene tubing.
   b. Acetate candy bags.
   c. ABS plastic sewer pipe.
   d. PVC plastic sewer pipe.
   e. Butt joints of ABS pipe (low pressure).
   f. Staking metal parts to plastic.
2. What types of tools may be used for cutting plastic sheets?
3. What type of saw blades are best suited to cutting thermoplastic sheet?
4. What is the difference between cohesive and adhesive bonding?
5. How does solvent cementing work?
6. How many types of bonding cements are there? Name them.
7. Which gas is used to hot gas weld PVC?
8. Which is probably the easiest plastic to hot gas weld in the laboratory?
9. Name the different gases used for hot gas welding.
10. What method is often used to cut thin plastic sheet instead of sawing? Why?
11. Name three types of die cutting. What are each used for?
12. What modification is necessary on a metal cutting drill bit for it to successfully drill plastic?
13. Describe the differences between heated tool, impulse, and ultrasonic heat sealing.
14. Name some uses of heat sealing.
15. Name some uses for friction welding.
16. Name some uses for hot gas welding.
17. How is Styrofoam usually cut? Why?
18. What substitute for a hot plate can be made in laboratory hot plate welding?
19. How does heated tool staking differ from ultrasonic staking?
20. How may a drilled hole in acrylic plastic be easily polished?
21. A dope cement is what kind of an adhesive?
22. Silicone bathtub seal is what kind of an adhesive?
23. Name several kinds of mechanical fasteners used for plastic parts.
24. Name three ways in which integral hinges may be made.

Coating and decorating may often be thought of as only adding eye appeal or beauty to a product. Actually plastic parts are coated or decorated for other reasons, too. The processes described in this chapter often improve the resistance to wearing, scratching, and marring in addition to adding beauty. These processes change the color, texture, contrast, and design of many products. Also, identification and product information are often placed on the product. Coating and decorating may be done either during or after the molding operation.

Successful surface finishes on plastic parts need clean surfaces that allow the finish to stick. Mold releases, dust, moisture, and excess surface plastizers may cause failure.

There are a number of different methods of decorating products but only the following will be discussed here: hot foil stamping, silk screen printing, painting, plating and vacuum metalizing, electrostatic powder coating, fluidized bed coating, in-the-mold decorating, two-color molding, and engraving.

Hot Foil Stamping

Hot foil stamping is a decorative coating for plastics as well as leather, paper and book covers. Although it has been in use a number of years, recent improvements have made it more popular in the plastics industry.

In the hot foil stamping process, a decorative coating is transferred from a carrier foil or film to the plastic part by heat and pressure. Heated dies or metal printing type are pressed against the uncoated side of the polyester or cellophane film backing (carrier foil). The coating from the foil sticks to the plastic and releases from the foil. A slight depression is made in the plastic which protects the stamped design from being rubbed off.

Many colors of foil, including real gold and silver leaf, are applied in this process. Aluminized polyester foils are used as lower cost substitutes for the real silver in many cases. Wood grain and other types of decorative patterned foils are also manufactured for use on plastics. Foils have different char-

COATING AND decorating

chapter 15

acteristics that make them suitable for certain plastics. This means, therefore, that care must be taken in selecting the right foil for a plastic product. Foils are supplied in rolls

Fig. 15-1. Electroplated plastic parts which add protection as well as eye appeal to auto parts.

318   Section V   FINISHING OPERATIONS

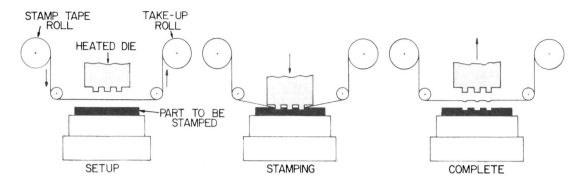

Fig. 15-2. This illustrates schematically the process of hot roll leaf stamping by using a heated metal die.

Fig. 15-3. Sample rolls of hot stamping foils.

of various widths and lengths, Fig. 15-3. Charts of foils and correct stamping temperatures are available from most foil manufacturers.

Dies for hot stamping are made from brass, steel, zinc, magnesium, or silicone rubber bonded to metal. Photo or chemically etched magnesium or zinc dies are used for short production runs on flat surfaces. Dies for long production runs should be engraved and cut deeply. They should be made from brass or steel. To make good impressions with metal dies, plastic surfaces must be flat or uniform. Die temperatures vary from about 250°F to 350°F (121°C to 177°C) depending on the plastic upon which it is stamped.

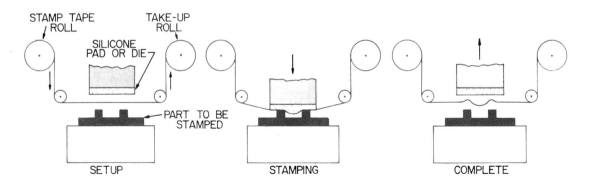

Fig. 15-5. This illustrates schematically the process of hot roll leaf stamping by using a hot silicone pad or die.

Coating and Decorating 319

A second type of hot foil stamping uses raised letters or designs on the plastic part. A heated metal plate with or without a silicone pad on it presses the foil against the raised decoration on the plastic. Silicone pads will produce good results on irregular surfaces. They need slightly higher die pad temperatures because they are good heat insulators. See Fig. 15-5.

Hot foil stamping machines, from hand-operated laboratory units to highly automated, multiple-station production units, are available. From a few to hundreds of impressions per hour are possible. Large machines have been developed which will transfer wood grained and other decorations to four or six surfaces of a boxlike object, such as a TV or radio cabinet, Fig. 15-7.

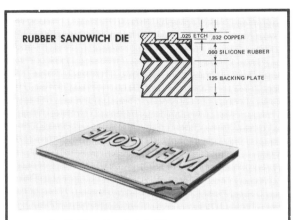

A. Rubber Sandwich Die

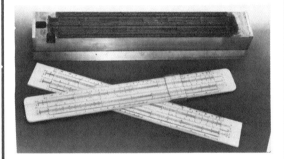

B. Engraved Hot Stamp Die and Products

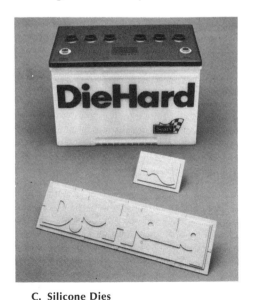

C. Silicone Dies

Fig. 15-4. Dies for Hot Stamping

Fig. 15-6. Small, Automatic-Production Hot Stamp Machine

Fig. 15-7. Automatic hot foil stamp machine for large flat surfaces. This machine will print 4- to 8-sided parts or 3-sided parts two at a time. Foil decoration is transferred by several heated silicone rollers.

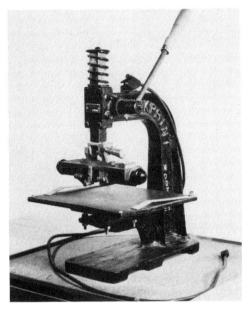

Fig. 15-9. Typical Laboratory Hot Foil Stamp Press

Fig. 15-8. Hot foil stamping an auto console.

Laboratory Equipment and Materials

Laboratory hot foil stamping may be done with a small manual machine, Fig. 15-9. It is the same kind that is often found in the graphic arts laboratory. A small unit may also be made which can be adapted to the drill press. Details are given in Fig. 14-40, Chapter 14.

Special type should be used for hot stamping. The use of lead type from the graphic arts laboratory is not advisable. The type deforms under the heat and pressure of hot stamping. If it is used, it should not be returned to the graphic arts area. Type made from other metals is available such as steel and brass, and that known as "service type." Service type looks like lead type but is made of a metal having a higher melting temperature. It is very economical while brass and steel type are more expensive. The service type is very satisfactory for general laboratory use.

Stamping foils for plastics commonly used in the laboratory may be obtained from several manufacturers who are listed in Appendix B. The foil width and the type of plastic upon which it is to be used must be specified.

Fixtures should be made to hold the various products. Two simple ones are shown in Fig. 15-10. They are made from wood and metal. They slide in a shuttle fixture which clamps to the table of the machine. See also Fig. 15-11.

Operation Requirements

Laboratory hot stamping is a simple process. However, some practice is needed for best results. When a manual machine such as the one shown in Fig. 15-11 is used, the operator must learn the "feel" of the right pressure to be applied. Correct timing of the impression is also important. The variables — time, temperature, and pressure — are accurately controlled on more expensive production machines such as those shown in Figs. 15-6 and 15-7. The variables can be adjusted and duplicated on them time and time again. This, however, must be learned with practice on manual machines.

Checklist of Materials and Equipment Needed

Assemble the following materials and equipment for hot foil stamping:

1. Hot foil stamping machine or drill press attachment.
2. Type, stamping die, or flat plate.
3. Stamping foil that will work on the plastic in use.
4. Pair of tweezers to remove hot type.
5. Scissors to cut foil if the machine does not have a roll feed.

Operation Sequence for Hot Foil Stamping

There are seven steps in performing the hot foil stamping process:

1. Choose the type and set it in the typeholder. It is set just as it is in graphic arts — upside down (nick up), face up and spelled from left to right. Lock up the type.

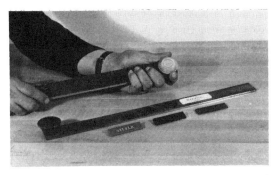

Fig. 15-10. Fixtures for hot stamping name tags and bottle caps.

Fig. 15-11. Fixtures mounted on typical laboratory hot foil stamp press.

2. Heat the typeholder, with the type in it, to about 225°F to 300°F (107°C to 149°C) depending upon the plastic to be stamped. Lower melt temperature plastics require lower stamping temperatures.
3. Choose the right foil for the plastic used.
4. Put the part to be stamped in the locating fixture.

5. Lower the die onto the plastic with the stamping foil in between them. Be sure the dull side (pigment side) of the foil is down against the plastic. Put a slight pressure on the lever until you "feel" the plastic give slightly. **Do not overimpress the plastic.**
6. Remove the pressure and take the foil off the plastic.
7. Remove the plastic part from the holding fixture.

Silk Screen Decorating

The silk screen process used by the graphic arts industry is also used on plastics. It imprints both designs and line work. Silk screen printing is an inexpensive method and makes a durable product. It may be done by hand or with automatic equipment. Production rates of 200 to 600 pieces per hour by hand and 2000 to 3000 per hour automatically are possible.

In the silk screen process, thick paint or ink is pushed through a stencil mounted on a fine screen. A rubber-faced **squeege** is used to push the ink or paint through the screen. The ink that comes through the stencil takes the shape of the openings in the stencil. The ink sticks to the plastic under the screen and dries. See Fig. 15-12.

Commercially, the "silk" screen may actually be nylon or stainless steel. A screen with 230 openings per inch is fairly standard. The squeege may be moved back and forth mechanically or by hand to push the ink through the screen. The proper type of paint or ink must be selected for each plastic used. Using the wrong kind of ink or paint will cause the plastic to craze (develop fine cracks), the ink not to stick, or other problems.

Photosensitive stencil materials are used for most industrial work. The blank stencils are first fastened to the screen. After exposure to the image to be printed, they are developed, washed out, and hardened. The stencils are then fixed and dried before use. Hand-cut stencils are ordinarily not used for production work. See Figs. 15-13 and 15-14.

Laboratory Silk Screen Decorating

Silk screen printing on plastics may be done in the school laboratory with supplies and equipment usually found in the graphic arts department. Either photosensitive or hand-cut stencils are suitable for use. Procedures for such work can be found in most detailed graphic arts textbooks.

Proper inks, paints, and stencils must be carefully selected. A stencil film base must be chosen that will work with the ink or paint to be used. Lacquer-based inks should be used with water-based stencils. Water- or oil-based inks should be used with lacquer-

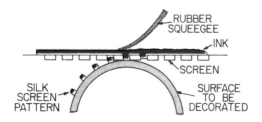

Fig. 15-12. Silk screening is a process of transferring paint through an open screen onto a flat or gently rounded surface.

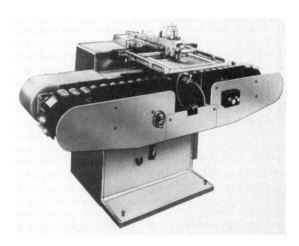

Fig. 15-13. Automatic Silk Screen Printing Machine

based stencils. Firms that supply the graphic arts industry can advise on and supply the right inks, paints, and stencil films.

Painting Plastics

Almost all plastics may be painted. It is difficult, however, to make paint stick on the waxy plastics like polyethylene, polypropylene, and fluorocarbon. Plastics are usually painted only for decorative effects and most often as a contrast to the molded-in color of the plastic part. Both enamel and lacquer base paints are used, depending on the plastic.

Enamels are made from thermosetting resins and a solvent. The thermosetting enamels are often oven-cured in production. The oven-cures can warp some plastics. Lacquers are thermoplastic resins that have been dissolved in a solvent. They are air-dried. The solvents in some can soften or craze certain plastics. Care must be taken to choose the right paint for the job.

Painting methods that may be used are (1) spray painting the entire surface, (2) mask spray painting, (3) flow coating, (4) dip coating, (5) roller coating, and (6) spray and wipe painting.

Spray Painting

Spray painting operations are used on automobile grills, stone shields, and dash panel parts; toys; home appliances; TV cabinets; and home accessories. Either the part

A. Starting the first pass.

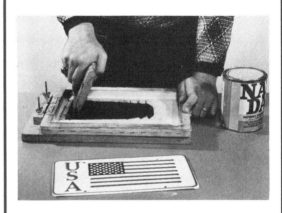

B. Making the second pass.

C. Removing the finished product.

Fig. 15-14. Products printed by the silk screen process.

Fig. 15-15. Silk Screen Printing in the Laboratory

324 Section V FINISHING OPERATIONS

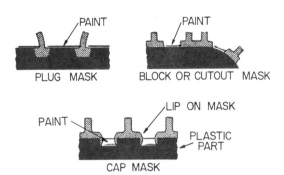

Fig. 15-16. The plug mask is used to fill and keep depressed areas clean while the surrounding area is painted. The block or cutout mask is used to confine paint to a shallow, recessed area that has no paint step.

Fig. 15-18. Automatic mask spray painting of taillight lenses.

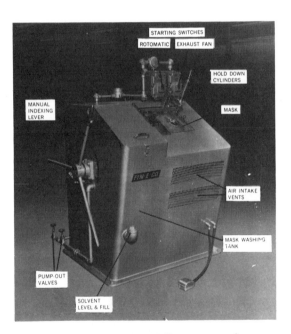

Fig. 15-17. An upspray rotating-gun mask spray painter with an automatic mask washer.

or the spray gun may be moved or rotated as the part is sprayed.

The entire surface of a part can be spray painted. Automotive parts — molded from one color plastic — usually dark, are often painted to match the rest of the car. Better color matches and lower costs are the main reasons the automotive industry uses this method. Dash panel, interior parts, and stone shields are examples of entire surface painting.

Mask Spray Painting

Parts such as automobile grills, instrument panel lenses and molded emblems are quite often mask spray-painted. This is a selective painting operation. Only part of the surface is painted. It gives background and/or decorative contrast to the part. Paint masks are designed to give sharp lines between the painted and non-painted areas. To achieve this sharpness, the part has a molded-in step or depression. The mask fits into the depression or over the lip formed by the part.

Three kinds of spray painting masks are shown in Fig. 15-16. The plug masks and cap masks give the sharpest paint lines of the three. Good masks eliminate wiping operations. Molded parts for mask spraying must be very uniform in size and shape, or paint will blow by the mask.

Flow Coating and Dip Coating

Flow coating and dip coating are methods in which the entire part is coated with paint. In flow coating, the part is flooded with paint and rotated as it is dried. Dip coating in-

Coating and Decorating 325

Fig. 15-19. Wheels being dip-coated.

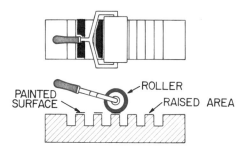

Fig. 15-20. Small Raised Areas Such as Letters, Figures, and Designs May Be Decorated by Roller Coating

Fig. 15-21. License plate being roller coated on a platen press.

Fig. 15-22. For roller coating, the part is held by a jig, a device that keeps the part in the correct position. The jig is locked in a chase, or frame.

volves dipping the part into a paint tank, Fig. 15-19. It is removed at the same rate as the paint drains. This makes the coating the same thickness at the bottom as at the top.

Roller Coating

Roller coating is often used where raised designs or letters are molded or formed into a part, Fig. 15-20. A good example is the popular thermoformed magnetic signs which are used on many cars and trucks. This is a quick, economical way to selectively paint a product. Usually only a paint roller or brayer is needed. It is done much like inking cold type for printing. In fact, using a platen press, Fig. 15-21, is a semiautomatic way of roller coating. Automatic and hand-fed roller coating machines are used in industry.

Spray and Wipe Painting

Spray and wipe painting involves the use of molded-in or indented areas in the plastic part. Paint is sprayed over the surface and into the depressions. The paint on the surface is wiped off with a cloth leaving the paint in the depressions. Sharp lines are possible when sharp angles have been molded between the recesses and the surface of the part. See Fig. 15-23.

Hand or spray painting of plastics may be done in the school laboratory with spray guns or aerosol spray cans. Products such as duck decoys and thermoformed wall placques can be hand-painted with a brush. Paint selected should be suited to the plastic used. Enamels work well on polystyrene while lacquer thinners soften it. Lacquers may be used on many other plastics. Latex paints may be required for some plastics. Some of the paint should be tried on a scrap of the plastic first before the whole part is painted. Several tests of different paints may be necessary.

Plating and Vacuum Metalizing

A bright metal coating may be applied to plastic parts. It is put on either the front (first surface) or back (second surface) of the part. First-surface metal coating gives protection as well as beauty to the surface of the part. It is most often done on opaque plastics. Second-surface metal coating is done on clear plastics so it can show through to the first surface. In this case the coating gives the beauty but the plastic provides the protection. Second-surface coatings give a deep shine because the coating is below the surface.

Two types of bright metal coatings are generally used:
1. Electroplating and
2. Vacuum metalizing.

Of the two, electroplating is the most durable, but vacuum metalizing is less expensive. Plating is usually done only on the first surface while vacuum metalizing may be done on either surface.

Electroplating

Plastics may be electroplated in the same way as metal. The only difference is that plastics are nonconductors of (will not transmit) electricity. In order to plate plastics, a conductive surface must be applied to the plastic. An electrically conductive coating of copper is chemically bonded to the cleaned plastic surface. From this point on, the electroplating process is almost the same as for metals. Bright chrome, silver, nickel, gold, and other metals are electrically deposited on the copper base coat.

Electroplating provides a sturdy, bright finish that protects the plastic part. The light weight of the plastic when compared to metal parts, and the bright metal finish produce many unique automotive and plumbing

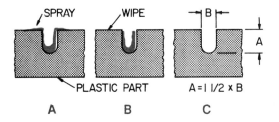

Fig. 15-23. Spray and Wipe Paint Process
   A. Paint is sprayed on.
   B. Paint is wiped from the surface.
   C. The depth of the recesses should be one and one-half times the width.

Fig. 15-24. Inspecting an electroplated automobile grill.

industry products. See Fig. 15-24. Others are being developed rapidly.

Electroplating may be done in the laboratory if the right equipment is available. Several equipment suppliers offer laboratory electroplating sets which may be used on plastics.

Vacuum Metalizing

Vacuum metalizing is an economical way to put a bright metal finish on plastics. The finish is not as durable as electroplating, however. It is used for plating items such as plastic model kit parts, automobile decorative parts, and household items, Fig. 15-25. Plastic films may also be vacuum metalized. Examples of metalized plastic film are the Echo I and Echo II balloon satellites and the sun shield for Skylab I. Metalized films were used on the Apollo moon flights. See Fig. 15-26.

Vacuum metalizing is a simple process but needs rather expensive equipment. The plastic is first cleaned and then sprayed with a base coat of lacquer. The dried base coat later bonds the metallic coat to the plastic. The parts are put on rotating racks which move into a large steel vacuum chamber, Fig. 15-27. The center of the rack has tungsten wire filaments which hold small metal clips. The clips are usually aluminum but may also be silver, gold, nickel, or chromium. The filled rack is pushed inside the vacuum

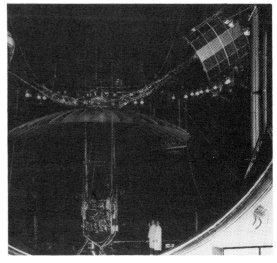

Fig. 15-26. Vacuum Metalized Space Hardware

Fig. 15-25. Vacuum Metalized Products

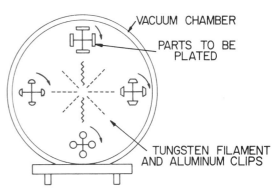

Fig. 15-27. An open-end view of a vacuum metalizing chamber. The plastic parts are rotated to ensure complete metal coverage.

chamber and the door is closed, Fig. 15-28. A near-perfect vacuum is pulled on the closed steel vacuum chamber. The tungsten filaments are then electrically heated to about 1200°F (649°C). At this temperature the metal clips melt and the metal spreads over the filament. The temperature is then raised to about 1800°F (982°C) and the metal on the filament "flashes" or evaporates off the filament. The **flashed** metal travels in a straight line and sticks to any surface it hits. Rotating the parts on the racks and flashing several filaments in sequence covers all the surfaces of each part. The air is then let back into the chamber, the door opened, and the parts removed.

Parts with patterned, crowned, or textured surfaces with rounded edges can be vacuum metalized. Parts should not have large flat surfaces or sharp corners. An optically flat surface is very hard to mold. If it is not perfectly flat, the distortion will show even more after it is metalized.

Vacuum metalizing is seldom done in the school laboratory. Equipment for the school is not presently available at a suitable cost.

Fluidized Bed Coating

Fluidized bed coating is a process that uses dry powdered plastic materials. Polyethylene is very popular for this process. The dry plastic powder is put in a chamber called a **fluidized bed**, Fig. 15-29. Air is blown up through a screenlike porous plate in the bottom of the chamber, setting the dry powdered plastic in motion. The air moving up through the chamber tends to float the

A. Moving rack into vacuum chamber.

B. Inside the chamber, showing the tungsten filaments in the center.

Fig. 15-28. Vacuum Metalizing Chamber

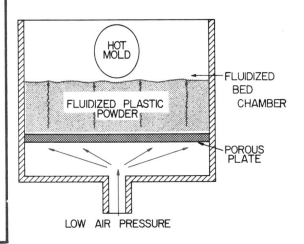

Fig. 15-29. Fluidized Bed Coating

dry powder and makes it feel like a liquid. It acts like a fluid (liquid), thus the name "fluidized."

A heated mold or object to be coated is lowered into the fluidized powder. The heat from the mold or object melts the powder on the surface. The plastic fuses to the surface of the mold or object. When a layer of plastic has been built up on the surface of the object, it is put back in the oven. The heat smooths out the surface of the plastic. It is then taken from the oven and cooled.

Importance

Fluidized bed coating is a part of the general category, "Coating and Decorating," in marketing information. This information shows that coating and decorating processes are presently about 6% of the total plastics market. Fluidized bed coating promises to increase its share of the market in the next 10 years. Typical examples of products with fluidized bed coatings are tool handles, dish drying racks, strainers, steel pipes, and steel window sashes. Also see Fig. 15-30.

Advantages

Fluidized bed coating can be used to build uniform coatings on parts from .005" (0.127 mm) to .080" (2 mm) in thickness. Coatings may be made on aluminum, carbon steel, brass, expanded metal, and nonmetallic materials such as glass and ceramics. The coatings are flexible and wear well. They are also an excellent electrical insulator. They do not break or chip like paint. The process is simple and inexpensive. An added advantage is that its application is nearly pollution free.

Disadvantages

Thin materials are difficult to coat. They do not hold enough heat to build up a coat. Materials of different thicknesses pick up coats of different thicknesses. In production coating, thicknesses may vary with the season. In warm months, the coating may be thicker and in cold months thinner due to plant temperature changes. Most plants are now temperature controlled to avoid this

Fig. 15-30. Fluidized bed-coated products in use at Knott's Berry Farm.

problem. Mechanization of fluidized bed coating is difficult because uniform coatings cannot be built up on a part that is in motion. Most fluidized bed coating is done manually.

Materials Used in Industry

Fluidized bed coating may be done with either thermoplastic or thermosetting plastic powders. The main limitation on materials is whether it can be ground into a fine powder. Some plastics heat up and melt when they are being ground into a powder. They have to be frozen to very cold temperatures to be ground. Plastics such as polyethylene, polypropylene, polyvinyl chloride, nylon, and cellulosics are now being used. The thermosetting plastic, epoxy, is also used for fluidized bed coating. A special coating grade of polyethylene is made for this process. It is the plastic most often used.

Industrial Equipment

Some large, mechanized industrial fluidized bed coating machines have been built. Recently machines 6' high x 5' wide x 12' long (1.8 m x 1.5 m x 3.65 m) have been made. One machine, built for the General Electric Company, holds 18,000 pounds (8165 kg) of plastic powder. Smaller manually operated units have been in use since the early 1950's. A unit like this is shown in Fig. 15-31.

Laboratory Equipment

Several laboratory fluidized bed coating machines are available for use, Fig. 15-32.

Fig. 15-31. Industrial Fluidized Bed Coating Machine

Fig. 15-32. Commercially Made Laboratory Fluidized Bed Coater

They range in size from 6" (150 mm) to 18" (450 mm) across. A simple fluidized bed coater may also be built for the laboratory. A wooden box may be built with two firebricks over an air chamber, Fig. 15-33. The firebricks should be tightly glued into place with silicone RTV bathtub seal. The box should be high enough to keep the plastic powder from flying around or twice as high as the depth of the powder needed. A depth of about 18" (450 mm) works well. An air line should be connected to the air chamber below the firebricks. An air pressure of about 3 to 5 pounds per square inch (20.7 to 34.5 kPa) is needed. Use an air regulator to control the air flow.

Operation Sequence for Fluidized Bed Coating

Fluidized bed coating may be used as a substitute for dip coating of tools. It may also be used as a durable finish for outdoor projects such as post lamps, tables, and chairs. Often the part requires a primer coating. Check the plastics manufacturer's instructions for this and other information. Study the operation sequence in Table 15-1.

Checklist for Materials and Equipment Needed

Assemble all the materials that are needed for fluidized bed coating.

1. Oven, electric stove, or rotational molder oven.
2. Fluidized bed coating machine.
3. Powdered plastic — coating grade polyethylene works very well. A molding grade may be substituted.
4. Primer — if needed.
5. Part to be coated.
6. Short piece of wire.
7. Insulated gloves and long-nosed pliers.
8. Dry color pigments — optional.

Coating and Decorating 331

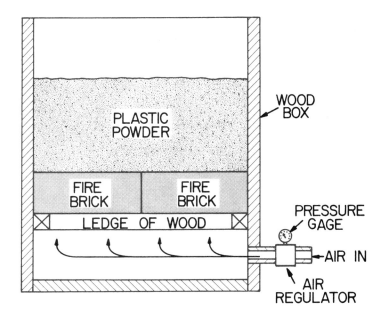

Fig. 15-33. Laboratory-Built Fluidized Bed Coater

Fig. 15-34. Materials and equipment ready for fluidized bed coating process.

TABLE 15-1
Operation Sequence for Fluidized Bed Coating

Sequence/Action	Reason	Troubleshooting
❶ Hang the part on a wire. Bend the other end of the wire to form a hook. Hang the part in the oven and preheat it to about 100°F (55°C) above the melting point of the plastic being used, Fig. 15-35. About 10 to 15 minutes at 350°F (177°C) for polyethylene will work.	The metal part to be coated must soak up enough heat to melt the plastic and make it stick.	Coating too thin: (a) Heat the metal longer. (b) Get the metal hotter.
❷ (a) Fill the fluidizer about 1/2 to 2/3 full of plastic powder. Add a dry color pigment if it is desired. (b) Turn on enough air pressure to **float** the plastic powder. About 3 to 5 psi (20.7 to 34.5 kPa) is enough.	(a) The chamber should not be too full of plastic powder. The plastic will spill over as it begins to float on the air. (b) Too much air pressure will make the plastic fly around.	Coating too thick: (a) Part too hot. (b) Part dipped too long.
❸ Put on insulated gloves. Take the part or tool from the oven with a pair of pliers and dip it in the fluidized plastic **immediately**, Fig. 15-36.	Safety first: Protect your hands with gloves when handling hot objects. If the part is not dipped right away, it will cool off.	Coating too thin: (a) Part cooled off before it was dipped. (b) Part not dipped long enough. (c) Part not heated long enough.
❹ When the part stops picking up plastic powder put it back in the oven, Fig. 15-37. Fuse it for about 2 to 4 minutes.	The plastic will soften and smooth out as it is reheated. Thicker parts sometimes will hold enough heat to completely fuse the plastic.	Coating not smooth: Fuse in the oven longer. Coating turns dark color: Coating fused too long or is too hot.
❺ Extra coats of plastic may be added by the same method, Fig. 15-38.	More plastic will be picked up if the hot part is dipped back into the fluidized bed.	

Coating and Decorating 333

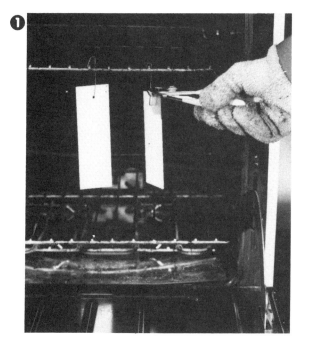

Fig. 15-35. Hang part on a wire in oven to heat.

Fig. 15-37. Remove part from fluidized bed to oven.

Fig. 15-36. Dip in fluidized bed.

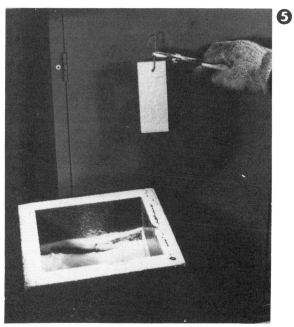

Fig. 15-38. Repeat for extra coats if desired.

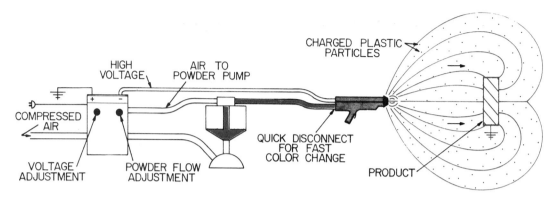

Fig. 15-39. Operation of an electrostatic powder coating system.

Electrostatic Powder Coating

Electrostatic powder coating is a process in which negatively charged plastic powder particles are sprayed onto a positively charged (grounded) product. The plastic particles are attracted to the grounded product much as a metal object is attracted to a magnet. The powder does not follow a straight line to the product because of the electrically charged field surrounding the product, Fig. 15-39. Plastic particles will actually be evenly deposited on the back side of a round object. The coated product is put in an oven and the plastic powder is fused to the product.

Importance and Advantages

Electrostatic powder coating gives a good surface covering with very low material loss. In fact, any powder that falls down can be collected and reused in the system. A 99% material efficiency is claimed for the process.

Electrostatic powder coating systems eliminate the air pollution problem. The plastic powder does not contain solvents as paint does. The plastics are 100% solids which melt on the product surface. No solvents are given off into the air during application and curing.

Powder coatings developed by this method are hard, tough, and attractive. They do not crack, chip, or peel like paint does. Products coated with plastics may be bent and the coating will flex with the product. Uniform coatings of from .003" to .030" (0.076 mm to 0.762 mm) thick may be made by electrostatic powder coating.

Materials Used in Industry

Almost any plastic powder may be applied to a part by the electrostatic powder coating method. Both thermoplastic and thermosetting plastics will work well. Polyethylene, polypropylene, thermoplastic polyester, and epoxy resins are common examples.

Industrial Equipment

Several manufacturers make industrial electrostatic powder coating equipment. The power supply usually operates on 115-volt, 60 Hz current. A DC voltage of from about 30 to 90 kilovolts (30,000 to 90,000 volts) and a current of 100 to 200 microamps is generated by the power supply. From about 15 to 25 cubic feet per minute (0.42 m^3 to 0.71 m^3) dry, clean compressed air at 50 to 100 psi (345 to 690 kPa) are required, depending on the manufacturer and size of the unit.

Laboratory Equipment

A laboratory sized electrostatic powder coating unit is currently available. It is, how-

Coating and Decorating 335

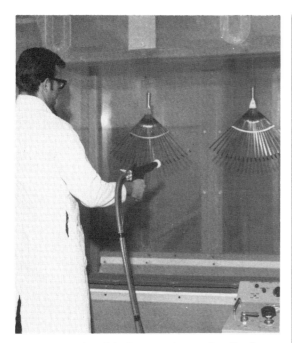

Fig. 15-40. Industrial Electrostatic Powder Coating

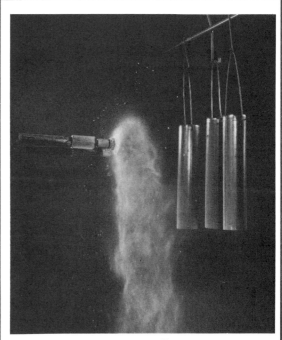

A. Current off.

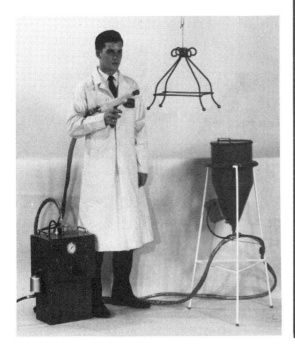

Fig. 15-41. Laboratory Powder Coating Unit

B. Current on.

Fig. 15-42. Electrostatic Powder Coating

ever, similar in cost to production units. Its main advantage is its small size.

In-the-Mold Decorating

Some thermosetting plastics and thermoplastics may be decorated in the mold. Products decorated by this method usually have hard, durable, scratch-free surfaces. Applications include melamine dinnerware (Fig. 15-43), wall plates for electric light switches, clock faces, kitchen utensil handles, and control knobs.

A decorative foil or overlay, the sheet which contains the decoration for the product, is placed in the mold **during** the molding process. Sheets used for in-the-mold decorating are printed by regular methods.

The printed overlay is made of rayon or cellulose paper about .002″ to .003″ (0.051 mm to 0.076 mm) thick. The printed paper is soaked in a melamine resin solution and dried. In thermoset in-the-mold decoration of melamine dinnerware, the part is partly cured in the mold first. The mold is opened and the printed overlay sheet is placed in the mold. The mold is closed and the cure is completed. The overlay should be put in the mold as soon as the part is cured enough to hold together. The sooner it is placed in the mold, the stronger the bond between the overlay and the part.

In-the-mold decorating may also be done during injection molding. The printed overlay sheet is put in the mold **before** injection. Often a static electric charge is applied to the overlay sheet to hold it in place during the injection molding process.

Two-Color Molding

Many products in use daily are two-color molded, such as typewriter keys, adding machine buttons, and telephone dial parts, Fig. 15-44. Two-color molding provides a very durable, wear-free way of making parts which need letters or numbers on them. These parts are made by first injecting some of the part in one mold and then the rest in another mold. Usually the main part is molded first, with openings left for the letters and numbers. One half of the mold is rotated with the part(s) in it. The second color is injected into the back, filling the openings left in the first "shot."

Engraving Plastics

An engraving machine is a popular and relatively inexpensive tool for making dur-

Fig. 15-43. In-the-Mold Decorated Melamine Dinnerware

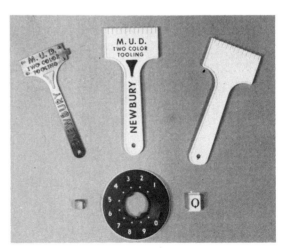

Fig. 15-44. Two-Color Injection Molded Parts

Coating and Decorating 337

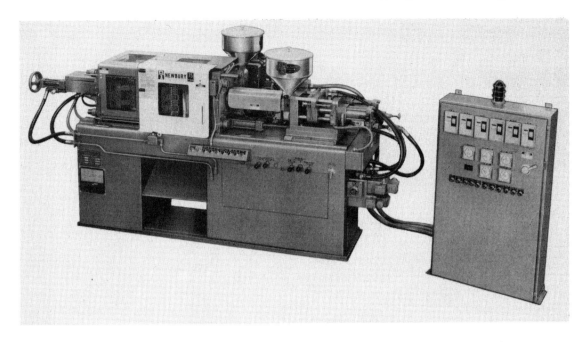

Fig. 15-45. Two-Color Injection Molding Machine
Note that there are two injection molding cylinders.

A. Cavity View
B. Plate View

Fig. 15-46. Two-Color Injection Molding Die
The circular plate rotates the parts 180° for the second shot.

Fig. 15-47. Engraving plastics in the laboratory.

able plastic signs, nameplates, tags, and dial plates. It may be used to engrave letters into a mold or other plastic products. Either forward or reverse lettering may be purchased for this process. The engraver is available in various sizes.

Actually, the engraver is a small pantograph machine which permits the copying of letters, Fig. 15-47. Letters of different sizes may be engraved by varying the ratio on the machine.

When cutting plastic signs or tags the engraver cuts through the top layer of plastic into a contrasting underlayer. This makes the sign show up very well.

STUDENT ACTIVITIES

1. Perform the hot foil stamping process on plastic products made by such processes as thermoforming and injection, rotational, and blow molding.
2. Set up a silk screen for the decoration of thermoformed and blow-molded products.
3. Use the roller coating method to paint raised letters and designs on such products as license plates and wall plaques.
   a. Experiment with different types of inks and paints.
   b. Set up manual or semiautomatic roller coating system.
   c. Use a platen press or proof press to roller-coat flat thermoformed products.
4. Engrave nameplates with a pantograph engraver.
5. Build a fluidized bed coating machine and coat products.
6. Experiment with different grades of powdered polyethylene to determine which gives the best coating and sticks the best.
7. Color some polyethylene powder for fluidized bed coating.

Questions Relating to Coating and Decorating

1. Besides eye appeal, what does coating or decorating often add to a plastic product?

2. Which coating and decorating technique is considered a part of the processing of a product?
3. How are painting and silk screening of plastics similar?
4. How are the silk screening and hot foil stamping processes similar?
5. What type of screen material is used in industry in place of silk screen printing?
6. How can spray painting techniques speed up the decorating of plastics?
7. How can selective areas of a part be painted by spray painting?
8. Name another painting technique besides spray painting and roller coating.
9. What kind of a finish can be applied to plastics by plating and vacuum metalizing?
10. Which is the most durable — plating or vaccum metalizing? Which is the least expensive?
11. What are the main differences between two-color molding and in-the-mold decorating?
12. What are the similarities between two-color molding and engraving?
13. How can electrostatic powder coating reduce air pollution?
14. What are the advantages of electrostatic powder coating over painting?

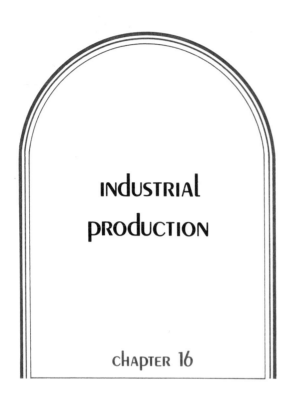

industrial production

CHAPTER 16

Mass Production

Industrial production is a system in which products are produced in large quantities. Companies may produce several models of a product with only slight changes. Parts for these products are **standardized** (all made alike). Many of these parts will fit into other models of the same product, such as in an automobile.

This system is often called **mass production** because large numbers of parts are made alike. Mass production contributed greatly to making the United States an industrial giant. It has caused the standard of living to constantly increase. It has increased the amount of goods people can buy in terms of hours, weeks, or months of work.

Interchangeability

The basis of industrial production is the machine. Machines mass-produce parts or whole products and make them all alike. They are made so much alike that it often takes very high precision instruments to

Reprinted by permission from the Ford Motor Company.

Fig. 16-1. Mass production of the automobile.

measure the difference between them. In many cases, differences between parts of .001" (0.0254 mm) to as little as .0001" (0.00254 mm) are maintained. Out of a box of 1000 identical parts, at least 970 will fit perfectly. This is a 3% reject rate. The production of identical parts provides the basis for what is called **interchangeability,** meaning any one of like parts can be used in the completion of a product.

Before interchangeable parts and mass production were first developed by Eli Whitney, each product was different. In 1798, the U. S. government needed 10,000 muskets (guns). At that time guns were all made by hand. Expert gunsmiths made and fitted each piece. If a part broke, a new one had to be made to fit in its place. The work took hours because each piece was different. To hasten the manufacture of muskets, Whitney worked several years to develop and build machines, jigs, and fixtures so that his gun parts could all be made alike. He demonstrated that individual parts could be taken at random from piles of 10 parts and a musket assembled, Fig. 16-2. By showing that the gun would still work when a part was taken out and replaced with a like part, Whitney established the value of mass production and interchangeability.

With mass production and interchangeable parts came the use of less skilled labor. Before this, each part or product was often made entirely by one person. He or she was highly skilled in making one product. Much of this piece work was done in or near the home. Mass production brought factories to the early towns and cities. With mass production, a number of machines could be set up by a master or skilled craftsman. They could then be run by unskilled workers. If adjustments or repairs were needed, the skilled craftsman could make them and keep many machines running.

The Assembly Line

A little over 100 years after Eli Whitney developed interchangeable parts, Henry Ford developed the **moving assembly line.** He put Whitney's interchangeable parts idea together with the conveyor system which he found in the meat packinghouses

Reprinted by permission from the Ford Motor Company.

Fig. 16-2. Eli Whitney demonstrating interchangeability.

of Chicago, Figs. 16-3 and 16-4. He brought the work to the workers. The workers stood still and the work moved in front of them. Workers became highly skilled in just one small assembly job. Before this, a worker often had to assemble a product from start to finish. Workers no longer had to learn a sequence of complicated operations. Now they learned only a few. **Labor was divided up into small pieces.** Each worker had a share, Fig. 16-5. A big job like building a car was divided into many small jobs, each done by a different worker. This system of assembly-line mass production speeded up production. More products were made at less cost. It also became more boring for the worker.

At about the same time that Henry Ford developed the assembly line, Fredrick W. Taylor tried to streamline the workers' efforts. He is called "the original efficiency expert." He wanted to find the speed at which a worker could best work. He also wanted to **eliminate waste motion.** His purpose was to speed production and keep the worker from becoming as tired. Taylor's work was often met with scorn by labor who felt he was trying to work them harder.

In summary, then, mass or industrial production is based on (1) interchangeable parts, (2) the moving assembly line, (3) the division of labor, and (4) the elimination of waste motion.

Fig. 16-3. Packinghouse Conveyor System

Reprinted by permission from the Ford Motor Company.

Fig. 16-4. First Ford Assembly Line

Reprinted by permission from the Ford Motor Company.

Fig. 16-5. Bringing the work to the worker.

Types of Manufacturing Systems

Several types of manufacturing systems are used by industry. They are classified by ownership and the services they perform. They are (1) captive operations, (2) subcontractors, (3) job shops, (4) consulting engineering firms, and (5) complete manufacturing corporations. In many cases it is hard to classify them because the systems are overlapping. Sometimes they may be classified in more than one way, depending on the product made or to whom it is sold.

Captive Operation

The captive operation is generally thought of as a unit of a larger company or corporation to which it supplies materials or parts. It may be the plastics manufacturing division of a large corporation. It may be the transmission or battery manufacturing division of an automobile manufacturing corporation. Most of the products made in a captive operation are used by the corporation that owns it. Some of its products may also be sold directly to other customers.

Captive operations of large corporations are usually supplied by the corporation with the overall design work for the total product to be made. The captive division then usually does the design work on the specific part or parts it will make for the parent corporation. Thus, a design department is often a part of a captive operation. In many cases all of the design work is done by the parent corporation. The jigs, fixtures, molds, and machines may be designed, built, and set up in the plant by the parent corporation, or sometimes by the captive division.

Captive operations may be started as new divisions of a corporation, or they may be small companies which are bought up by the larger corporations.

Subcontractors

Large corporations often have contracts with other corporations or smaller companies for the manufacture of parts or assemblies to be used in a product being produced. These corporations or smaller companies, known as subcontractors, are completely independent of the large corporation. By subcontracting with these companies for the manufacture and supply of certain parts or assemblies, the large corporation often saves money. The smaller companies or subcontractors may be able to offer lower prices because they often **specialize** in certain types of manufacturing operations. Thus, they can invest in special equipment and offer savings to large corporations who desire a certain type of part or product.

Most of the design and engineering work is done for the subcontractor by the larger (contracting) corporation. The subcontractor is usually just responsible for setting up the job in its own factory. Even the molds and dies may be made and owned by the contracting corporation.

Large corporations usually have at least two prime sources for a part or material which they buy. This is so a strike at one of the subcontractors will not completely shut down the large corporation's operation.

In addition to supplying the needs of other corporations, a subcontractor may also have a line of products which it makes and sells through its own organization.

Job Shops

A job shop is similar to a subcontracting operation. It usually makes the same types of items but is often thought of as smaller in size. Engineering and design work is often supplied to the job shop by the contracting corporation. It is usually responsible only for setting up and manufacturing the contracted products.

Job shops may make long or short production runs or one-of-a-kind prototype parts. Job shops often specialize in certain types of work.

Consulting Engineering Firms

Consulting engineering firms are usually thought of as groups of designers and engineers who are hired by a company or corporation. They may design or engineer a product, or they may just serve as a consultant to the design or engineering department of another company or corporation.

Consulting engineering firms may also have the facilities for the manufacturing of certain products. They may act as designers, engineers, and subcontractors for a product or whole line of products for another corporation. They may also act as an independent manufacturer of a product or line of products.

Complete Manufacturing Corporation

The complete manufacturing corporation is a firm which makes and markets a product, line, or several lines of products. The large complete manufacturing corporation often controls or owns much of the whole operation from the raw materials to the finished products. It sometimes even owns its distribution and sales outlet system. It usually has its own design, manufacturing, and marketing departments. These corporations obtain or buy raw materials and parts or whole products from captive operations, subcontractors, job shops, consulting engineering firms with manufacturing capability, other corporations, or some of its own divisions.

Two or more divisions of large corporations may even bid against each other for the contract to produce a product for the parent corporation. As a result, this open competition within a corporation saves the customer money.

Fig. 16-5A. Some companies hire outside companies to design or engineer their products.

Fig. 16-6. When a bid is prepared, the company gives prices, terms, and promises of delivery of a product.

Production and Plastics

Plastics products are true mass production items. Many of the plastics processes are high-speed production methods. Injection molding, compression molding, extrusion, thermoforming, and other processes produce parts at a rapid rate. Most plastics processes have now been **automated** to achieve a more uniform product with a minimum of labor. New **computer control systems** have further developed the quality control of plastics products. Computers are lowering the cost of products by reducing the number of rejected parts.

The production of both plastics materials and products is becoming highly automated. Many of the products are almost exclusively mass-produced. Dies and molds are designed to fit into high-speed production machines to turn out hundreds or thousands of parts per hour.

The Plastics Industry

Today, there are over 14,000 plastics processing plants in the United States. They are divided into three large categories which sometimes overlap:

1. The plastics materials manufacturers who produce the basic plastic resins or compounds;
2. the primary processors who convert plastics into solid shapes; and
3. the secondary processors who further fashion and decorate the plastics.[1]

Plastics Materials Manufacturers

The primary function of the materials companies is the formulation of plastics from basic chemicals. These plastic compounds are sold in the form of granules, powder,

---

[1]Data on the breakdown of the plastic industry is derived from the following sources:
   The Story of the Plastics Industry, The Society of the Plastics Industry, Inc.
   Directory and Buyers' Guide of Members, 1973-74, The Society of the Plastics Industry, Inc.
   "Plastics Processing USA 1974, Manufacturing Plant Census #6," by **Plastics Technology Magazine.**

pellets, flakes, and liquid resins or solutions for processing into finished products. Some plastics materials companies may go a step further and form the resin into sheets, rods, tubes, or film.

The membership of the Society of the Plastics Industry, Inc. lists 275 companies as manufacturers of plastics materials in the United States. A majority of these are chemicals manufacturing companies. Some purchase chemicals from which they formulate the plastic resins and compounds. Others only make the compound after purchasing the resin.

Primary Plastics Processors

Primary plastics processors are those firms that convert plastics materials into usable shapes or products. These firms can be divided into the following classifications according to the processes they perform:

1. **Molders.** Molders produce finished products by forming the plastic in or on a mold of the desired shape. There are over 5400 injection molding plants in the United States. In addition, 2722 plants do thermoset molding, 1131 do blow molding, 591 do rotational molding, and 2269 process foam plastic products. Over 2800 plants thermoform plastic sheet into products.
2. **Extruders.** Extruder plants are divided into two groups. The first group manufactures sheets, film, rods, tubing, profile shapes, pipe and coat wire. The second group includes the producers of threadlike plastic filaments which, when woven into cloth, are used for clothing, seat covers for cars, buses, trains, airplanes, and furniture. The filaments are also used to manufacture insect and industrial screening. About 3065 plants in the USA extrude plastics.
3. **Film and Sheeting Processors.** These processors make vinyl film and sheeting by calendering, casting, and extruding. Over 675 plants manufacture these materials in this country.
4. **High-Pressure Laminators.** These plants form sheets, rods and tubes from paper, cloth, and wood impregnated with plastic resins. Data on the number of these plants is combined with reinforced plastics manufacturers.
5. **Reinforced Plastics Manufacturers.** Reinforced plastics include processes that combine liquid polyesters, epoxies, phenolics, and silicones with reinforcements such as glass fibers, asbestos, synthetic fibers, and sisal to form strong, rigid structural plastics and plastics products by molding and forming. There are 2009 such plants in the United States.
6. **Coaters.** Coating plants make use of calendering, spread coating, preimpregnating, dipping, and vacuum deposition to coat fabric, metal, and paper with plastics. There are nearly 3100 coating plants.

The total number of primary processing plants in the USA is about 13,500. To be included in these data, a plant must use at least $100,000 worth of raw plastics materials per year. There are duplications in the figures showing the number of plants, by type, as many plants carry out more than one process.

Secondary Processors

The secondary processing operations of the plastics industry are those in which plastics products such as sheets, rods, tubes and molded parts are converted into saleable products. Very few plants, about 660, are solely engaged in secondary processing. Most plants listed (about 10,500) do both primary and secondary plastics processing at one location. Secondary plastics processing is divided into the following classifications.

1. **Decorating.** Decorating includes printing, marking, electroplating, and vacuum metallizing. About 3600 plants are involved in decorating plastics.
2. **Bonding, Welding, and Heat Sealing.** In these operations, plastics sheet and

film are converted into usable products such as shower curtains, rainwear, inflatable toys, upholstery, luggage, and closet accessories. Over 8700 plants are engaged in these operations.
3. **Machining.** Plastics sheets, rods, tubes and special shapes may be machined, using all types of machine tools, to produce industrial parts, jewelry, and consumer products. Plastics are machined in more than 4750 plants in the USA.
4. **Fabricating and Finishing.** Plastics products are fabricated from standard shapes, sheets, rods and tubes by welding and bonding. About 5400 plants currently fabricate parts from plastics. Finishing operations such as buffing, ashing, and deflashing are done in 3911 plants.

Mold Makers

Mold making is done in at least 5358 plastics plants in the United States. Many thousands of independent mold makers also supply the plastics industry with molds.

The Five Divisions of Industrial Organization

Most large industrial organizations are made up of five divisions: (1) research and development, (2) manufacturing, (3) marketing, (4) industrial relations, and (5) finance and control. The success of a company or corporation is dependent on the operation of these five divisions. In some small companies, several divisions may be combined because of the size of the total operation.

Research and Development

Research and development (R & D) is the division of a corporation or company that is in charge of researching and developing new products, Fig. 16-7. Its main job is to work with ideas for creating new products which can be manufactured and later sold to consumers. It is also responsible for developing new knowledge about the material or processes with which the company works.

A. In research centers, new applications of natural and synthetic materials are developed.

B. Research is carried on to find new knowledge that can be used to create new products.

C. One of the first steps in the development of a new product idea is creating its design.

Fig. 16-7. Research and Development

Fig. 16-8. Management in the manufacturing division plans, organizes, and controls production of the company's products.

A. Establishing goals for the sale of plastics products is one responsibility of the marketing division.

B. Marketing methods may include use of new materials that improve packaging efficiency.

Fig. 16-8A. Marketing

Manufacturing

The manufacturing division of the company produces the product or line of products, Fig. 16-8. It is responsible for setting up machines and obtaining all the materials needed for production. This division also plans, organizes, and controls production of the products and maintains the production facilities.

Marketing

Marketing is the division which not only sells the products but also looks for new product ideas and new markets where the products can be sold. Its operation includes marketing research, sales planning, product promotion, sales, and distribution.

Marketing research seeks to find where the market is, what the public wants to buy, and how effective the advertising and sales promotions are. **Sales planning** is the overall planning of sales objectives and goals. **Product promotion** is the actual advertising of the product and includes the design of the package. Both advertising and packaging are a way of bringing a **sales** message to the public. Finally, the marketing division is in charge of the **distribution** of the product to the consumer.

Industrial Relations

Industrial relations is that division of a company which deals with the hiring, training, retaining, job satisfaction, and firing of its employees. It works within a framework of policies, or regulations, established by the company. Day-to-day contact with the employees and consideration of their wages,

Industrial Production

Fig. 16-9. During a work week, employees spend about one-half of their waking hours at their jobs. Being satisfied with the work they do is important.

hours, benefits, and training are part of the responsibilities of this division. It is also responsible for the operation of the office staff which supports the manufacturing or work force.

Finance and Control

The finance and control division is responsible for the business and financial operation of the company. Its main activity is the handling of the company's money. It obtains and controls the money needed to operate the business. It is also responsible for the purchase of all supplies and materials necessary for that business.

This division also organizes the company in one of several ways: (1) proprietorship, (2) partnership, (3) cooperative, or (4) a corporation. Each of these methods of organization has its own advantages and disadvantages. These will be discussed later in this chapter.

How Divisions of the Organization Are Related

Three of the divisions of the industrial organization provide a direct line from the idea to the product which is sold to consumers. These divisions are research and development (R & D), production, and marketing. The other two divisions — industrial relations and finance and control — support the total effort of the company. Figure 16-10 shows their relationship.

Organizing for Industrial Production

Industrial production is usually organized on the basis of certain needs. These may include a need to develop a new product, a need to sell a new product, a need to realize a profit and/or a need to produce more products. Often a combination of these needs is present when a company is organized for production. Industrial production meets the needs of the consumers by pro-

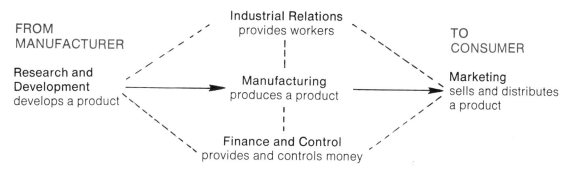

Fig. 16-10. Relationship of the corporate divisions.

viding the products they want and the needs of the company which must make a profit in order to stay in operation.

Plastics is the world's fastest growing industry, and many new companies have been organized to produce plastics products. Plastics production continues to provide excellent opportunities for new companies, new products, and new jobs.

To investigate the organization of a plastics processing company provides a valuable opportunity for students to learn about the world of industry in general. The rest of this chapter discusses how companies are organized and operated to produce their products. It offers suggestions of how students can form a simulated corporation for the production of plastics products. This group activity provides a valuable experience in the design, production, and marketing of products; the organization and management of the company; and the management of workers.

Starting a Manufacturing Business

To simulate a manufacturing business, a class or group should gather together for the purpose of sharing ideas. A temporary chairman may be elected just to keep order and guide the discussion. A recorder may also be needed. Everyone in the group should be invited to suggest ideas for a product to manufacture, Fig. 16-11. No idea should be ruled out or criticized at this point. This type of discussion is called **brainstorming.** It is a method of sharing and recording ideas. No idea offered should be disregarded. The longer the list the better.

After ideas are recorded, they may be reviewed and each item discussed. The group should consider methods that could be used to make each of the products listed. Products requiring equipment or materials other than those available should be eliminated from the list.

Narrow the list of ideas down until at least one or as many as five good ideas remain. In a class situation where 20 to 25 students are participating, it is suggested that the ideas be narrowed to five. The group may then be reorganized into five subgroups. Each subgroup can act as a research and development team, taking one of the five ideas and analyzing its possibilities.

The research and development teams should analyze the ideas from several standpoints. They should:

Fig. 16-11. Brainstorming session in a simulated manufacturing business.

Fig. 16-11A. Developing a design idea is part of the work of getting started to produce a product.

1. Develop several proposed designs for the product;
2. Suggest several ways to produce the product;
3. Suggest materials appropriate for each design;
4. Do a limited market research to determine competition, prices, and demand;
5. Estimate manufacturing costs; and
6. Determine a suggested selling price.

After analyzing its product idea from these several standpoints, each R & D team should prepare a report which can be given before the entire group. At this point, each may choose to present one or several designs of the product idea. The team may then make the presentation showing drawings of the idea. The presentation may be divided among several persons who give design data, proposed manufacturing methods, market research, cost estimates, and a suggested selling price.

After hearing all the presentations, the entire group should select the product idea it decides will most likely succeed. Selection should be based on:

1. Amount of time required for both tooling up and producing the product. It should not exceed the time available.
2. The tooling required. Students should have skills necessary to build the tools, jigs, fixtures, or dies needed to make the product.
3. The equipment and materials available. The capacity of the laboratory plastics equipment should be checked before the group decides on molds or dies to be made.
4. Number of available workers. The production or assembly line for a product should not require more workers than are available in the group. Extra workers might be hired to help a small group, however. If only a few production workers are needed, others in the group may supervise the work, keep records, or alternate with production workers.

5. Selling price of the product. Product price should be kept as near $2 as possible to ensure success of the student enterprise. For this reason, a product should be selected that is inexpensive to make. Normally, a product priced from $1.49 to $1.98 **or less** will sell best. The product should be sold at not less than twice its cost (100% markup).
6. Simplicity and usefulness of product. It should be as small, simple, and useful as possible. It should be easy and fast to produce in a school laboratory and have as few pieces as possible to assemble. The product should lend itself to continuous line production.

Simple, easy to manufacture, well-designed products at the right price will produce a profit. Products which appeal to the peer group of the class often sell faster than those which do not. Remember, the purpose of this simulated industrial venture is not only to produce a good product, but to make a profit as well.

After a product is chosen, the research and development divisions of a classroom group may become inactive. The students may then be assigned to other activities. In an industrial situation, the R & D department would actively look for new challenges once a new product is put into production.

Forming the Organization

Once a product has been chosen, the company, or organization, will produce it. Companies are organized on the basis of one of four types of ownership: (1) proprietorship, (2) partnership, (3) cooperative, and (4) corporation.

The Proprietorship

The oldest and simplest type of business organization is the proprietorship. It is **owned by one person** who is completely responsible for its operation. About 80% of all businesses in the United States **are** proprietorships. They receive about 25% of the United States business income. The owner

Fig. 16-12. Types of ownership.

gets all the profits and is free to manage the company as he or she desires. The business is limited by the ability of the owner to raise money. The owner is also liable (responsible) for all debts, and his or her personal property can be taken to satisfy these debts.

The Partnership

The partnership is an association of two or more people in a business. They are **co-owners.** About 10% of all U. S. businesses are partnerships. A partnership has all the advantages of an individual proprietorship, with the partners sharing in decisions, profits, and losses. In addition, more money can be raised for operation. All the partners are liable for the debts of the business, and their personal property can also be taken to satisfy any debts.

The Cooperative

The cooperative is an **association of consumers** who group together to collectively buy, manufacture, and sell many goods and/or services. The collective buying often saves money. The group pays for the goods the same as the owners would in any other type of business. Usually, all of the profits from the business are paid back to the consumer members in the form of stock certificates or dividend payments. This, in effect, lowers the price of the goods purchased. Cooperatives are not subject to as many taxes as other businesses are, further reducing costs.

The Corporation

A corporation is a business formed by one or more persons. It is a legal entity in itself. That is, its rights and responsibilities are defined by law in a **charter** (written permission to come into existence). The corporation can own property; buy, manufacture, and sell products; and hire or fire people. The people who form a corporation buy **stock** (an amount of ownership) in it. Stock may also be sold to other people. In this way, very large sums of money can be raised for the operations of the corporation. People **invest** (buy shares of stock) in the corporation because they expect the corporation will operate at a profit. When the corporation makes a profit, it pays its **stockholders** or **investors,** dividends (money) according to the amount of stock they own. The corporation is liable for its debts but its stockholders are not. All the stockholders can lose if the corporation fails is the money they have invested in it. An investor's personal property cannot be taken to settle the debts of the corporation. About 10% of the business in the United States are corporations. However, they do about 65% of the dollar volume of business.

A corporation is managed by a **board of directors** which sets the policies (rules for management). The members of this board are elected by all of the stockholders. The

Fig. 16-12A. Management often seeks legal advice from an attorney when applying for a corporate charter.

Fig. 16-12B. If a corporation makes more products than it can sell, it soon runs out of money to pay its bills. Without money from sales, the corporation cannot continue to operate.

board of directors elects the officers of the corporation. If a majority of the stockholders do not like the way the business is being run, it can elect a new board of directors and, in turn, new officers. The stockholders have the final control of a corporation.

Establishing the Corporation

The corporation is probably the best type of business organization for a class or school group to form. It is representative of the largest segment of U. S. business. It may be formed simply by the group gathering together and electing a board of directors. The directors then elect the following officers of the corporation and a chairperson:

1. Chairperson — who usually conducts the meetings of the board. In a school group the teacher may be elected to this office and may then start the election of the officers.
2. President — who can conduct the rest of the election. This officer will act for the chairperson when so delegated.
3. Vice president of finance and control — who also acts as treasurer.
4. Vice president of research and development — (If research and development of product has been completed and no other products are planned, this office may be omitted.)
5. Vice president of manufacturing.
6. Vice president of marketing.
7. Vice president of industrial relations — who also serves as secretary of the board.

The duty of each vice president will be to head his or her respective division, each controlling that function of the corporation.

Application for Incorporation and Bylaws

After a group of people decides to start a corporation, it must apply for incorporation in the state in which the company is formed. The Articles of Incorporation are drawn up and submitted in the application for incorporation. This statement includes the purpose of the corporation, its name, home office,

ARTICLES OF INCORPORATION OF
Plasti-Vac Corporation

The undersigned incorporator or incorporators, desiring to form a corporation (hereinafter referred to as the "Corporation") pursuant to the provisions of the Indiana General Corporation Act (Medical Professional Corporation Act/Dental Professional Corporation Act/Professional Corporation Act of 1965), as amended (hereinafter referred to as the "Act", execute the following Articles of Incorporation.

IED-561   Class

ARTICLE I
Name

The name of the Corporation is   Plasti-Vac Corporation

ARTICLE II
Purposes

The purposes for which the Corporation is formed are:

To purchase, manufacture, fabricate, assemble, sell, both wholesale and retail products made from plastics and other materials; To buy, own, sell and lease materials and equipment necessary for carrying out the purposes of this corporation; To do all things necessary for making a profit.

ARTICLE III
Period of Existence

The period during which the Corporation shall continue is   Winter Quarter, 1976

ARTICLE IV
Resident Agent and Principal Office

Section 1. Resident Agent. The name and address of the Resident Agent in charge of the Corporation's principal office is   William T. Sargent   131 Practical Arts   Ball State University, Muncie, Indiana   47306
   (City)   (State)   (Zip Code)

Section 2. Principal Office. The post office address of the principal office of the Corporation is

134 Practical Arts, BSU   Muncie   Indiana   47306
(Address)   (City)   (State)   (Zip Code)

ARTICLE V
Shares

Section 1. Number.

A. The total number of shares which the Corporation has authority to issue is ....
B. The number of shares which the corporation designates as having par value is   1,000   with a par value of $ 1.00
C. The number of shares which the corporation designates as without par value is ....

Section 2. Terms.

The stock herein authorized shall be common stock and nonassessable. Additional stock shall be sold at such a price as the Board of Directors may determine.

ARTICLE VI
Requirements Prior To Doing Business

The Corporation will not commence business until consideration of the value of at least $1,000.00 (one thousand dollars) has been received for the issuance of shares.

Fig. 16-13. Sample articles of incorporation for class corporation.

(Continued on next page)

Fig. 16-13 (Cont.)

ARTICLE VII
Director(s)

Section 1. Number of Directors. The initial Board of Directors is composed of ..20.. member(s). The number of directors may be from time to time fixed by the By-Laws of the Corporation at any number. In the absence of a By-Law fixing the number of directors, the number shall be ..20..

Section 2. Names and Post Office Addresses of the Director(s). The name(s) and post office address(es) of the initial Board of Director(s) of the Corporation is (are):

Name	Number and Street or Building	City	State	Zip Code
Each class member enrolled in IED-561 Winter Quarter 1976				

Section 3. Qualifications of Directors. (If Any)

NONE

ARTICLE VIII
Incorporator(s)

The name(s) and post office address(es) of the incorporator(s) of the Corporation is (are):

Name	Number and Street or Building	City	State	Zip Code
Frederick Collins,	Ball State Univ.,	Muncie,	Ind.	47306
Gerald L. Steele,	Ball State Univ.,	Muncie,	Ind.	47306

ARTICLE IX
Provisions for Regulation of Business and Conduct of Affairs of Corporation

The Board of Directors shall have full authority with reference to the conduct of the business of the corporation.

IN WITNESS WHEREOF, the undersigned, being the incorporator(s) designated in Article VIII, execute these Articles of Incorporation and certify to the truth of the facts herein stated, this ..2.. day of ..March.., 19..76..

Gerald L. Steele
(Written Signature)

Frederick Collins
(Written Signature)

GERALD L. STEELE
(Printed Signature)

FREDERICK COLLINS
(Printed Signature)

STATE OF INDIANA ss:
COUNTY OF ..Delaware..

I, the undersigned, a Notary Public duly commissioned to take acknowledgements and administer oaths in the State of Indiana, certify that ..Gerald L. Steele.., being of the incorporator(s) referred to in Article VIII of the foregoing Articles of Incorporation, personally appeared before me; acknowledged the execution thereof; and swore to the truth of the facts therein stated.

Witness my hand and Notarial Seal this ..2.. day of ..March.., 19..76..

Shirley Hanna
(Written Signature)

SHIRLEY HANNA
(Printed Signature)

Notary Public

My Commission Expires:

January 31, 1977

This instrument was prepared by ..Edgar S. Wagner.., Attorney at Law,

135 Practical Arts, BSU, Muncie, Indiana 47306
(Number and Street) (City) (State) (Zip Code)

Corporate Certificate No. 151
(Sept. 1969)

STATE OF INDIANA
OFFICE OF THE SECRETARY OF STATE

CERTIFICATE OF INCORPORATION

OF

PLASTI-VAC CORPORATION

I, LARRY A. CONRAD, Secretary of State of the State of Indiana, hereby certify that Articles of Incorporation of the above Corporation, in the form prescribed by my office, prepared and signed in duplicate by the incorporator(s), and acknowledged and verified by the same before a Notary Public, have been presented to me at my office accompanied by the fees prescribed by law; that I have found such Articles conform to law; that I have endorsed my approval upon the duplicate copies of such Articles; that all fees have been paid as required by law; that one copy of such Articles has been filed in my office; and that the remaining copy of such Articles bearing the endorsement of my approval and filing has been returned by me to the incorporator(s) or his (their) representatives; all as prescribed by the provisions of the Indiana General Corporation Act, as amended.

Wherefore, I hereby issue to such Corporation this Certificate of Incorporation, and further certify that its corporate existence has begun.

In Witness Whereof, I have hereunto set my hand and affixed the seal of the State of Indiana, at the City of Indianapolis,

this .... third .... day of .... March ...., 19 76

LARRY A. CONRAD, *Secretary of State*

By..........
Deputy

Fig. 16-14. Certificate of incorporation.

> **BYLAWS**
> **of**
> **Plastic-Vac Corporation**
>
> ARTICLE I — Ownership and Control
> 1. Ownership of this corporation shall be vested in the members of IED 561 American Industries Practicum, Ball State University, Winter 1976.
> 2. To retain class enrollment, each member of IED 561, Winter 1976, must own at least one share of stock in Plastic-Vac Corporation.
> 3. Corporate control shall be vested in a Board of Directors consisting of the stockholders of this corporation with the course instructor acting as chairman.
> 4. Stockholders shall constitute a quorum when nine-tenths of the outstanding stock of the corporation is represented either in person or by proxy. Stockholders meetings may be called by presenting the chairman of the board with a written petition representing one-half of the outstanding stock of the company.
>
> ARTICLE II — Board of Directors
> 1. The business of this corporation shall be guided by a Board of Directors consisting of the membership of this corporation and the chairman. The President shall act for the Chairman when so delegated.
> 2. Meetings shall be called by the Chairman or President when needed or by the shareholders as provided in Article I, No. 4.
> 3. A Quorum shall consist of at least nine-tenths of the Board of Directors. The Chairman of the Board shall cast a vote only in the event of a tie.
> 4. Vacancies in the board will be filled by the chairman and approved by a two-thirds majority of the stockholders.
>
> ARTICLE III — Officers
> 1. The elected officers of the Corporation shall be President, Vice President for Production, Vice President for Marketing, Vice President for Finance and Control, Vice President for Industrial Relations, and Vice President for Research and Development.
> 2. Duties of said officers are those described in **Job Descriptions for Production Classes,** which is to be regarded as a part of this document.
> 3. Officers of this Corporation shall be elected for the duration of this course and Plastic-Vac Corporation.
> 4. A two-thirds vote is necessary by the Board of Directors to remove and replace an Officer.
> 5. Nominations for President shall be made by the Board of Directors. Vice Presidents shall be selected by the members of each functioning department respectively.
>
> ARTICLE IV — Compensation
> 1. The President, after consultation with the other Officers, shall recommend a compensation schedule to the Board of Directors. The Board of Directors shall determine the wages, salaries, and commissions.
>
> ARTICLE V — Liquidation
> 1. The Corporation is organized with the intention of operating for the duration of the Winter Quarter.
> 2. A two-thirds vote of the Board of Directors called to consider termination of activities may dissolve this Corporation.
>
> ARTICLE VI — Amendments
> 1. These Bylaws may be amended by a two-thirds vote of the Board of Directors at any regular meeting.
>
> Adopted by the Board of Directors
> Plasti-Vac Corporation
>
> on the 3 day of February , 1976.
>
> *Frederick C. Collins*

Fig. 16-15. Sample bylaws for class corporation.

amounts and types of stock, and names of the original incorporators. The Secretary of State usually issues a Certificate of Incorporation to the new corporation if all the laws have been obeyed. These laws vary from state to state.

At the formal organizational meeting, the board of directors establishes the bylaws of the corporation. The bylaws state the regulations that concern ownership and control of the corporation, duties of the board of directors, rules for selling stock, regular meeting of board of directors and of shareholders, and payment of dividends.

The board of directors of the student manufacturing group may meet to establish the Articles of Incorporation. Discussion should lead to the following decisions: (See example in Fig. 16-13.)

Article I — Name of the Corporation
Artcile II — Purposes
Article III — Period of Existence
Article IV — Resident Agent and Principal Office
Article V — Shares
Article VI — Requirements Prior to Doing Business
Article VII — Directors
Article VIII — Incorporators
Article IX — Provision for Regulation of Business and Conduct of Affairs of Corporation

The secretary of the board of directors will prepare the Articles of Incorporation which may be submitted to the school administrative office for consideration and approval. The principal of the school may act as Secretary of State in issuing a Certificate of Incorporation, Fig. 16-14.

Bylaws of the corporation should be submitted along with the Articles of Incorporation. See sample bylaws in Fig. 16-15.

Financing a Corporation

Determining Capital Needs

A business needs **capital** (money) with which to operate. Before raising capital, officers of the corporation must decide how much money is needed. They need to decide:

1. How much money will be needed for materials and supplies before money from sales come in.

   In the class simulated corporation, usually enough money is needed to cover the cost of the entire production run.

2. How much money is needed to buy or rent equipment and property.

   Whether or not the school charges the group for equipment is up to the individual situation. Paying a symbolic charge makes the operation more realistic.

3. How much money is needed for continued research and development of the product, such as building prototypes (models). Often one or more prototypes must be developed in order to find ways of making a product and develop jigs and fixtures. Prototypes are also used by

Fig. 16-15A. Corporation officers must decide how much money will be needed for a place in which to operate as well as for machines, materials, and salaries of people who will do the work.

Fig. 16-16. A corporation acquires the capital for operation by borrowing money, selling stocks, and/or reinvesting its profits.

the marketing division to set up a sales campaign.
4. How much money is needed to pay salaries and wages before money from sales comes in.

Usually in school manufacturing groups only token wages or no wages at all are paid. Often payment for labor is by receiving the products made or dividends on stock investments. **This policy should be decided, with the approval of the school when the simulated corporation is formed.**

Raising Capital

Captital may be raised by several methods:
1. Borrowing money,
2. Selling stocks, and
3. Reinvesting profits.

Money may be **borrowed** from a bank or any person who is willing to lend it to the corporation. Usually, the lender wants some type of **security** for a loan. Security is an assurance that the loan will be repaid. The security may be the right to take some of the borrower's property if the loan cannot be repaid. In addition, the lender wants to earn money (interest) on the loan or investment. The lender is guaranteed a certain percentage of interest by the person or corporation receiving the loan. Proprietorships, partnerships, cooperatives, and corporations may all apply for loans.

Corporations may raise capital by **selling stock** to investors. The corporation may then use this money for operation of its business. Stock certificates, such as the one shown in Fig. 16-17, give the investors part ownership in the corporation. Each share of stock gives the investor one vote in the election of directors and any other business brought before a stockholders' meeting.

Most corporations finance their operations by selling stock. Additional capital may be raised through loans, reinvestment, and/or another issue of stock. The price of a share of stock, called the **par value**, should be low enough to attract the most investors. A prospectus, or statement of the corporation and its expectations for growth, is usually issued to help sell the stock. See Fig. 16-18.

A business can also finance its operation by **reinvesting** its available capital. This is possible only after the business is under way and money is being received from sales. Part of the profits may be kept in the business, or reinvested, to help pay for new equipment, property, materials, research and development, or salaries and wages.

A school group or class should plan to sell stock at about 50¢ to $1.00 per share. (Industrial corporate stocks are often priced at the outset at $10 per share par value.) One share per person in the class may be sold to finance the simulated company. If more money is needed, additional shares may be sold to class members as well as individuals outside of the class or group. The amount of capital raised should about equal the immediate needs of the simulated company. A slight surplus is desirable.

Once the capital is raised it should be turned over to the vice president of finance and control who acts as the company trea-

Fig. 16-16A. Money to operate a business is often obtained through a bank loan officer.

Fig. 16-17. Stock certificate which will show the corporation's name, the stock owner's name, and number of shares issued.

PROSPECTUS
Plasti-Vac Corporation

Plasti-Vac Corporation is a confident and enlightened group of students and educators responding to a need in contemporary society. Plasti-Vac Corporation was incorporated in Muncie, Indiana on February 3, 1976. We need investors who are interested in using their capital for personal and financial growth.

We are offering stocks which provide unsurpassed opportunity for capital growth. Corporation profits will be distributed at the termination of this course in the form of dividends (not to exceed 10%) and grants of social and educational significance.

Income derived from our initial stock issue will be used to finance research and to start production of a line of vacuum-formed plastic products. Our major product line will include a series of personalized number plate emblems. Plasti-Vac feels that this is a timely product for those individuals seeking identity in our homogenized, computerized society. Our products are:

    Durable
    Attractive
    Personalized
    Inexpensive
    Patriotic

All shares are purchased through the Vice President of Finance and Control or a designated representative.

Fig. 16-18. Sample prospectus for class corporation.

surer. He or she should then make arrangements for the safekeeping of the money. A wise practice is to set up a checking account in a bank with the teacher (chairman of the board) as cosigner for all checks.

Organizing the Operating Divisions

After the corporation has been organized and capital raised, the following five operating divisions should be set up:

1. Research and Development
2. Manufacturing
3. Marketing
4. Industrial Relations
5. Finance and Control

Vice presidents who were elected or selected earlier by the board of directors will head each of these divisions. Remaining members of the class should then become members of one of the operating divisions, being selected according to their interests. The vice president and members of each operating division will be responsible for the duties of their division as presented here.

Study the organization chart, Fig. 16-19, to see how the various responsibilities are divided up. If there are not enough people for each position indicated, have one person handle two or more of the light responsibilities. Areas of heavy, medium and light responsibility are indicated on the organization chart. Positions of heavy responsibility usually require two or more people, and medium responsibility one person. One, two,

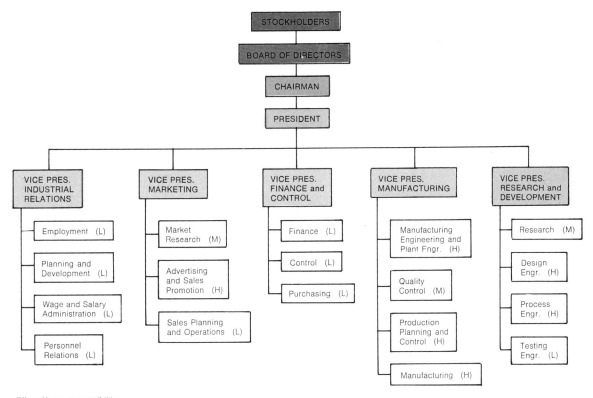

(H) = Heavy responsibility
(M) = Medium responsibility
(L) = Light responsibility

Fig. 16-19. Organization Chart of a Simulated Corporation

or even three areas of light responsibility may be handled by one person. All the people within a division must work as a team to accomplish the goals of that division.

Research and Development

The main functions of the research and development division have been discussed earlier. In industrial situations, the functions include (1) research, (2) design engineering, (3) process engineering, and (4) testing engineering.

Research may be either pure or applied. **Pure research** is research for the sake of finding new knowledge. **Applied research** is directed at a certain problem. Design engineering is that which is associated with the design of a new product. Process engineering is associated with the processes necessary to produce products. In testing engineering, the products produced by the company or by competitors are tested.

At this point, the product idea has been selected by the simulated corporation. Design of the product should now be assigned to members of the research and development team. A set of working drawings and a prototype (model) should be made for the manufacturing division which will use them to set up production procedures. When the drawings and prototype are completed, the research and development division may be dissolved and members can take up assigned duties elsewhere in the class corporation.

Manufacturing

The manufacturing division can be divided into two sections: (1) production and (2) service. The production section is responsible for making and assembling the product to be manufactured. The service section is often divided into five areas: (a) plant and tool engineering, (b) production planning and control, (c) quality control, (d) industrial engineering and (e) maintenance.

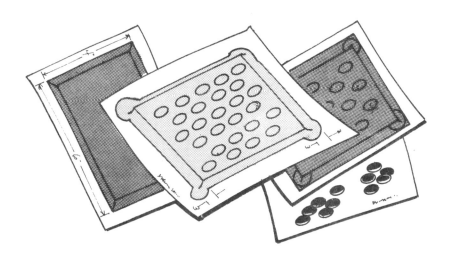

Fig. 16-20. Ideas for a product are developed.

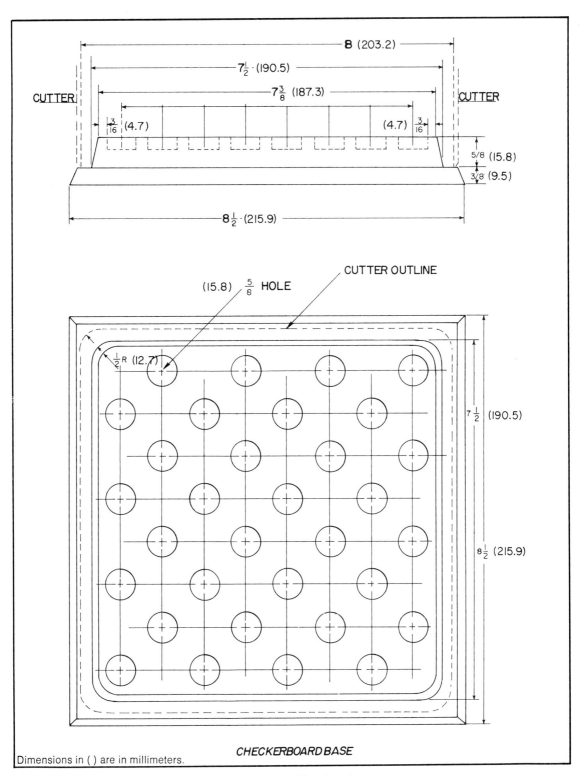

Fig. 16-21. Working Drawing

364   Section VI   PRODUCTION

Fig. 16-22. Prototype

Fig. 16-23. An example of simple tooling which includes molds and cutting location jig for checkerboard.

Fig. 16-23A. There are many jobs for people to do in managing a corporation. The person who most often becomes a manager is eager, can get along well with others, and can carry out responsibilities.

The manufacturing division of a student company will probably be the largest of all the divisions. Actual production of the product will likely require most of the students in the group. They may all have to work on the production line, including the officers.

Plant and tool engineering is the first function of the manufacturing division. The plant and tool engineers develop and make, have made, buy, or rent the jigs, fixtures, molds, and tools needed for the manufacture of the product. This operation is one of the most important functions of the whole manufacturing division. A good product depends on good tooling (jigs, fixtures, molds, etc.) and in the end will determine the financial success of the company.

Plant and tool engineering will probably take as long as any of the company functions. The vice president should assign several people to work in this area. Jigs, fixtures, tools, and molds should be tried out and perfected well in advance of the expected production date. It is sometimes necessary to revise them several times before actual production begins. Tooling should be simple, sturdy, accurate, and dependable enough that almost anyone can be taught to use it.

Production planning and control is set up after tooling has been perfected. A production plan must be made that describes the operations to be performed and the route that parts will take.

Several types of charts or diagrams are used in industry to plan and control the manufacture of goods. Four which are often used are (1) operation analysis sheets, (2) operation process charts, (3) flow process charts, and (4) flow diagrams. Examples of these can be found in several books such as the **Industrial Engineering Handbook**.[1] These charts use the standard ASME (American Society of Mechanical Engineers) symbols. The charts used in industry vary from company to company, depending on their needs, Fig. 16-24.

[1] Industrial Engineering Handbook, Second Ed. Maynard, H. B. Ed., McGraw-Hill, Inc., 1963.

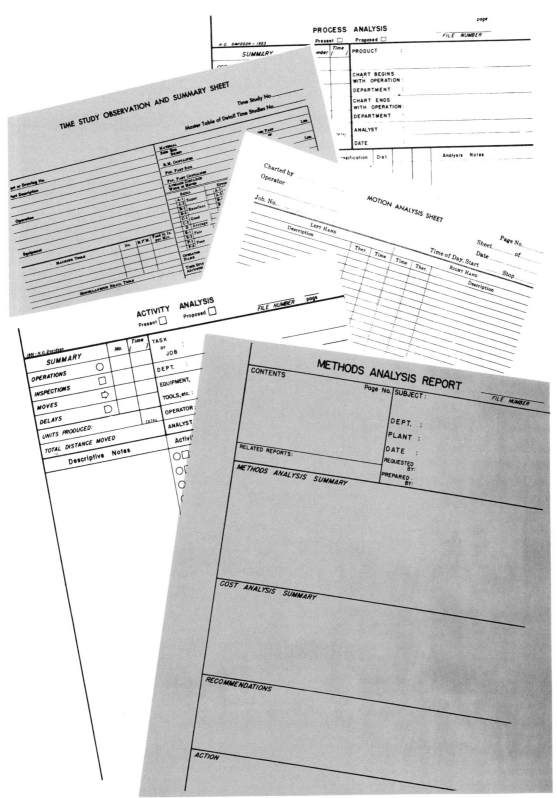

Fig. 16-24. Industry uses various types of charts for production planning and control.

OPERATION ANALYSIS SHEET
Job or Part Name: Checkerboard Top

Oper. No.	Operation Description	Tools and Equipment	Machine	Materials
1	Select and inspect the material.	Ruler Micrometer		.040 High Impact Polystyrene
2	Cut the plastic sheets into 12" (30 cm) strips.	Plastic Blade	Circular Saw	.040 High Impact Polystyrene
3	Cut the 12" (30 cm) strips into 12" (30 cm) squares.		Squaring Shear	.040 High Impact Polystyrene
4	Thermoform checkerboard top.		AAA Thermoformer	.040 High Impact Polystyrene
5	Inspect for flaws — visual.			.040 High Impact Polystyrene
6	Hot die-cut tops to outline shape.		Hot Die Cutter	.040 High Impact Polystyrene
7	Inspect for flaws — visual and clean trimmed edge.	Knife		.040 High Impact Polystyrene

Fig. 16-25. Typical operation analysis sheet showing manufacturing operations for a simple product.

Planning begins with the **operation analysis sheet**. It is like a plan of procedure often used in the school laboratory. It should include at least the following information for simple manufacturing operations:

1. Description of basic operations.
2. Sequence in which operations are performed.
3. The facility or machine required.
4. Location of the facility or machine.
5. Jigs, fixtures or special tooling needed.
6. Inspection equipment needed.

Information on material size, description and specifications may also be given, as well as a sketch.

An operation analysis sheet should be completed for each part of the product. It should establish the procedure and operations in making the part. Two important points must be remembered: (1) use as few operations as possible and (2) locate as many or all the operations from one point on the part. Locating operations from one point on the part will eliminate many human and machinery errors.

After the operation analysis sheets have been completed for all parts, an **operation process chart** can be made. It will show how each part flows into the final assembly of the product. "An operation process chart is a graphic representation of the points at which materials are introduced into the process, and of the sequence of inspections and all operations except those involved in materials handling."[2] See Figs. 16-26 and 16-27.

[2] Ibid., p. 2-21.

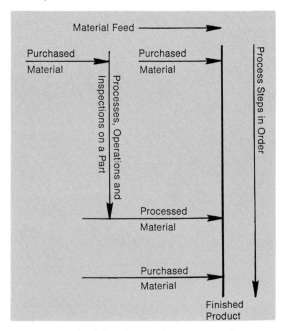

Fig. 16-26. Principle of operation process chart construction.

Subject Charted: _____

Charted By : _____ Date: _____

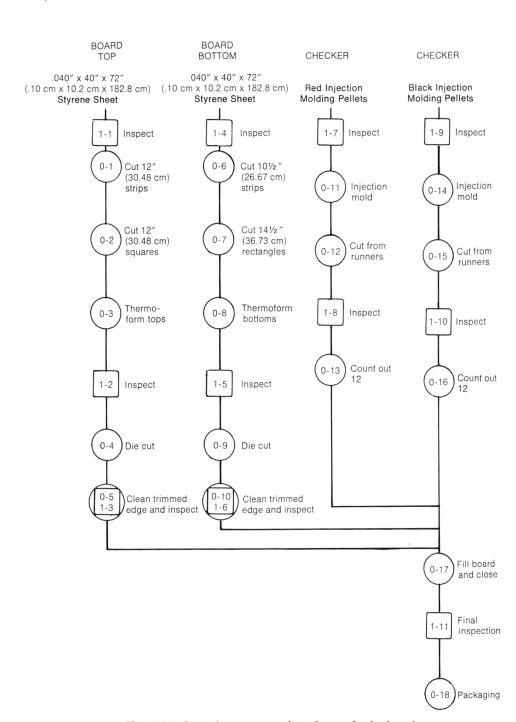

Fig. 16-27. Operation process chart for a checkerboard.

SUMMARY

	Present		Proposed	
	No.	Time	No.	Time
Operations	5	4½		
Transportation	5			
Inspections	3	4		
Delays	2			
Storages	0			
Distance Traveled	65 Ft. (19.812 m)		(Ft. m)	

Flow Process Chart

Part or Product: Checkerboard Bottom
Present Method ☒   Proposed Method ☐
Charted by: Gerald L. Steele
Date: March 9, 1976   Dept.: Plastics
Chart: 2 of 3

Oper. No.	Symbol	Process Description	Time (Min.)	Dist. Moved	Changes Recommended
1	☐	Inspect the sheet material.	3		
2	⬠	Move stock to circular saw.	10	10′ (3.0 m)	Store close to saw.
3	○	Cut stock into 10½″ (26.67 cm) strips.	5		Get 11″ (27.9 cm) stock instead.
4	⬠	Move strips to squaring shear.	1	15′ (4.5 m)	Bring sq. shear closer to saw.
5	○	Cut strips into 14½″ (36.73 cm) lengths.	½		
6	⬠	Move blanks to thermoformer.	½	30′ (9.0 m)	
7	D	Store blanks near thermoformer.			
8	○	Thermoform bottoms.	2		Decrease cooling time w/comp. air.
9	⬠	Move to die cutter.		8′ (2.4 m)	Move thermoformer closer.
10	☐	Inspect bottoms.	½		
11	○	Die cut bottoms.	½		Combine w/die cutting of tops.
12	☐	Inspect part for warp and clean trimmed edge.	½		
13	⬠	Move to final assembly.		2′ (0.6 m)	
14	D	Store until parts can be assembled.			Not needed if parts flow smoothly.

Fig. 16-28. Flow process chart for checkerboard bottom.

A **flow process chart** may be made using information from the operation analysis sheets and the operation process chart. It is a "graphic representation of the sequence of all operations, transportations, inspections, delays, and storages occurring during a process or procedure. It includes information considered desirable for analysis, such as time required and distance moved."[3] A simple version of the flow process chart is shown in Fig. 16-28. In very simple manu-

[3]Ibid., p. 2-25.

facturing operations, it may be possible to omit the operation process chart and use the flow process chart in its place. To do so may require a more complicated flow process chart, however.

A flow process chart may be made for each process, part, or product. The one shown in Fig. 16-28 is for making one part of a checkerboard. In this case a flow process chart is needed for each of two other parts.

Flow process charts may be used to find better or easier ways to make a part or product or to perform an operation. Each step in the making of a part can be evaluated by asking the following questions:

1. Can an operation be omitted?
2. Can two of the operations be combined?
3. Can the sequence of production be changed?
4. Can a movement in production be shortened or omitted?
5. Can an operation be simplified?
6. Can inspection be simplified?
7. Can a delay or storage be omitted?
8. Can the product be redesigned or a different material used to omit a step in production?

Several symbols are used on operation process and flow process charts to show what action is being done. These symbols and their definitions are shown in Fig. 16-29. They are also used on the flow diagrams described later. Small templates like the one shown in Fig. 16-30 are often used to draw in the symbols.

The **flow diagram,** when used with the operation analysis sheet, may be the most valuable graphic illustration of the production process. It shows how the part(s) flow

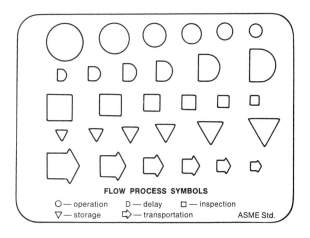

Fig. 16-30. Flow chart template for ASME standard symbols.

Symbol	Name	Action
○	OPERATION	Modification of an object at one workplace.
⇨	TRANSPORTATION	Changing the location of the object.
□	INSPECTION	A check on the quality or quantity of the object(s).
D	DELAY	Holding of an object or waiting for another operation.
▽	STORAGE	Stockpiling objects in a protected location.
⌂	COMBINED ACTIVITY	When two activities are performed at one station or by one operator.

Fig. 16-29. Symbols used for process charting.

from work station to work station. It shows what type of activity is performed at each station and where the part goes from there, Fig. 16-31. It also shows the relationship of one part to another when they are assembled. The flow diagram is made on a floor plan of the manufacturing area. It is like a map of the whole manufacturing operation.

Another advantage of the flow diagram is that it can show how equipment or operations may be shifted for greater efficiency. It will show where materials would be traveling long distances from station to station.

Manufacturing can be most efficient if successive operations are located next to each other, eliminating transportation. Some manufacturing plants and school laboratories have fixed equipment locations. If this is the case, the material will have to be moved. However, when equipment may be moved, a more efficient operation may be set up. The flow diagram, along with the operation process chart, may be used to set up a compact and efficient manufacturing operation. Workers may hand parts from station to station. Conveyors may also be used to move parts to the next station.

Quality control is another function of the manufacturing division. The quality control group inspects products or parts for flaws and correct size, Fig. 16-33. Often special checking fixtures are set up to make these inspections. Sometimes simple measuring tools such as calipers, squares, etc. may be used.

Parts not acceptable for use may sometimes be repaired, salvaged, or recycled for use. The manufacturing division usually sets up a repair area to salvage as many defective parts as possible.

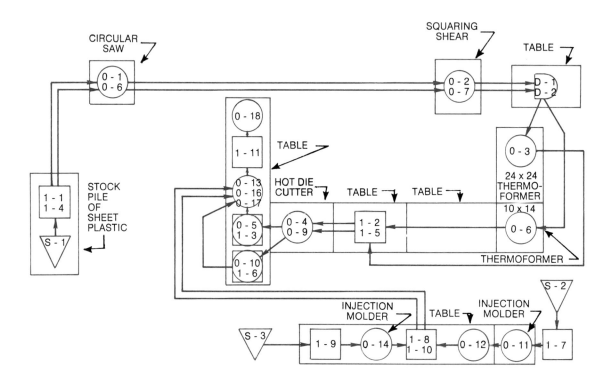

Fig. 16-31. Flow diagram of checkerboard production.

Fig. 16-32. Checkerboard production in the school laboratory.

A. Checkerboard parts are checked for fit before the next production step begins.

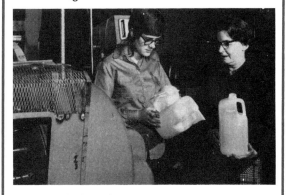

B. Final products are also checked. Here, quality control supervisors examine polyethylene containers.

Fig. 16-33. Quality control is a check to make sure the product will meet the standards set for it.

The **industrial engineering** section plans the use of people, facilities, tools, jigs, and fixtures to develop the desired quality product at lowest cost. This section includes methods study, work measurement, plant layout, and materials handling study. The industrial engineers can be described as the efficiency experts of the operation. They are responsible for reducing the cost and improving the efficiency of the manufacturing system. Industrial engineers use analysis sheets, flow charts, and motion study charts to help determine ways to improve production. In a simulated industry this function may be combined with plant and tool engineering as shown in Fig. 16-19.

The **maintenance** section of the manufacturing division keeps the equipment and plant in top operating condition. It makes sure all jigs, fixtures, and machines are working properly at all times. It also keeps the plant clean so that accidents do not happen.

Starting Production

Once the all important setup procedure is completed, production may begin. It should be done in three phases: (1) the pilot run, (2) the trial run, and (3) the production run.

The **pilot run** should be done by the plant and tool engineering and the production planning and control sections. In this run, a few pieces should be made to be sure the production line is set up properly. Only a few people are needed, and the product may be produced one segment or part at a time. When a few of each part are completed, they should be assembled. This will let the plant and tool engineering group check the tools and equipment. It will also let the production planning and control group check and analyze the production run for delays, shortages, stoppages, etc. Quality control can check the control system at this time, also.

The **trial run** is a training session. It is run after workers have been hired for the various jobs. Usually the industrial relations division of a large company is in charge of the training program. However, in a simulated enterprise such as this, those who have perfected the product, operations, and sequences are best suited to train the workers.

Fig. 16-34. Preparing for final assembly.

Job Description

Job Title: Machine Operator

Principal Duties:
1. Operate correctly and safely a wide variety of machines common to the company's operation.
2. Recognize and report malfunctions in the operation of assigned machines.
3. Make routine inspection checks of workpieces to assure quality production.

Materials or Products Worked On:
Various types of rough and semi-finished parts associated with a variety of products.

Typical Measuring Instruments and Tools:
Rule

Typical Equipment Used:
Belt sander
Drill press
Thermoformer
Hot die cutter

Fig. 16-35. Job Description

Job descriptions are written for each job on the basis of information provided in the operations and flow process charts. This is a function of the industrial relations division.

Using the job descriptions, the people in charge of worker training can show each worker how to do his or her job. During this time, a few pieces are produced. As soon as these pieces begin to flow down the production line, other workers are trained to further process the parts. No speed is expected during the trial run. When all production workers have learned their jobs, a regular production run may be scheduled.

The regular **production run** should be a well organized operation with close supervision and quality control. Care must be taken to produce good quality products that will sell.

Usually, the number of products to be produced is controlled. In a class situation, the goal may be a minimum of one product per class member and a maximum of 100. Sometimes sales are so good that more than 100 products will need to be produced. By having a close working relationship with the marketing division, the production department will be aware of how many parts to make. A check with marketing on a day-to-day basis is important just before the production run.

Marketing

The marketing division is in charge of not only selling the product but its package design, pricing, sales force, and distribution. There is an overlap between marketing and manufacturing in product packaging. Product packaging is performed as a part of the production process. Since it is designed by the marketing division, however, it will be considered here.

Packaging serves several purposes. It must, of course, hold and protect the product. It must also attract buyers, show the product name and brand name, and help convince people to buy the product. The package becomes a salesman, especially if the product is sold off the shelf.

Industrial Production 373

Fig. 16-34A. Marketing may arrange exhibits which attract interest and provide information.

A number of different kinds of packages may be used, depending on the product and how it is sold. Products manufactured by school enterprises may use boxes, sacks, heat-sealed plastic tubes, or skin and blister packages. See Fig. 16-36. An instruction sheet for use and/or care of the product should be included in or on the package. The instruction sheet may be dittoed, mimeographed, or printed in some other way. The package design should be completed in advance of the production run. Procedures for several of these packaging operations are contained in earlier chapters.

Pricing the product early is necessary so that sales may begin before production. Three important factors must be considered in pricing the product: (1) competition, (2) costs, and (3) profit margin.

The marketing division must find out what the **competition** is, that is, what other similar products are available to customers. One or several members of the division should check the prices that several other companies are charging for similar products. If there is no competition, costs and profit margin will determine the price. Most businesses will charge as much as possible for a product and still sell it easily. This means that if there is little or no competition, a company is able to charge more for its

Fig. 16-36. Packaging for different products.

Fig. 16-37. Packaging products as part of the production process.

Selling Tips

Memo to Salespeople of Plasti-Vac Corp.

PRODUCTS: Plastic products

Make sure you understand the following before attempting to sell our product.

1. Our methods of manufacturing our product and those of competitors.
2. Relative importance of each company in in-industry.
3. Sales policies of companies in industry.
4. Product's qualities, such as raw material used, styling, durability, and appearance.
5. Performance details, speed, accuracy, and research tests.
6. Price details, means of figuring, and practices of computation.
7. Shipping information, means, times, and costs.
8. Packaging.

To obtain the information of above, study the working drawings of the products, contact your head supervisor, and/or the executive board. Also study films, charts, and slides pertaining to these products.

Technique of Selling

Capture Attention. Make a comparison of our product to one with which the prospect is familiar.

Arouse Interest. Stress benefits of using our product or service, not the product or service itself.

Convince Prospect. Use "yes-but" approach; don't argue; and give considertion to prospect's point of view.

Create Desire. Point out in detail what product will do for him.

Secure Order. Probe for signs of acceptance, remove any doubts, and urge prospect to act now.

Note: To be a **good** salesman you **must**:

1. Be willing to work.
2. Be eager to get ahead, to increase wages and status.
3. Have a deep respect, interest, and liking for sales work.
4. Have a positive and enthusiastic attitude.

Fig. 16-38. Example Sales Kit

It may include sample or illustration of product, selling tips, sales record, letter of introduction, and sales invoice.

---

product. On the other hand, if competition is great, it must price its product so the price will attract customers.

Before a price is set, an accurate estimate of the cost of the product must be made. In industrial production, this cost **must include** materials, labor, use of equipment, utilities, buildings, office help, and management. For the purpose of the student enterprise, some of these factors may not be considered in fixing the product cost. Remember that waste, scrap and rejected parts must be considered, too. These often add 10% to 25% to the cost per item. Study this additional expense carefully.

Profit is the last to be considered in pricing. In a student enterprise, the profit margin is usually 100% of the **direct material and labor costs.** These costs include the actual amount of **material** and the actual amount of **labor** used in making the part as well as **waste allowances.** Industry operates on a 500% to 700% markup over direct material and labor costs. The 500% to 700% markup is divided many ways. It pays for the business overhead such as buildings, equipment, facilities, management, and office help. It also allows markups for distributors and dealers. Freight costs often are taken from the markup. Last, but not least, the manufacturer needs to make a profit. This often is as low as 5% to 10% after all the above expenses are paid.

If the list price of the item is too high when a reasonable markup is added, a reduction in the cost of production should be considered. The cost may be lowered by changing materials, increasing production efficiency, or changing the design.

Sales Force and Territory. The marketing division is responsible for setting up a sales force and outlining its territory. In industry, some companies use direct sales (sell directly to the customers) while others make their sales through sales representatives, jobbers (wholesalers), and dealers (retailers).

Student enterprises often use direct sales. In order to experience sales activity and making direct contact with customers, the entire group should become salespeople. This will give the widest sales coverage possible in the shortest time.

The marketing division may choose to set up a territory for each salesperson or let each sell where desired. Each salesperson should be given a sales kit which includes at least:

1. A picture, diagram, or sample of the product.
2. Some selling tips.
3. Information on the product.
4. Order pad.
5. Order record sheet.

The simulated enterprise should decide whether or not to allow a sales commission. If sales commissions are not paid, it will mean higher earnings on the stock later, if a profit is made.

Even if sales commissions are not paid, accurate sales records must be kept by the marketing division. The manufacturing division will want to know exactly how many products are sold. The finance and control division will want to know how many products have been **paid** for and delivered and how many have been sold but **not paid** for or delivered.

Advertising. Advertising is also an important responsibility of the marketing division. It should be aimed at those who are **most likely** to buy the product, otherwise the effort is wasted. Publishers of school materials, for instance, advertise in magazines for educators, not in family magazines. If the product is one a student will buy, display posters that advertise it in the halls of the schools. If it is a product parents will buy, try some posters in store windows (local merchants will usually cooperate). Be sure to let the public know where the product can be purchased, when, and from whom.

Distribution. Industry uses many types of distribution systems. Some of them are complicated, and produtcs may be shipped and stored several times before they reach the consumer.

The student enterprise should use a simple method of distributing its products. The one which often works the best is to have each salesperson personally deliver and take payment for each product he or she sells. When products are picked up for delivery, a record is kept of how many each salesperson takes. When sales money is turned in, it is checked off against the number of products charged out to the salesperson.

Fig. 16-39. Companies advertise their products in many ways.

When the products are delivered, the salesperson should check for customer satisfaction, making sure the product works. If it does not, the marketing division is responsible for repairing or replacing it.

Finance and Control

The finance and control division has three main responsibilities:
1. Finance (obtaining and allotting money).
2. Control (accounting for the money).
3. Purchasing (buying whatever is necessary).

In a corporation, much of the money for operation is obtained by selling stock, as discussed before. The finance and control group must keep accurate records of all stock sold, date of stockholder's purchase, to whom it is sold, and the number and cost of shares.

A corporation may obtain additional capital to buy certain things like materials and tools by using **short-term credit.** This means the corporation is granted a short period (sometimes a month or more) to pay for the purchase. In the simulated corporation, some materials may be bought on short-term credit (or charged) if the production run and sales are made within 30 days. The income from the sales can then be used to pay the creditors. If materials must be bought for cash,

Fig. 16-40. Finance and Control

Plasti-Vac Corporation

General Ledger

Date	Description	Expenses	Income	Balance
2-3	Sale of stock		$20.00	$ 20.00
2-5	Material for prototype	$ 3.25		16.75
2-6	Materials for molds	4.50		12.25
2-12	Income from advance sales		37.00	49.25
2-20	Plastic sheet stock	37.50		11.75
2-25	Income from sales		80.00	91.75
2-28	Plastic sheet stock	20.50		71.25
2-29	Plastic pellets	10.80		60.45
2-30	Freight bill	5.25		55.20
3-3	Income from sales		92.00	147.20
3-4	Sales commissions	28.00		119.20
3-4	Gift to scholarship fund	75.00		44.20
3-4	Phone calls	2.20		42.00
3-4	Liquidation of stock @ $1.00	20.00		22.00
3-4	Stock dividend @ $1.10	22.00		00.00
				00.00

Fig. 16-41. General ledger for a simulated corporation.

however, operating capital from sale of stock must be used.

Finance and control must also **account** for the corporation's money by keeping accurate records of all income and expenses. A simple ledger can be used to show purchases of materials and equipment, sales, dates, amounts, and accountings. Note the example ledger in Fig. 16-41.

Other records that should also be kept by the finance and control group include:

1. Materials purchased.
2. Supplies purchased.
3. Labor costs (if any).
4. Sales income (from marketing).
5. Equipment, tools, facilities, and utilities.
6. Inventory.

By keeping separate lists (ledgers) of these items used and their costs, it will be simple to prepare a financial statement at the close of the business.

In industry, a **financial statement**, often called a **profit and loss statement**, is prepared at the close of each business period. This may be monthly, quarterly, semiannually, or yearly. This statement is used as an analysis tool to compare with previous business periods, to show where improvements can be made and profit or loss can be expected.

In the simulated corporation, the financial statement is prepared at the close of business. If accurate records have been kept, it should be easy to complete. An example statement is shown in Fig. 16-42.

The finance and control division is also responsible for **purchasing.** Each purchase of material, supplies, equipment, etc. for the student enterprise should be cleared through this group. A **purchase requisition** should be filled out by the person requesting a purchase and submitted to the vice president of finance and control or someone designated by him or her. A **purchase order** is then issued for the purchase of the goods. The teacher of a student enterprise should be directly involved with all purchases, having final authority and co-signing each purchase order.

```
              Plasti-Vac Corporation
             Profit and Loss Statement
                (First Quarter, 1975)

Income
  Product sales                                    209.00
  Total income                                    $209.00
Expenses
  Cost of goods
    Inventory (beginning)              $00.00
    Raw materials purchased             76.55
    Inventory (end of production)       00.00
    Total cost of goods                             76.55
  Direct labor                                      00.00
  Total Expenses                                   $76.55
Gross Profit (income minus expenses)              132.45
Overhead                                           35.45
Net Profit (gross profit minus
  operating expenses)                             $97.00
```

Fig. 16-42. Simplified profit and loss statement for a simulated corporation.

Closing the Business

Most business corporations are set up as perpetual corporation (to continue forever). In the classroom, however, this is usually not practiced. At the end of the term or a specified period of time, the student corporation is terminated (closed). After all the products have been manufactured and sold, all business activity stops and the business is closed. There are four steps in closing the business:

1. Paying all bills and obligations.
2. Preparing a profit and loss statement.
3. Distributing profit or collecting for losses.
4. Liquidating stock.

All bills and other obligations should be paid so that the financial statement, Fig. 16-42, may be completed. The total costs involved in research and development, manufacturing, marketing, industrial relations, and finance and control can then be figured and the profit and loss statement completed.

When a corporation operates at a profit, it can use the surplus capital in either of two ways — (1) reinvest it in the business or (2) declare a dividend on stock. Profit that is reinvested or put back in the business is often called **plowback.** It is a method of financing new research and development and general expansion which all healthy businesses use. Dividends which are a return to the shareholders on money they invested

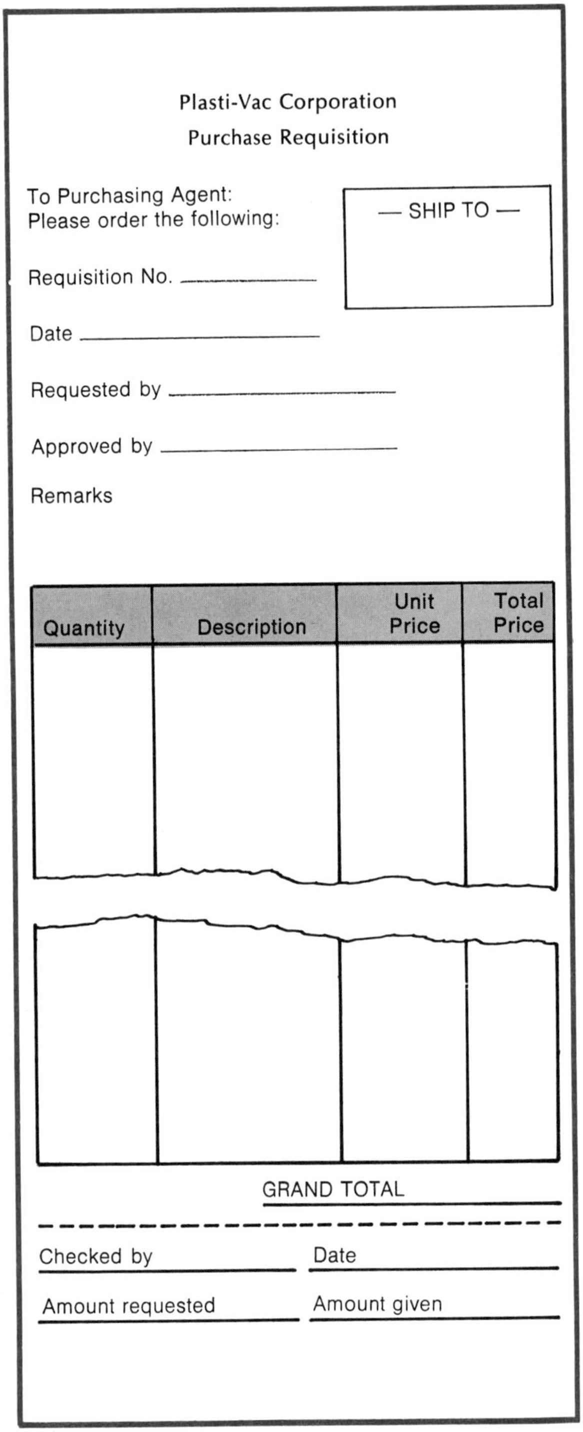

Plasti-Vac Corporation
Purchase Requisition

To Purchasing Agent:
Please order the following:

— SHIP TO —

Requisition No. ______

Date ______

Requested by ______

Approved by ______

Remarks

Quantity	Description	Unit Price	Total Price

GRAND TOTAL ______

Checked by ______ Date ______

Amount requested ______ Amount given ______

Fig. 16-43. Purchase Requisition

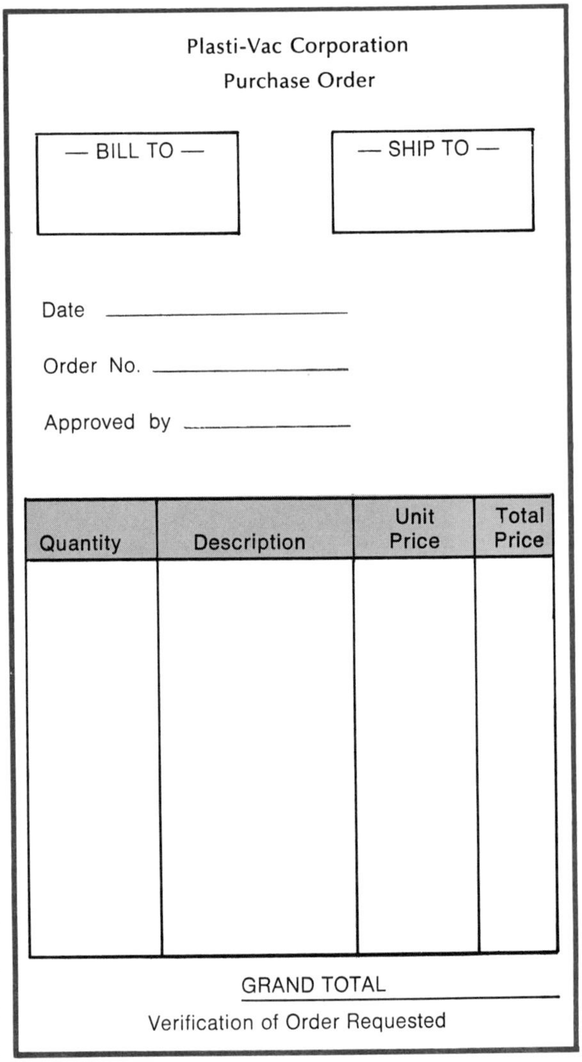

Plasti-Vac Corporation
Purchase Order

— BILL TO — — SHIP TO —

Date ______

Order No. ______

Approved by ______

Quantity	Description	Unit Price	Total Price

GRAND TOTAL ______

Verification of Order Requested

Fig. 16-44. Purchase Order

in the company may be paid on a quarterly, semiannual, or annual basis.

If a profit is made by the student corporation, it can be divided among the shareholders on a per share basis at the close of business. Each shareholder may first be repaid for his or her stock investment by selling it back to the corporation. This is known as cashing in stock. The surplus capital, after all the stock has been cashed in, can be divided among the shareholders. Such a return on stock investment is called a **dividend.** It repays the shareholder for the use of his or her money. It is like collecting interest if a loan were made.

Any loss by the company must, however, be deducted from the value of each share of stock before it is cashed in. In this case each share will be worth less than the original price. At the end of a simulated enterprise, the financial statement should show no money or other credits on hand. Any property owned by the simulated enterprise should be sold or given to the school before the business is closed.

STUDENT ACTIVITIES

1. Design plastics products which can be mass produced easily.
2. Design molds for producing plastic products.
3. Set up a small simulated industry to mass produce and market your product(s).
4. Invite representatives of industry, lawyers, unions, and government to come into the classroom to explain in more detail the various aspects of operating a business. One or more of the students' parents may be able to provide this information.
5. Learn more about stocks by watching the stock market and reporting on certain stocks. Keep a record of a given stock for a period of time. Learn why stock values climb and fall.
6. Form a union to bargain with the business that has been set up. Find out more about unions from a local union source.

Questions Relating to Mass Production

1. What was the first and probably the most important development in the concept of mass production?
2. Why was precision measurement necessary before mass production could be fully developed?
3. What is an allowable reject rate for mass production?
4. How did the meat packing industry influence early mass production?
5. Who was the original efficiency expert?
6. Compare the different types of manufacturing systems in use by industry.
   a. How does a consulting engineering firm differ from a complete manufacturing firm?
   b. How are job shops different from captive operations?
   c. What other type of manufacturing firm is a subcontractor like?
7. Name the three categories of plastics firms.
8. How are the five divisions of large industrial organizations related? Which ones form a direct line to the consumer? Why?
9. How does the relationship to the product differ with industrial relations and production?
10. What are the rules for brainstorming?
11. What does a certificate of stock entitle its owner to?
12. What is a stock prospectus used for?
13. What is the difference between a flow process chart and a flow diagram?
14. Why is quality control necessary?
15. Packaging performs at least two functions. What are they?
16. Name several ways in which products may be distributed.
17. What is the purpose of a financial statement?
18. Name several ways of financing a business enterprise.

Appendix A

References

Plastics Materials Guide

Generic Name and (Common Name)	Abbreviation*	Typical Trade Names	Approx. % of Market (1975)	Relative Cost	Characteristics
Acrylonitrile Butadiene Styrene (ABS) Thermoplastic	ABS	Cycolac Kralastic Absinol	2.8%	Medium	Good tensile strength, flexural rigidity, and impact resistance over a wide temperature range. Excellent electrical insulation properties. Stain and chemical resistance to most inorganic solutions.
Polyacetal (Acetal) Thermoplastic	POM	Delrin Celcon Dielux	0.25%	Medium high	Rigid, resilient, low coefficient of friction, good impact strength, good heat resistance, solvent resistant, low creep, platable. High specific gravity. Some processing difficulty. Poor cementability. Degradation point close to melt point.
Methacrylate (Acrylic) Thermoplastic	PMMA	Plexiglas Lucite Elvacite Acrylite	1.8%	Medium	Outstanding weather resistance. Excellent optical properties. Rigid, good tensile strength. Fair heat resistance. Poor impact strength except special grades. Resistant to most acids and alkalis. Attacked by alcohol and hydrocarbon solvents.
Alkyds Thermosetting		Durez Dyal Glaskyd	2.7%	Low	Generally used in protective coatings. Low moisture absorption, good electrical properties, heat resistance and dimensional stability. Also available in premix form for reinforced plastics.
Cellulosics (Acetate) (Butyrate) (Propionate) Thermoplastic	CA CAB CAP	Tenite Forticel Arnel	0.5%	Medium	A variety of plasticizers compatable with cellulosics result in a wide variety of properties. Outstanding clarity. Excellent gloss. Slow burning. Good impact strength. Resistant to aliphatic solvent but dissolved by ketones. Medium moisture absorption, low shrinkage. Good processability.
Epoxy Thermosetting	EP	Hysol Epon Epikote	0.8%	Medium	Primarily used in protective coatings and reinforced plastics (much as polyester is). Excellent adhesive characteristics. High tensile strength, tough, good chemical resistance. Good electrical properties. Can be cured with little or no external heat. Low shrinkage.
Fluorocarbon Thermoplastic	PTFE PCTFE	Teflon Kel-F	1%	Very high	Outstanding weathering, outstanding chemical resistance, excellent heat resistance, and excellent electrical properties. Lowest coefficient of friction. Expensive, difficult to process. Very high melting point. High specific gravity. Family includes PTFE, FEP, TFE, PVF_2, etc.
Polyamide (Nylon) Thermoplastic	PA	Zytel Ultramid Xylon	0.5%	Medium high	Rigid, resilient, low coefficient of friction. Good impact strength, heat resistant, abrasion resistant, and self-extinguishing. Solvent resistant but high moisture absorption. Often filled with glass fibers and MOS_2. Nylon 6, 6/6, and 610 refer to the number of carbon atoms in the principal reactive components.

*American Society for Testing and Materials — D — 1600 — 71A

Generic Name and (Common Name)	Abbreviation	Typical Trade Names	Approx. % of Market (1975)	Relative Cost	Characteristics
Polycarbonate Thermoplastic	PC	Lexan Merlon	0.4%	High	High impact strength, high tensile strength, and heat resistant. Good dimensional stability and electrical properties. Poor chemical resistance. High moisture absorption and some difficulty in processing.
Phenol formaldehyde (Phenolic) Thermosetting	PF	Bakelite Varcum Plenco	4.0%	Low	High elastic modulus, excellent heat, chemical and moisture resistance. Excellent electrical properties, low heat conductivity, flame resistant and non-dripping. Retains its properties over wide temperature and humidity range. Low cost but color limited to dark colors.
Polyester Thermosetting and Thermoplastic		Stypol Valox Polylite Pleogen Laminac	3.1%	Low	Primarily used in reinforced plastics with glass and other fibers. Many grades are available. Rigid, heat resistant, chemical resistant and has good tensile strength. It can be cured either with or without external heat or pressure. Has good electrical properties, and it can be modified with various fillers and/or reactive monomers.
Polyethylene, high density Thermoplastic	HDPE	Alathon Hi-Fax Marlex	9.9%	Low	Lightweight (0.946 to 0.965 density), rigid, easily processed, excellent chemical resistance and electrical properties. Translucent, easily colored, nontoxic and an excellent food container material. Poor weatherability and low heat resistance.
Polyethylene, low density Thermoplastic	LDPE	Petrothene Poly-Eth Norchem Dylan	20.2%	Low	Lightweight (0.910 to 0.945 density), flexible, easily processed, excellent chemical resistance and electrical properties. Translucent to clear in thin films, nontoxic, odorless, and tasteless. Poor weatherability, low heat resistance, and stress cracks easily.
Polypropylene (Propylene) (Olefin) Thermoplastic	PP	Profax El Rexene Avison Marlex	8.3%	Low	Lightest weight of all plastics (0.890 to 0.900 density), boilable, tough, more rigid than polyethylene, stress crack and chemical resistant. Good electrical properties and clarity in films. Unlimited flex life (living hinge), poor low-temperature impact properties except if oriented.
Polystyrene (Styrene) Thermoplastic	PS	Styron Dylene Dylite Evenglo	14.6%	Low	Easily processed, high optical clarity but brittle if not modified such as high-impact polystyrene. Poor weatherability, poor chemical and solvent resistance. Burns easily. Available as a foam (styrofoam, etc.). One of the lowest cost plastics. Colors easily.

(continued)

Generic Name and (Common Name)	Abbreviation	Typical Trade Names	Approx. % of Market (1975)	Relative Cost	Characteristics
Polystyrene copolymer (Hi-impact styrene) (Impact PS) Thermoplastic	PS	Styron Bakelite Dylene Kraton	Included in the styrene market	Low	Wide versatility, the properties can be varied greatly by "tailoring" it with BUTADIENE. Translucent to opaque with good colorability. Easily processed with fast cycles. Used extensively for sheet plastics for thermoforming. Medium to high impact strength, lower tensile strength than polystyrene.
Polyphenylene oxide Thermoplastic	PPO	Noryl Alphalux	1.0%	High	High heat resistance, good moisture resistance, self extinguishing, low creep, good dimensional stability, and good physical properties. Some processing difficulties, degradation point very close to melt point. Modified PPO, Noryl, is lower in cost and easier to process.
Polyvinyl chloride, flexible (Vinyl) Thermoplastic	PVC	Geon Marvinol Opalon Exon	16.5%	Low	Readily modified with plasticizers and other additives to produce a wide variety of properties. Flexible to rigid, good weathering, self-extinguishing, tough, and abrasion resistant. Also available as a liquid in plastisol form. Gives off HCl gas if overheated.
Polyvinyl chloride, rigid (Vinyl) Thermoplastic	PVC	Airco Intramix Geon	3.2%	Low	Relatively low in cost compared to engineering thermoplastics, flame resistant, and has good chemical resistance to most acids, alkalis, oils and greases. One of the highest available combinations of stiffness and impact resistance. Very good weatherability, good scratch and abrasion resistance. Difficulty in processing due to poor heat stability. Gives off HCl gas when it degrades.
Polyurethane (Urethane) Thermosetting, thermoplastic and elastomer	PUR	Texin Flexane Isonate Mista Foam	5.75%	High	Excellent abrasion resistance, good tear strength, good solvent resistance, and good flexibility. Difficult to process. Available in cellular foam, either rigid or flexible. Fatigue resistant, good insulation value and cushioning.
Silicone (RTV) Thermosetting elastomer	SI	GE Silicone Silastic Eccosil	1.0%	High	Liquid or solid materials with good flexibility and surface release properties. Low coefficient of friction. May be used in high temperature applications as a wax ingredient or rubber substitute. May also be used as a glue or sealer. Good weatherability, tear strength, and arc resistance.
Styrene acrylonitrile Thermoplastic	SAN	Tyril Lustran Piccflex	.35%	Low	Less easily processed than regular polystyrene. Increased heat distortion and chemical resistance over polystyrene. Better weatherability, increased strength, and higher water absorption than polystyrene. Clear and easily colored.

Generic Name and (Common Name)	Abbreviation	Typical Trade Names	Approx. % of Market (1973)	Relative Cost	Characteristics
Urea and melamine formaldehyde (Urea) and (Melamine) Thermosetting	UF MF	Plaskon Diaron Premelite	3.0%	Low	Hardness and mar resistance unsurpassed by other plastics, especially melamine. Excellent colorability and gloss. May be made any color or hue. Outstanding resistance to oils, solvents, and greases. Superior flame resistance and good electrical properties. Good strength and high heat distortion. Melamine has better heat and acid resistance while urea costs less. Used in place of phenolic where color and high heat resistance are necessary.

Note: This list contains only those families of plastics which are in most common use. Usage figures are derived from statistics published in **Modern Plastics** magazine, January 1976. Up-to-date statistics for the previous year are published in that source each January. Most materials can be obtained in many forms — coatings, adhesives, foams, molding compounds, powders, sheet, rod and tubes, etc. Characteristics are those for molding compounds unless otherwise specified.

chronology of plastics.*

DATE	MATERIAL	EXAMPLE
1868	Cellulose Nitrate	Eyeglass Frames
1909	Phenol-Formaldehyde	Telephone Handset
1909	Cold Molded	Knobs and Handles
1919	Casein	Knitting Needles
1926	Alkyd	Electrical Bases
1926	Analine-Formaldehyde	Terminal Boards
1927	Cellulose Acetate	Toothbrushes, Packaging
1927	Polyvinyl Chloride	Raincoats
1929	Urea-Formaldehyde	Lighting Fixtures
1935	Ethyl Cellulose	Flashlight Cases
1936	Acrylic	Brush Backs, Displays
1936	Polyvinyl Acetate	Flash Bulb Lining
1938	Cellulose Acetate Butyrate	Irrigation Pipe
1938	Polystyrene or Styrene	Kitchen Housewares
1938	Nylon (Polyamide)	Gears
1938	Polyvinyl Acetal	Safety Glass Interlayer
1939	Polyvinylidene Chloride	Auto Seat Covers
1939	Melamine-Formaldehyde	Tableware
1942	Polyester	Boat Hulls
1942	Polyethylene	Squeezable Bottles
1943	Fluorocarbon	Industrial Gaskets
1943	Silicone	Motor Insulation
1945	Cellulose Propionate	Automatic Pens and Pencils
1947	Epoxy	Tools and Jigs
1948	Acrylonitrile-Butadiene-Styrene	Luggage
1949	Allylic	Electrical Connectors
1954	Polyurethane or Urethane	Foam Cushions
1956	Acetal	Automotive Parts
1957	Polypropylene	Safety Helmets
1957	Polycarbonate	Appliance Parts
1959	Chlorinated Polyether	Valves and Fittings
1962	Phenoxy	Bottles
1962	Polyallomer	Typewriter Cases
1964	Ionomer	Skin Packages
1964	Polyphenylene Oxide	Battery Cases
1964	Polyimide	Bearings
1964	Ethylene-Vinyl Acetate	Heavy Gauge Flexible Sheeting
1965	Parylene	Insulating Coatings
1965	Polysulfone	Electrical/Electronic Parts
1966	Polyphenyl Oxide (PPO)	Electrical/Electronic and Appliance Parts

* From the booklet *The Story of the Plastics Industry* (SPI).

Identification Tests for

Resin		Burning	Heating in a Test Tube for Odor	Melting Point °F (°C)
Acetal		Blue flame; no smoke; drippings may burn	Formaldehyde	347 (175)
Acrylate & Methacrylate		Blue flame; yellow top	Fruit-like	374* (190)
Acrylic - rubber modified		Yellow flame; spurts	Characteristic [3]	279 (137)
Acrylonitrile - Butadiene - Styrene		Yellow flame; black smoke; drips	Characteristic [3]	—
Acrylonitrile - Styrene		Yellow flame; black smoke; clumps of carbon in air	Illuminating gas and acrylonitrile	268 (131)
Cellulose Acetate		Yellow flame; sparks; drippings may burn	Acetic Acid	[4] 446 (230)
Cellulose Acetate Butyrate		Blue flame; yellow tip; sparks; drippings may burn	Rancid butter	[4] 356 (180)
Cellulose Nitrate		White flame; very rapid	Sharp	Decomposes
Cellulose Propionate	plastic burns — not self-extinguishing	Blue flame; yellow tip; sparks; drippings may burn	Fragrant	[4] 456 (236)
Cellulose Triacetate		Yellow flame; drips	Acetic Acid	572 (300)
Ethyl Cellulose		Yellow flame; blue top; drippings may burn	Burned sugar	—
Ethylene		Blue flame; yellow top; drippings may burn	Paraffin	221 (105) low density / 248 (120) high density
Methylstyrene		Yellow flame; black smoke; clumps of carbon in air; softens	Illuminating Gas	349 (176)
Polyester Films		Yellow flame; smokes; drips	Characteristic [3]	482 (250)
Propylene		Blue flame; yellow top, swells and drips	Sweet	334 (168)
Styrene		Yellow flame; dense smoke; clumps of carbon in air	Illuminating Gas	374 (190)
Styrene - Methylmethacrylate		Yellow flame; smoke	Characteristic [3]	351 (177)
Vinyl Acetate		Yellow flame; smoke	Acetic Acid	140 to 190 (60 to 88)
Vinyl Alcohol		Yellow flame; smoke	Characteristic [3] Initially soapy	446 (230) Decomposes
Vinyl Butyral		Blue flame; yellow top; melts and drips; drippings may burn	Rancid Butter	[4] 345 (174)
Vinyl Chloride - Vinyl Acetate	plastic [1] burns	Yellow flame with green	Hydrochloric Acid [2]	261 (127)
Carbonate		Decomposes	—	430 (221)
Chlorinated Ether		Sputtering; bottom green; top yellow; black smoke; carbon in air	Characteristic [3]	358 (181)
Nylon 66 nylon ZYTEL® 101	plastic burns — self-extinguishing	Blue flame; yellow top; melts and drips	Burned wool	489 (254)
ZYTEL 42				489 (254)
610 nylon ZYTEL 31				415 (213)
6 nylon ZYTEL 211				405 (207)
11 nylon 'Rilsan'				351 (177)
alcohol soluble nylon: ZYTEL 63				311 (155)
ZYTEL 69				293 (145)
Vinyl Chloride Polymers[2]		Yellow flame; green at edges; softens and chars	Hydrochloric Acid	302 (150)
Vinylidene Chloride Polymers		Yellow flame; ignites with difficulty; green spurts	Hydrochloric Acid	313 (156)
Fluorocarbons FEP TFE	plastic does not burn	Deforms	—	554 (290) 621 (327)
Fluorohydrocarbons Vinylidene Fluoride Polymers		—	—	340 (171)
Fluorochlorocarbons		Deforms; slight melting; drips	Weak acetic acid	383 (195)

footnotes:

[1] Some compositions burn slowly, others are self-extinguishing

* Will vary between grades 6, 7 or 8.

[2] Plasticizers will usually lower specific gravity and/or may mask odor.

[3] Odors are difficult to describe, but recognizable — use controls. (Known sample)

[4] Wide range of melting points, depending upon composition.

Reproduced by permission of E. I. du Pont de Nemours & Co. (Inc.), Plastics Department, Wilmington, Delaware 19898

Thermoplastic Materials

References

Cold CCl₄	Ace-tone	Warm Ethanol	Ben-zene	CHCl₃	Cold 4.2M HCl	Boiling 4.2M HCl	90% Formic Acid	Special Solvents	Specific Gravity @ 23°C.	Examples of Materials
I	I	I	I	I	—	—	—	—	1.42 -1.43	DELRIN®[1]
I	S	I	S	S	—	—	—	Toluene	1.18 -1.19	LUCITE®[1]
I	PS	I	S	PS	—	—	—	Partially soluble in esters	1.12 -1.16	"IMPLEX"[2]
PS	—	I	PS	S	—	—	—	Esters & ethylene dichloride	1.04	"CYCOLAC"[3]
I	S	—	—	—	—	—	—	Methyl ethyl ketone	1.08 -1.06	"TYRIL"[5], "FOSTACRYL"[6] "LUSTRAN A"[18]
—	S	—	I	S	—	—	—	Acetic Acid	1.25 -1.35	
PS	S	I	—	S	—	—	—	—	1.17 -1.24	"TENITE" BUTYRATE[9]
I	S	—	—	PS to I	—	—	—	Esters & aliphatic hydrocarbons	1.34 -1.40	"XYLONITE"[10]
S	S	—	—	S	—	—	—	Esters	1.20 -1.24	"FORTIFLEX"[11]
—	S	—	—	—	—	—	—	Dioxane	1.27 -1.28	"B X TRIACETATE"[10]
—	S	S	I	S	—	—	—	Some hot aromatics Benzene - Toluene	1.10 -1.16	"ETHOCEL"[5]
I	I	I	I	I	I	I	I	Some hot aromatics	.914- .930	ALATHON®[1]
I	I	I	I	I	I	I	I	Benzene - Toluene	.94 - .96	
—	S	—	S	S	—	—	—	Esters & some chlorinated hydrocarbons	1.03	"CYMAC" 400[4]
I	I	I	I	I	I	I	—	Metacresol and hot nitrobenzene	1.38 -1.40	MYLAR®[1]
I	I	I	I	I	I	I	I	Some hot aromatics Benzene - Toluene	.85 - .91	ALATHON®[1]
—	S	—	S	S	—	—	—	Esters	1.04 -1.08	"DYLENE"[12]
—	S	—	S	S	—	—	—	—	1.14	"ZERLON"[5]
—	S	—	S	S	—	—	—	Cyclohexanol	1.18 -1.20	ELVACET®[1]
I	I	—	I	I	—	—	—	Hot water	1.27 -1.31	ELVANOL®[1]
I	I	S	I	I	—	—	—	Esters, PS	1.07 -1.20	BUTACITE®[1]
—	S	I	—	S	—	—	—	Cyclohexanone	1.34 -1.37[2]	"VELON"[16]
PS	PS	I	PS	S	—	—	—	—	1.2	"LEXAN"[7]
I	I	I	I	I	I	I	I	—	1.4	"PENTON"[8]
I	I	I	I	I	I	S	S	—	1.14 -1.15	ZYTEL®[1]
I	I	I	I	I	I	S	S	—	1.14	
I	I	I	I	I	I	I	S	—	1.09	
I	I	I	I	S	S	S	S	—	1.13	
I	I	I	I	I	I	I	I	—	1.04	
I	I	S	I	I	S	S	S	Hot 80% aqueous ethanol	1.08	
I	I	S	I	I	S	S	S		1.08	
I	—	I	I	—	—	—	—	Cyclohexanone	1.30 -1.60	"EXON"[16] "VIPLA"[14]
—	PS	I	—	—	—	—	—	—	1.63 -1.73	"SARAN"[5]
I	I	I	I	I	I	I	I	—	2.10 -2.19	TEFLON®[1]
I	I	I	I	I	I	I	I	—	2.14 -2.17	TEFLON[1]
—	S	—	—	—	—	—	—	Dimethyl acetamide	1.76	"KYNAR"[17]
I	I	I	I	I	I	I	I	Dimethyl acetamide	2.1	"KEL-F"[13]

[5] S — Soluble
I — Insoluble
PS — Partially Soluble

1. E. I. du Pont de Nemours & Co. (Inc.)
2. Rohm and Haas Co.
3. Marbon Chemical Div., Borg-Warner Corp.
4. American Cyanamid Co.
5. The Dow Chemical Co.
6. Foster Grant Co.
7. General Electric Co.
8. Hercules Powder Co.
9. Eastman Chemical Products, Inc.
10. BX Plastics Ltd.
11. Celanese Plastics Co.
12. Koppers Co.
13. Minnesota Mining and Manufacturing Co.
14. Chemore Corp.
15. Dayton Industrial Products Co.
16. Firestone Plastics Co.
17. Pennsalt Chemical Co.
18. Monsanto Company

Polymer Chemistry

TABLE 1
Polymers Based on the Ethylene Chain

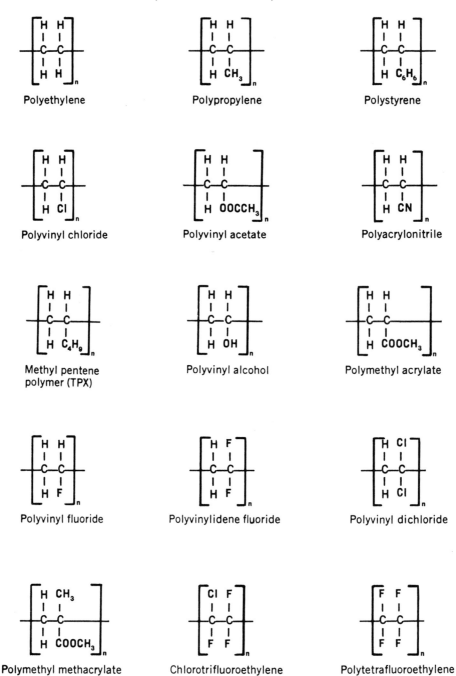

TABLE 2

Carbon "Backbone" Copolymers

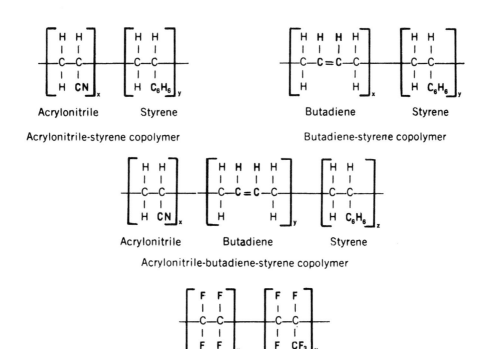

TABLE 3
Polymers with Oxygen in the Chain

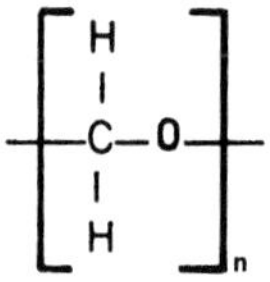

Polyacetal resin

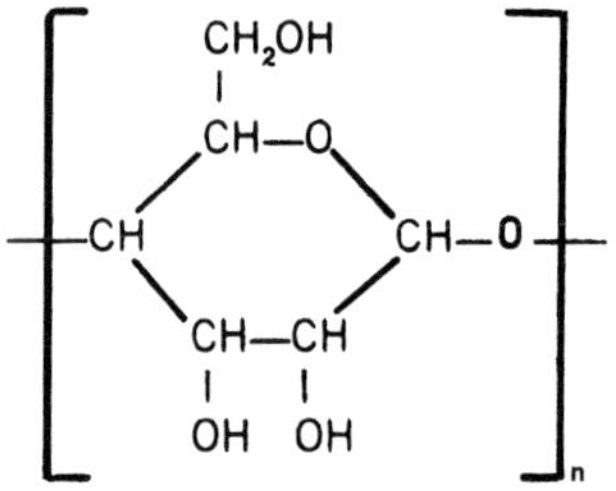

Cellulose (natural polymer)

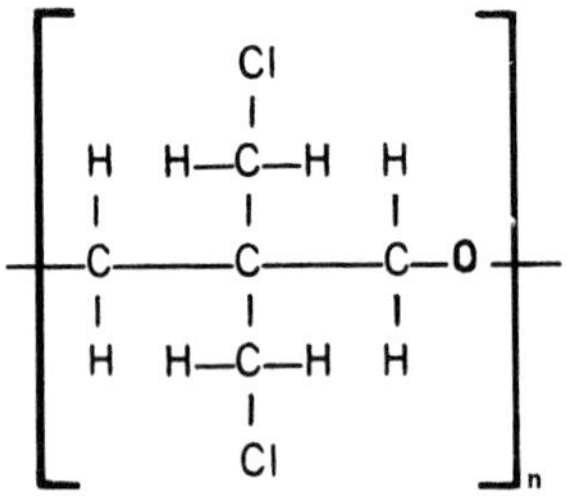

Chlorinated polyether

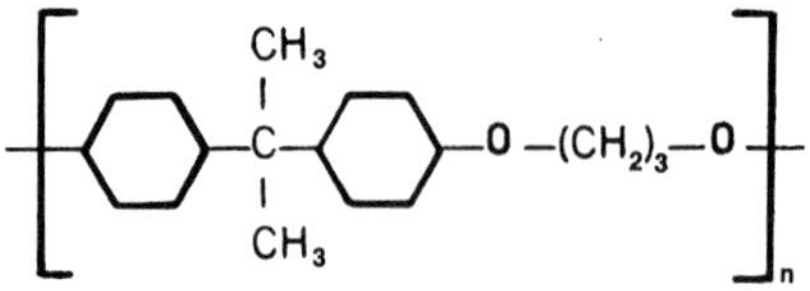

Phenoxy resin (polyhydroxyether)

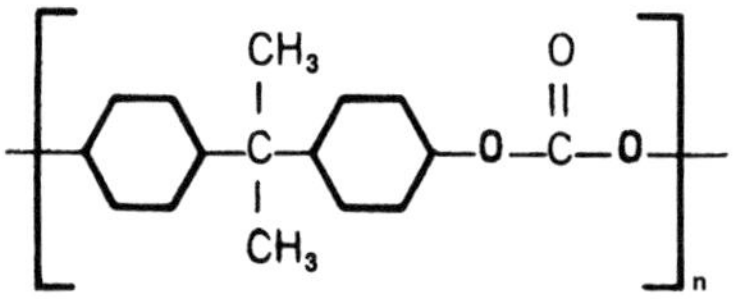

Polycarbonate

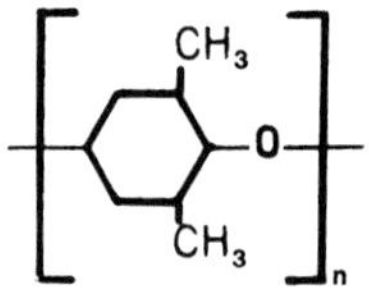

Polyphenylene oxide

TABLE 4

Polymers with Nitrogen in the Chain

$$\left[-\underset{\underset{H}{|}}{N}-(CH_2)_5-\overset{\overset{O}{\|}}{C}-\right]_n$$

Nylon 6

$$\left[-\underset{\underset{H}{|}}{N}-(CH_2)_6-\underset{\underset{H}{|}}{N}-\overset{\overset{O}{\|}}{C}-(CH_2)_4-\overset{\overset{O}{\|}}{C}-\right]_n$$

Nylon 6/6

$$\left[-\underset{\underset{H}{|}}{N}-(CH_2)_6-\underset{\underset{H}{|}}{N}-\overset{\overset{O}{\|}}{C}-(CH_2)_8-\overset{\overset{O}{\|}}{C}-\right]_n$$

Nylon 6/10

$$\left[-\underset{\underset{H}{|}}{N}-(CH_2)_{10}-\overset{\overset{O}{\|}}{C}-\right]_n$$

Nylon 11

$$\left[-O-(CH_2)_x-O-\overset{\overset{O}{\|}}{C}-\underset{\underset{H}{|}}{N}-(CH_2)_y-\underset{\underset{H}{|}}{N}-\overset{\overset{O}{\|}}{C}-\right]_n$$

Polyurethanes

TABLE 5
Thermoset Structures

Phenol-formaldehyde resin

Urea-formaldehyde

Melamine-formaldehyde

Sample Log Sheets for Recording Experimentation

To record their experimentation with plastics processes, students may prepare log sheets like the samples shown here.

(Chapter 4)
Injection Molding Log Sheet

Material					Machine — Mold					Parts	
Exp. No.	Name Manufacturer Code	Density or Specific Gravity	Flow # or Melt Index	Cost Per Lb.	Mold Type	Injection Forward Time	Molding Temp.	Gauge Pressure	Nozzle Pressure	No. of Parts	Results

(Chapter 5)
Extrusion Log Sheet

Material					Machine — Die					Product		
Exp. No.	Name Manufacturer Code	Density or Specific Gravity	Flow # or Melt Index	Cost Per Lb.	Rear Zone Temp.	Front Zone Temp.	Die Temp.	Screw Speed	Die Type	Takeoff Speed	Feet Made	Results

(Chapter 6)
Blow Molding Log Sheet

Material					Machine — Die			Product			
Exp. No.	Name Manufacturer Code	Density or Specific Gravity	Flow # or Melt Index	Cost Per Lb.	Barrel Temp.	Die Temp.	Die Type	Mold	Blowing Time	No. Made	Results

(continued)

(Chapter 7)
Compression Molding Log Sheet

Material					Machine — Mold					Part	
Exp. No.	Name Manufacturer Code	Density or Specific Gravity	Flow # or Plasticity	Cost Per Lb.	Mold Type	Weight of Charge	Molding Temp.	Molding Pressure	Cure Time	No. of Parts	Results

(Chapter 8)
Thermoforming Log Sheet

Material				Machine — Mold				Part	
Exp. No.	Name Distributor	Thickness	Cost Per Sq. Ft.	Machine	Mold Type	Molding Temp.	Heating Time	No. of Parts	Results

(Chapter 13)
Rotational Molding Log Sheet

Material					Machine — Mold					Part	
Exp. No.	Name Manufacturer Code	Density or Specific Gravity	Flow # or Melt Index	Cost Per Lb.	Mold Type	Weight of Charge	Molding Temp.	Rotation Speed	Fusion Time	No. of Parts	Results

TEMPERATURE CONVERSION TABLE

Common Fahrenheit temperatures with corresponding Celsius temperatures used in plastics processing.

°F	°C	°F	°C
100	37.7	475	246.1
125	51.6	500	260.0
150	65.5	525	273.8
175	79.4	550	287.7
200	93.3	575	301.6
225	107.2	600	315.5
250	121.1	625	329.4
275	135.0	650	343.3
300	148.8	675	357.2
325	162.7	700	371.1
350	176.6	725	385.0
375	190.5	750	398.8
400	204.4	775	412.7
425	218.3	800	426.6
450	232.2		

Trade Names of Selected Plastics Materials

Trade names are special names given to their products by manufacturers, wholesalers, and dealers. They are used to identify the products from those of their competitors (other companies). The following is a list of some trade names common to the plastics industry. A complete list may be found in the directory section of a current issue of *Modern Plastics Encyclopedia*.

Trade Name	Plastic Material	Manufacturer
Absaglas	Fiberglass reinforced ABS	Fiberfil Div., Dart Industries, Inc.
Abson	ABS resins and compounds	B. F. Goodrich Chemical Co.
Acroleaf	Hot stamping foil	Acromark Co.
Acrilan	Acrylic, modacrylic, nylon, and polyester fibers	Monsanto Co.
Acrylite	Acrylic molding compounds; cast acrylic sheet	American Cyanamid Co.
Arnel	Cellulose (triacetate fiber)	Celanese Corp.
Aropol	Polyester resins	Ashland Chemical Co.
Atlac	Polyester resin	Atlas Chemical Industries, Inc.
Avisun	Polypropylene	Avisun Corp.
Bakelite	Polyethylene, ethylene co-polymers, epoxy, phenolic, polystyrene, phenoxy, ABS, and vinyl resins and compounds	Union Carbide Corp.
Beetle	Urea molding compounds	American Cyanamid Co.
Blapol	Polyethylene compounds	Reichhold Chemicals, Inc.
Blendex	ABS resin	Marbon Chemical Co.
Boltaron	ABS or PVC rigid plastic sheets	General Tire and Rubber Co.
Cadco	Plastic rod, sheet, tubing, film; resins, glass fibers, glass strand	Cadillac Plastic and Chemical Co.
Capran	Nylon film	Allied Chemical Corp.
Carbaglas	Fiberglass reinforced polycarbonate	Fiberfil Div., Dart Chemical Co.
Castcor	Cast polyolefin films	Mobil Chemical Corp.
Castethane	Castable molding urethane	Upjohn Co.
Celcon	Acetal copolymer resins	Celanese Plastics Co.
Celpak	Rigid polyurethane foam	Decar Chemical Products Co.
Chemfluor	Fluorocarbon plastics	Chemplast, Inc.
Chem-o-sol	PVC plastisol	Chemical Products Corp.
Cordo	PVC foam and films	Ferro Corp.
Cycolac	ABS resins	Borg-Warner Chemicals
Cymel	Melamine molding compounds	American Cyanamid Co.
Dacovin	PVC compounds	Diamond Shamrock Corp.
Delrin	Acetal resin	E. I. du Pont de Nemours and Co., Inc.
Diaron	Melamine resins	Reichhold Chemicals, Inc.
Dow Corning	Silicones	Dow Corning Corp.
Duco	Lacquers	E. I. du Pont de Nemours and Co., Inc.
Durethene	Polyethylene film	Arco/Polymers, Inc.
Dyfoam	Expanded polystyrene	W. R. Grace and Co.
Dylan	Low- and medium-density polyethylene	Arco/Polymers, Inc.
Dylel	ABS plastics	Arco/Polymers, Inc.
Dylene	Polystyrene resin and oriented sheet	Arco/Polymers, Inc.
Dylite	Expandable polystyrene beads	Arco/Polymers, Inc.

(Continued on next page)

Trade Name	Plastic Material	Manufacturer
Eccosil	Silicone resins	Emerson and Cuming, Inc.
El Rexene	Polyethylene, polypropylene, polystyrene and ABS resins	Rexene Polymers Co.
Elvacite	Acrylic resins	E. I. du Pont de Nemours and Co., Inc.
Epolite	Epoxy compounds	Rezolin, Inc.
Epon	Epoxy resin	Shell Chemical Co.
Epotuf	Epoxy resins	Reichhold Chemicals, Inc.
Estane	Polyurethane resins and compounds	B. F. Goodrich Chemical Co.
Ethafoam	Polyethylene foam	Dow Chemical Co.
Evenglo	Polystyrene resin	Arco/Polymers, Inc.
Flexane	Urethanes	Devcon Corp.
Fluorosint	TFE - fluorocarbon base composition	Polymer Corp.
Foamthane	Rigid polyurethane foam	Pittsburgh Corning Corp.
Formica	High-pressure laminate	American Cyanamid Co.
Fortiflex	Polyethylene resins	Celanese Plastics Co.
Fosta	Molding and extrusion grade nylon	Foster Grant Co.
Fostacryl	Thermoplastic polystyrene resins	Foster Grant Co.
Fostafoam	Expandable polystyrene beads	Foster Grant Co.
Genthane	Polyurethane	General Tire and Rubber Co.
Geon	Vinyl resins	B. F. Goodrich Chemical Co.
Glaskyd	Alkyd molding compounds	American Cyanamid Co.
Halon	TFE molding compounds	Allied Chemical Corp.
Hetron	Fire-retardant polyester resins	Durez Div., Hooker Chemical Corp.
Hex-One	H. D. polyethylene	Gulf Oil Co.
Hi-Fax	Polyethylene	Hercules Inc.
Implex	Acrylic molding powder	Rohm and Haas Co.
Isonate	Urethane systems	Upjohn Co.
Kodacel	Cellulosic film	Eastman Chemical Products, Inc.
Korad	Acrylic film	Rohm and Haas Co.
Kralastic	ABS high-impact resins	Uniroyal, Inc.
Kraton	Styrene-butadiene copolymers	Shell Chemical Co.
Krene	Plastic film and sheeting	Union Carbide Corp.
Kydex	Acrylic/PVC sheets	Rohm and Haas Co.
Laminac	Polyester resins	American Cyanamid Co.
Lexan	Polycarbonate resin, film, and sheet	General Electric Co.
Lucite	Acrylic resins	E. I. du Pont de Nemours and Co., Inc.
Lustran	SAN and ABS molding and extrusion resins	Monsanto Co.
Marafoam	Polyurethane foam resin	Marblette Co.
Marlex	Polyethylenes, polypropylenes, other polyolefin plastics	Phillips Petroleum Co.
Merlon	Polycarbonate	Mobay Chemical Co.
Micarta	Thermosetting laminates	Westinghouse Electric Corp.
Miccrosol	Vinyl plastisol	Michigan Chrome and Chemical Co.
Microthene	Powdered polyolefins	U. S. Industrial Chemicals Co.
Mini-Vaps	Expanded polyethylene	Agile Div./Nalge Co.
Mirrex	Calendered rigid PVC	Tenneco Chemicals, Inc.
Mista Foam	Urethane foam	M. R. Plastics and Coatings, Inc.
Monocast	Nylon	Polymer Corp.
Moplen	Polypropylene	Novamont Corp.
Multrathane	Urethane elastomers	Mobay Chemical Co.

Trade Name	Plastic Material	Manufacturer
Multron	Polyesters	Mobay Chemical Co.
Mylar	Polyester film	E. I. du Pont de Nemours & Co., Inc.
Nalgon	Plasticized PVC tubing	Nalge Co.
Naugahyde	Vinyl coated fabrics	Uniroyal, Inc.
Nimbus	Polyurethane foam	General Tire and Rubber Co.
Norchem	L. D polyethylene	Northern Petrochemical Co.
Noryl	Polyphenylene oxide	General Electric Co.
Nylafil	Fiberglass reinforced nylon	Fiberfil Div., Dart Industries, Inc.
Nylatron	Filled nylons	Polymer Corp.
Olefane	Polypropylene film	Avisun Corp.
Oleflo	Polypropylene resin	Avisun Corp.
Orlon	Acrylic fiber	E. I. du Pont de Nemours & Co., Inc.
Paraplex	Polyester resins	Rohm & Haas Co.
Pelaspan-Pac	Expandable polystyrene	Dow Chemical Co.
Penton	Chlorinated polyether	Hercules Inc.
Permelite	Melamine molding compounds	Melamine Plastics, Inc.
Petrothene	Low-, medium- and high-density polyethylene	U. S. Industrial Chemicals Co.
Plaskon	Melamine, urea, nylon	Allied Chemical Corp.
Plastic Steel	Epoxy tooling and repair materials	Devcon Corp.
Plenco	Phenolics	Plastics Engineering Co.
Pleogen	Polyester resins	Mol-Rez Div., American Petrochemical Corp.
Plexiglas	Acrylic sheets and molding powders	Rohm & Haas Co.
Pliobond	Adhesive	Goodyear Tire and Rubber Co.
Pliovic	PVC resins	Goodyear Tire and Rubber Co.
Plyophen	Phenolic resins	Reichhold Chemicals, Inc.
Polycarbafil	Fiberglass reinforced polycarbonate	Fiberfil Div., Dart Industries, Inc.
Polyfoam	Polyurethane foam	General Tire and Rubber Co.
Poly-Eth	Low-density polyethylene	Gulf Oil Corp.
Polylite	Polyester resins	Reichhold Chemicals, Inc.
PPO	Polyphenylene oxide	General Electric Co.
Pro-fax	Polypropylene	Hercules Inc.
Resinol	Polyolefins	Allied Resinous Products, Inc.
Restfoam	Urethane foam	Stauffer Chemical Co.
Richlite	Phenolic fibre laminate	The Henry Co.
Rolox	Two-part epoxy compounds	Hardman, Inc.
Royalite	Thermoplastic sheeting	Uniroyal, Inc.
Rucoblend	Vinyl compounds	Ruco Div., Hooker Chemical Corp.
Saran	Polyvinylidene chloride	Dow Chemical Co.
Scotchpak	Heat-sealable polyester film	3M Co.
Selectrofoam	Urethane foam	PPG Industries, Inc.
Silastic RTV	Silicone rubber	Dow Corning Corp.
Structoform	Resin-reinforced fiberglass sheets	Melamine Plastics, Inc.
Styrafil	Fiberglass reinforced polystyrene	Fiberfil Div., Dart Industries, Inc.
Styrofoam	Expanded polystyrene	Dow Chemical Co.
Styron	Polystyrene	Dow Chemical Co.
Super Dylan	High-density polyethylene	Arco/Polymers, Inc.
Surlyn	Ionomer resin	E. I. du Pont de Nemours & Co., Inc.
Teflon	TFE fluorocarbon resins	E. I. du Pont de Nemours & Co., Inc.
Tenite	Polyolefins and cellulosics	Eastman Chemical Products, Inc.
Texin	Urethane elastomer molding compound	Mobay Chemical Co.

Trade Name	Plastic Material	Manufacturer
Textolite	Industrial laminates	General Electric Co.
Tuftane	Polyurethane film and sheet	B. F. Goodrich Chemical Co.
Tynex	Nylon filaments	E. I. du Pont de Nemours & Co., Inc.
Tyril	Styrene acrylonitrile	Dow Chemical Co.
Ultramid	Nylon	BASF Corp.
Ultrapas	Melamine formaldehyde compounds	Dynamit Nobel Sales Corp.
Ultrathene	Ethylene-vinyl acetate resins and copolymers	U. S. Industrial Chemicals Co.
Uraglas	Fiberglass reinforced polyurethane	Fiberfil Div., Dart Industries, Inc.
Uralite	Urethane compounds	Rezolin, Inc.
Uvex	Cellulose acetate butyrate sheet	Eastman Chemical Products, Inc.
Varcum	Phenolic resins	Reichhold Chemicals, Inc.
Vectra	Polypropylene fibers	Fibers Div., Chevron Chemical Co.
Velon	PVC film and sheeting	Firestone Plastics Co.
Vibrathane	Polyurethane elastomer	Uniroyal, Inc.
Vitel	Polyester resin	Goodyear Tire and Rubber Co.
Vynaloy	Vinyl sheet	B. F. Goodrich Chemical Corp.
Vyram	Rigid PVC materials	Monsanto Co.
Whirlclad	PVC plastic coatings	Polymer Corp.
Wilflex	Vinyl plastisols	Flexible Products Co.
Xylon	Nylon	Fiberfil Div., Dart Industries, Inc.
Zendel	Polyethylene	Union Carbide Corp.
Zerlon	Copolymer of acrylic and styrene	Dow Chemical Co.
Zetafin	Polyolefin copolymers	Dow Chemical Co.
Zytel	Nylon resins	E. I. du Pont de Nemours & Co., Inc.

Appendix B
Resources

Plastics Industry Organizations and Their Publications

**The Society of the
Plastics Industry, Inc.
355 Lexington Ave.
New York, NY 10017**

This is an industry-oriented organization financed by dues to large and small plastics or plastics related firms. Dues structure is in direct proportion to the gross sales in plastics. It represents the industry in legislative and other capacities. Individual membership is also possible. The SPI has 54 operating units — councils, divisions and committees: It is divided into three sections geographically: Midwest, New England, and Western. The SPI is also very active in Canada with many operating divisions.

SPI Publications
(1) The Story of the Plastics Industry
(2) The Need for Plastics Education
(3) Plastics Education Guide
(4) Finishing and Decorating Plastics
(5) Injection Molding
(6) Compression and Transfer Molding
(7) An Introduction to Extrusion
(8) Thermoset Molding
(9) Recycling Plastics
(10) Answers to Questions You are Asking About Plastics and the Environment
(11) The Plastics Industry and Solid Waste Management
(12) Directory and Buyers Guide of Members
(13) SPI Film Catalog
(14) Plastics and the Human Body

**The Society of
Plastics Engineers, Inc.
656 West Putnam Ave.
Greenwich, CT 06830**

This is a people-oriented organization dedicated to promote the scientific and engineering knowledge relating to plastics. It is done through local monthly meetings, occasional Regional Technical Conferences (RETEC), National Technical Conferences (NATEC), and the society's annual International Conference (ANTEC). The society offers its members:

1. The latest scientific and technical data at society meetings.
2. Several plastics publications.
3. Close personal contacts with leaders in the field.
4. An opportunity to express one's ideas and accomplishments.

Both members and nonmembers are welcome at all its meetings. It is made up of 68 sections in the USA, Canada, Japan, and Europe.

SPE Publications
(1) **Plastics Engineering** — the journal of the Society, monthly, free to members.
(2) **Polymer Engineering and Science** — a journal of record, published bimonthly.
(3) **SPE Newsletter,** semiannual survey report and Society news.
(4) **ANTEC and NATEC Reprint Volumes** — papers presented at ANTEC and NATEC conferences.
(5) **RETEC Reprint Books** — papers presented at regional technical conferences.
(6) **SPE Plastics Science and Engineering Series** — a series of books sponsored by SPE, written by outstanding authorities. Sold at a discount to SPE members. Write for a list.

The Plastics Education Foundation
1913 Central Ave.
Albany, NY 12205

The Plastics Education Foundation was formed and chartered by the SPI and SPE to serve as a focal point for all educational activities relating to the plastics industry and to mobilize available management and engineering manpower into the development of plastics education. It was also chartered to plan and coordinate the development of teaching materials and assist local SPE chapters in developing plastics education programs. The PEF will assist educational institutions and instructors in providing effective plastics programs. It also provides the plastics industry with appropriate in-service training programs.

PEF Publications
(1) **PEF Small Equipment Purchasing Guide.**
(2) **100 Years Young** — A history of the plastics industry, 36 pp.
(3) **Jobs in the Plastics Industry** — 34 pp.
(4) **SPE Education Chairman Handbook,** 28 pp.
(5) **How to Buy Plastics Shop Equipment** — A reprint from March 1973 **School Shop,** 4 pp.
(6) **1959 Plastics Safety Handbook.**
(7) **Where Its At in Plastics.**
(8) **Job Descriptions for Plastics Manufacturing Plants,** 12 pp.
(9) **The Custom Molder** — The role of the custom molder in the plastics industry, 8 pp.
(10) **How to Infiltrate the Establishment Without Losing Your Identity.**
(11) **Better Living and a Better Environment with Plastics,** 4 pp.
(12) **Plastics Industry Workers,** SRA, 4 pp.
(13) **An Experimental Resource Unit in Plastics for Industrial Arts,** a 205-page curriculum guide.
(14) **Plastics Education Guide,** 47 pages.
(15) **Plastics for Industrial Arts,** State of Kansas curriculum guide, 41 pages.

A free copy of the current publications list is available.

Selected Plastics Periodicals

Plastics Technology
Bill Brothers Publications, Inc.;
630 Third Avenue; NY 10017
Many practical articles, some technical, on all phases of the plastics industry. Processing handbook issued annually.

Plastics World
Cahners Publishing Co., Inc.;
270 St. Paul St.; Denver CO 80206
General plastics industry articles, many practical tips.

Plastics Engineering
Society of Plastics Engineers;
656 W. Putnam Avenue;
Greenwich, CT 06830
Technical articles on plastics processing and polymer properties.

Modern Plastics
McGraw-Hill, Inc.; P. O. Box 430;
Hightstown, NJ 08520
Many good articles on plastics. Encyclopedia issue is bible of the industry, should be in every library.

Plastics Design & Processing
Lake Publishing Corp.;
Box 270; Libertyville, IN 60048
Industry news on design and processing of plastics.

Journal of Cellular Plastics
Technomic Publishing Co., Inc.;
750 Summer Street;
Stamford, CT 06901
Technical articles on plastic foams including urethanes, polyethylene, polystyrene, etc.

Machine Design
Plastics Reference Issue
Penton Publications; Penton Building;
Cleveland, OH 44113
Up-to-date articles and references on plastics.

Modern Packaging
McGraw-Hill, Inc.;
P. O. Box 430; Hightstown, NJ 06520
Many articles on plastic packaging. Encyclopedia issued once a year as part of the subscription price.

Canadian Plastics
1450 Don Mills Road;
Don Mills, Ontario, Canada
Current articles on Canadian plastics developments.

Trade Publications

These publications are available free from the firms who publish them. They contain many informative articles.

Durez Molder, Hooker Chemical Corp., Box 535 Walck Rd., North Tonawanda, NY 14120.

Industrial Research, Beverly Shores, IN 46301.

Polygram, Mobay Chemical Co., Penn - Lincoln Parkway West, Pittsburgh, PA 15205.

DuPont Magazine, Wilmington, DE 19898.

Rohm and Haas Reporter, Independence Mall West, Philadelphia, PA 19105.

Close-Up, Eastman Close Up, 260 Madison Ave., NY 10016.

Goodchemco News, 3135 Euclid Avenue, Cleveland, OH 44115.

The ABC's of Modern Plastics, Union Carbide Corp. Chemicals and Plastics Dept., 270 Park Ave., NY 10017.

Handbook of Polyolefin, Phillips Petroleum Co., Chemical Dept., Bartlesville, OK 74003.

Glossary of Plastics Terms, Phillips Petroleum Co., Chemical Dept., Bartlesville, OK 74003.

Bakelite Review, Union Carbide Corp., Chemicals and Plastics Dept., 270 Park Ave., NY 10017.

Monsanto Magazine, Monsanto Chemical Co., 800 N. Lindberg Blvd., St. Louis, MO 63166.

Section VII   APPENDICES

Selected Audiovisual Sources

Catalogs and brochures of films, filmstrips, overhead transparencies, and teaching aids may be obtained from the following sources. A review of other film catalogs may reveal additional sources. Many film libraries stock films on plastics which may be rented. Also, several film services distribute free-loan films for industrial concerns.

1. Association - Sterling Films
   512 Burlington Ave.
   LaGrange, IL 60525
2. DCA Educational Products, Inc.
   4865 Stenton Ave.
   Philadelphia, PA 19144
3. Educators Progress Service, Inc.
   Randolf, WI 53956
4. The Henry Co.
   Kitco Division
   6405 East Kellog St.
   Wichita, KA 67218
5. Howard W. Sams & Co., Inc.
   4300 W. 62nd St.
   Indianapolis, IN 46206
6. Modern Talking Picture Service
   1212 Avenue of The Americas
   New York, NY 10036
7. National Audio-Visual Center
   General Services Administration
   National Archives & Record Service
   Washington, DC 20409
8. Plastik-Labs, Inc.
   6336 Avon Ave.
   Kalamazoo, MI 49002
9. The Plastics Education Foundation
   1913 Central Ave.
   Albany, NY 12205
10. Scott Education Division
    Lower Westfield Road
    Holyoke, MA 01040
11. The Society of The Plastics Industry, Inc.
    355 Lexington Ave.
    New York, NY 10017
12. Sterling Educational Films
    241 East 34th Street
    New York, NY 10016
13. Universal Educational and Visual Arts
    221 Park Avenue South
    New York, NY 10003

Selected Plastics Laboratory Equipment and Supply Sources

AAA Plastic Equipment
P.O. Box 11512
Ft. Worth, TX 76110
(Thermoformers)

Acromark Company
60 Locust Avenue
Berkley, Heights, NJ 07922
(Hot stamp machines)

All Purpose Roll Leaf Corp.
37 W. Century Road
Paramus, NJ 07652
(Hot stamp foils)

American Handicrafts Company
1011 Foch Street
Ft. Worth, TX 76107
and cities throughout the USA
(General plastics craft supplies)

Boy Machines, Inc.
91 Commercial Street
Plainview, NY 11803
(Injection molders)

Binks Manufacturing Company
9201 West Belmont Avenue
Franklin Park, IL 60131
(Fiberglass equipment)

Boin Arts and Crafts Company
91 Morris Street
Morristown, NJ 07960
(Hobby craft supplies)

Brabender, C. W., Instruments, Inc
50 East Wesley Street
So. Hakensack, NJ 07606
(Extrusion equipment)

Brodhead-Garrett Company
4560 East 71st Street
Cleveland, OH 44105
(General supplies)

Brown Plastics Engr. Company, Inc.
1823 Holste Road
North Brook, IL 60062
(Extruders and take-off equipment)

Carver, Fred S., Inc.
W. 142 N. 9050 Fountain Blvd.
Menomonee Walls, WI 53051
(Compression and injection molders)

Cadillac Plastics and Chemical Co.
15111 Second Avenue
Detroit, MI 48203
and other major cities
(Sheet, rod and tube)

Chicago Crafts Service
615 North LaSalle Street
Chicago, IL 60610
(Crafts)

Comet Industries, Inc.
2500 York Road
Elk Grove Village, IL 60007
(Thermoformers)

Cope Plastics
111 West Delmar Avenue
Godfrey, IL 62035
(Sheet, rod and tube and equipment)

Commercial Plastics and Supply Corp.
1642 Woodhaven Drive
Cornwells Hts., PA 19020
(Sheet, rod and tube)

Craft Service
337 University Avenue
Rochester, NY 14607
(General crafts)

Day-Glo Color Corp.
4732 St. Clair Avenue
Cleveland, OH 44103
(Colorants)

D-M-E Company
29111 Stephenson Highway
Madison Hts., MI 48071
(Mold making supplies)

Dake Corporation
641 Robbins Road
Grand Haven, MI 49417
(Molding presses)

Dayton Plastics, Inc.
2554 Needmore Road
Dayton, OH 45414
(Sheet, rod and tube)

Di-Acro Division
Houdaille Industries, Inc.
578 - 8th Avenue
Lake City, MN 55041
(Thermoformers and injection/blowmolders)

Delvie's Plastics, Inc.
2320 S.W. Temple
Salt Lake City, UT 84110
(Sheet, rod and tube and equipment)

Dri-Print Foils
329 New Brunswick Avenue
Rahway, NJ 07065
(Hot stamp foils)

France Engineering Company
Route #5, Loader Road
Coshocton, OH 48312
(Heat sealing equipment)

Flex-O-Glass, Inc.
Flex-O-Film Division
1100 North Cicero Avenue
Chicago, IL 60651
(Thermoforming film and heat seal tubing)

Foster and Allen, Inc.
26 Commerce Street
Chatham, NJ 07928
(Extrusion take-off equipment)

Franklin Manufacturing Corp.
692 Pleasant Street
Norwood, MA 02062
(Hot stamp machinery)

Gane Brothers & Lane, Inc.
1335 W. Lake Street
Chicago, IL 60607
(Hot stamping type)

Haake, Inc.
244 Saddle River Road
Saddle Brook, NJ 07662
(Extrusion and take-off equipment)

Hastings Plastics, Inc.
1704 Colorado Avenue
Santa Monica, CA 90404
(General plastics supplies)

The Henry Company
P.O. Box 18166
Wichita, KS 67218
(Injection molders and supplies)

Hilliard Industries, Inc.
201 Wood Avenue
Middlesex, NJ 08846
(Injection molders)

Howard and Smith, Inc.
31270 Stephenson Highway
Royal Oak, MI 48068
(General laboratory equipment)

Hyaline Plastics, Inc.
1019 North Capital Avenue
Indianapolis, IN 46206
(Sheet, rod and tube)

IMS Company
24050 Commerce Park Road
Cleveland, OH 44122
(Molding supplies and scrap granulators)

Industrial Arts Supply Company
5724 West 36th Street
Minneapolis, MN 55416
(Plastics supplies and equip.)

Jaco Manufacturing Company
468 Geiger Street
Berea, OH 44017
(Injection molders)

Kamweld Products Company, Inc.
742 Providence Highway
Norwood, MA 02062
(Hot gas welders)

Kingsley Machine Company
850 Cahuenga Blvd.
Hollywood, CA 90038
(Hot stamp equipment)

Lawter Chemicals, Inc.
990 Skokie Blvd.
Northbrook, IL 60062
(Colorants)

McKilligan Industrial Supply Co.
494 Chenango Street
Binghampton, NY 13901
(Plastics supplies and equip.)

Meyers Materials, Inc.
5101 East 65th Street
Indianapolis, IN 46220
(Sheet, rod and tube; fiberglass and foam)

Michigan Chrome and Chemical Co.
8615 Grinnel Avenue
Detroit, MI 48213
(Fluidized bed equipment and plastics)

New Hermes Engraving Machine Corp.
20 Cooper Square
New York, NY 10003
(Engravers)

Orbit, Inc.
P.O. Box 351
Deming, NM 88030
(Thermoformers)

Packaging Aids Corp.
P. O. Box 77203
San Francisco, CA 94107
(Heat sealing equipment and supplies)

Polymer Corporation
2120 Fairmont Avenue
Reading, PA 19606
(Fluidize bed coaters and resins)

Plastik-Labs, Inc.
6336 Avon Avenue
Kalamazoo, MI 49002
(Teaching systems and equipment)

Quality Vacuum Forming Mach. Co., Inc.
12767 Saticoy Street
North Hollywood, CA 91605
(Thermoformers)

Rainville Company, Inc.
200 Clay Avenue
Middlesex, NJ 08846
(Injection, extrusion, granulators, etc.)

Ransburg Corporation
P.O. Box 88220
Indianapolis, IN 46208
(Powder spray, fluid bed, fiberglass equipment)

Ren Plastics
5656 South Cedar Street
Lansing, MI 48909
(Plastics, resins, and supplies)

The Satterlee Company
2200 East Franklin Avenue
Minneapolis, MN 55404
(General laboratory equipment)

The Simplomatic Manufacturing Co.
4416 West Chicago Avenue
Chicago, IL 60651
(Injection molders)

Testing Machines, Inc.
400 Bayview Avenue
Amityville, NY 11701
(Plastics testing equipment)

Vega Enterprises, Inc.
Route #3, Box 300-B
Decatur, IL 62521
(Rotational molders, laminating presses)

Wabash Metal Products, Inc.
P.O. Box 298
Wabash, IN 46992
(Hydraulic molding presses)

Wayne Machine and Die Co.
100 Furler Street
Totowa, NJ 07512
(Lab extruders and take-off equipment)

Bibliography

American Federation of Labor and Congress of Industrial Organizations, **Collective Bargaining - Democracy on the Job.** Washington, D. C.: AFL-CIO, 1965, Pp. 30.

This is the AFL-CIO. Washington, D. C.: AFL-CIO, 1972, Pp. 24.

Why Unions? Washington, D. C.: AFL-CIO, 1969, Pp. 16.

American Society for Testing and Materials Part 26. **Plastics Specifications with Closely Related Tests.** Philadelphia, PA: ASTM, annual.

Part 27. **Plastics - General Methods of Testing.** Philadelphia, PA: ASTM, annual.

Amrine, Harold T.; Ritchey, John A.; and Holley, Oliver S., **Manufacturing Organization and Management,** 2nd Ed. Englewood Cliffs, NJ: Prentice-Hall, Inc. 1966, Pp. 568.

Baird, Ronald J. **Industrial Plastics.** South Holland, IL: Goodheart-Willcox Co., Inc., 1971.

Barnes, Ralph M., **Motion and Time Study,** 6th Ed. New York: John Wiley & Sons, Inc., 1968, Pp. 799.

Barr, Robert A., **Controlled Melting Extrusion Screws.** New Brunswick, NJ: Midland-Ross Corp., 1972, Pp. 12.

Beck, Ronald D., **Plastic Product Design.** New York: Van Nostrand Reinhold Co., 1970, Pp. 471.

Benjamin, W. P., **Plastic Tooling Techniques and Applications.** New York: McGraw-Hill Book Co., 1972.

Benning, Calvin J., **Plastic Foams.** New York: Van Nostrand Reinhold Co., 1969.

Bernhardt, Ernest C., **Processing of Thermoplastic Materials.** New York: Reinhold Publishing Corp., 1959, Pp. 690.

Billimeyer, F. W. Jr., and Ford, Renee, "The Anatomy of Plastics." Reprint from **Science and Technology.** New York: Conover-Mast Publishing Co., January, 1968, Pp. 17.

Bockhoff, Frank S. and Neuman, J. A., **Welding of Plastics.** New York: Reinhold Publishing Corp., 1959, Pp. 270.

Borro, E. F. Sr., "Compression and Transfer Molding." Reprint from **Modern Plastics Encyclopedia,** 1964 Ed., Hooker Chemical Corp., North Tonawanda, NY

Buffa, Elwood S., **Modern Production Management,** 3rd Ed. New York: John Wiley & Sons, Inc., 1969, Pp. 793.

Butzko, Robert, **Plastic Sheet Forming.** New York: Reinhold Publishing Corp., 1958, Pp. 181.

Cherry, Raymond, **General Plastics.** Bloomington, IL: McKnight Publishing Co., 1964, Pp. 156.

Cobb, Boughton, **Fiberglass Boats, Construction and Maintenance.** New York: Vachting Publishing Corp., 1965, Pp. 178.

Cook, J. G., **The Miracle of Plastics.** New York: Dial Press, 1964, Pp. 272.

Daumas, Maurice, **A History of Technology and Invention Progress through the Ages.** New York: Crown Publishers, 1969, 2 Vol.

Dearle, Denis A., **Opportunities in Plastics Careers.** New York: Vocational Guidance Manuals, 1963, Pp. 144.

DePuy, Charles H. and Rinehart, Kenneth L. Jr., **Introduction to Organic Chemistry.** New York: John Wiley & Sons, Inc., 1967, Pp. 392.

Dietz, Albert G. H., **Plastics for Architects and Builders.** Cambridge, MA: The MIT Press, 1969, Pp. 129.

DuBois, John H., **Plastics History USA.** Boston, MA: Cahners Books, 1972, Pp. 447.

DuBois, John H. and John, F. S., **Plastics.** New York: Reinhold Publishing Corp., 1967, Pp. 342.

DuBois, John H. and Pribble, W. I., **Plastics Mold Engineering.** New York: Reinhold Publishing Corp., 1965, Pp. 672.

Duffin, Daniel, **Laminated Plastics.** New York: Reinhold Publishing Corp., 1966, Pp. 249.

Eldon, R. A. and Swan, A. D., **Calendering of Plastics.** New York: American Elsevier Publishing Co., 1971, Pp. 106.

Farkas, Robert, **Heat Sealing.** New York: Reinhold Publishing Corp., 1964, Pp. 179.

Fricker, Ronald H., **An Introduction to American Industry.** Philadelphia, PA: DCA Educational Products, Inc., 1972, Pp. 68.

George, Claude S. Jr., **Management in Industry,** 2nd Ed. Englewood Cliffs, NJ: Prentice Hall, 1964, 618 pp.

Glickstein, **Basic Ultrasonics.** New York: J. F. Rider, 1960, Pp. 137.

Griff, Allan L., **Plastics Extrusion Technology.** New York: Reinhold Publishing Corp., 1968, Pp. 352.

Hanlon, Joseph F., **Handbook of Package Engineering.** New York: McGraw-Hill Book Co., 1971.

Haslam, John, **Identification and Analysis of Plastics.** New York: Van Nostrand Reinhold Co., 1965, Pp. 483.

Jones, David A., **Blow Molding.** New York: Reinhold Publishing Corp., 1962, Pp. 210.

Keen, Martin L., **The How and Why Wonder Book of Chemistry.** New York: Wonder Books, 1961, Pp. 48.

Kell, R. M., and Stickney, P. B., "How Structure Determines Properties of Plastics," **Materials in Design Engineering,** May 1964, Pp. 79-82.

Lappin, Alvin R., **Plastics Projects and Techniques.** Bloomington, IL: McKnight Publishing Co., 1965.

Lindbeck, John R., **Designing Today's Manufactured Products.** Bloomington, IL: McKnight Publishing Co., 1972., Pp. 400.

Maynard, H. B., **Industrial Engineering Handbook,** 2nd Ed. New York: McGraw-Hill Book Co., 1963.

Milby, Robert V., **Plastics Technology.** New York: McGraw-Hill Book Co., 1973, Pp. 581.

Narcus, Harold, **Metallizing of Plastics,** Stamford, CT: Reinhold Publishing Corp., 1960, Pp. 208.

New York, the University of the State of, **An Experimental Resource Unit in Plastics for Industrial Arts.** Greenwich, CT: Plastics Education Foundation, 1966, Pp. 205.

Parry, Robert W., et al., **Chemistry.** Englewood Cliffs, NJ: Prentice Hall, Inc., 1970, Pp. 630.

"Plastics — Where, When, How to Use Them in your Plant," **Factory Magazine.** New York: McGraw-Hill Book Co., August 1963. Reprints available.

Prall, George, **Cooling Blown Tubular Film.** Reprint from **Modern Plastics,** April 1969.

———, **How to Build a Blown Polyethylene Film Line.** Reprint from **Plastics Technology,** July-August 1968.

———, **How to Handle Blown Film Trim and Scrap.** Reprint from **Modern Plastics,** March 1973.

———, **How to Solve Blown Film Problems.** Reprint from **Paper, Film and Foil Converter,** October 1972 to June 1973.

Redfarn, C. A. and Bedford, J., **Experimental Plastics,** 2nd Ed. New York: Inter-science Publishing Inc., 1960, Pp. 140.

Rexene Polymers Co., **Tubular Polyethylene Film Extrusion.** Paramus, NJ, Pp. 18.

Richardson, Terry A., **Modern Industrial Plastics.** Indianapolis, IN, 1974, Pp. 384.

Rosato, Dominick V., **Markets for Plastics.** New York: Van Nostrand Reinhold Co., 1969, Pp. 408.

Rubin, Irvin I., **Injection Molding Theory and Practice.** New York: John Wiley & Sons, Inc., 1972, Pp. 657.

Shubin, John A., **Business Management.** New York: Barnes & Noble, Inc., 1957, Pp. 372.

Simonds, H. R. and Church, J. M., **Concise Guide to Plastics,** 2nd Ed. New York: Reinhold Publishing Co., 1954, Pp. 244.

Skinner, W. and Rogers, D., **Manufacturing Policy in the Plastics Industry.** Homewood, IL: Richard D. Irwin, Inc., 1968, Pp. 336.

Society of the Plastics Industry, Inc., **Plastics Engineering Handbook,** 3rd Ed. New York: Reinhold Publishing Corp., 1960, Pp. 565.

Sonneborn, Ralph H., **Fiberglass Reinforced Plastics.** New York: Reinhold Publishing Corp., 1954, Pp. 244.

Steele, Gerald L., **Fiber Glass Projects and Procedures.** Bloomington, IL: McKnight Publishing Co., 1962, Pp. 159.

Stewart, Alan S., **A Glossary of Urethane Industry Terms.** Louisville, KY: The Martin Sweets Co., Inc., 1971, Pp. 133.

Swanson, Robert S., **Plastics Technology Basic Materials and Processes.** Bloomington, IL: McKnight Publishing Co., 1965, Pp. 232.

Whittington, Floyd R., **Whittington's Dictionary of Plastics.** Stamford, CT: Technomatic Publishing Co., 1968.

Zade, H. P., **Heat Sealing and High-Frequency Welding of Plastics.** New York: John Wiley & Sons, Inc., 1959.

Association and Industrial Publications

Boonton Molding Co., Inc., 300 Myrtle Ave., Boonton, NJ 07005. **A Ready Reference for Plastics,** 13th Ed., 1973.

Burlington Industries, Inc., Greensboro, NC. **Textile Fibers and Their Properties,** 1970, Pp. 63.

Fred S. Carver, Inc., 9050 Fountain Blvd., Menomonee Falls, WI 53051. "Embedding and Preserving Specimens in Transparent Plastic" by Sheldon Winkler. Reprint from **Dental Digest,** May 1969.

Celanese Plastics Co., 550 Broad St., Newark, NJ. **Celanese Plastics Magazine,** Vol. 23, No. 1, March 1965. Special issue on John Wesley Hyatt and Celluloid®.
Compounding Fortiflex®, P3D.
Extrusion of Fortiflex®, P3B.
Fortiflex® Finishing, P3F.
Fortiflex® Polyethylene Blow Molding, P3C.
Fortiflex® Properties and Applications, P1A.
Injection Molding Fortiflex®, P3A.
Standard Tests on Plastics, G1C.

Chemplex Company, Rolling Meadows, IL 60008, "Minimizing Weld Line Effects in Blown PE Film," by George Prall, reprint from **Modern Plastics,** Nov. 1970, Pp. 3.

Dow Chemical — USA, Midland, MI 48640.
Fundamentals of Injection Molding Technology.
Troubleshooting Injection Molding Technology.

DuPont de Nemours & Co., E. I., Plastics Department, Wilmington, DE 19893.
Blow Molding of Alathon® Polyofin Resins, Pp. 28.
Extrusion Coating, Pp. 16.
The Extrusion of Polyethylene Film, Pp. 32.

Ford Motor Co., Educational Affairs Department, Dearborn, MI 48121. **The Evolution of Mass Production,** Pp. 52.

Fry Plastics International, Inc., 8601 S. Figuerea St., Los Angeles, CA 90003.
The Story of Acrylics, 1963, Pp. 56.
The Story of Cellulose, 1963, Pp. 56.
The Story of Nylons, 1964, Pp. 44.
The Story of Polyethylene, 1963, Pp. 56.
The Story of Vinyls, 1963, Pp. 56.

Gem-O-Lite Plastics, North Hollywood, CA. **Polyester and Fiberglass** by Maurice Lannon, 1969, Pp. 121.

General Motors Corp., Detroit, MI. **American Battle for Abundance, A Story of Mass Production** by Charles F. Kettering and Allen Orth, 1947, Pp. 103.

Gulf Oil Company, Chemical Row, Orange, TX 77630. **A Short Course in Blow Molding Linear Polyethylene,** Pp. 31.

Hooker Chemical Corp., North Tonawanda, NY 14120. **Wood Particle Molding,** Bulletin No. 400, Pp. 8.

Kamweld Products Company, Norwood, MA. **Handbook for Welding and Fabricating Thermoplastic Materials** by S. J. Kaminsky and J. A. Williams, 1964, Pp. 176.

Laramy Products Company, Inc., Cohasset, MA. **Making Better Plastic Welds** by Donald W. Thomas, 1962, Pp. 64.

Man-Made Fiber Producers Association, Inc., 350 Fifth Ave., New York 10001. **Guide to Man-Made Fibers,** Pp. 16.

Modern Plastics, 1221 Avenue of the Americas, New York 10020. "Handbook of Plastics," reprint from current **Modern Plastics Encyclopedia.** "Modern Plastics Directory," reprint from current **Modern Plastics Encyclopedia.** "Modern Plastics Technical Review," reprint from current **Modern Plastics Encyclopedia.** "Plastics in the 1980's — A 15-Year Outlook," reprint from **Modern Plastics,** Nov. 1968, Vol. 45, No. 15.

Monsanto Company, 800 N. Lindberg Blvd., St. Louis, MO 63166. **The Fundamentals of Blow Molding,** Bulletin No. 5023B, Pp. 29.

Morgan Industries, Inc., 3311 E. 59th St., Long Beach, CA 90805. **Cutting Costs in Short-Run Plastics Injection Molding,** revised edition, 1973, Pp. 105.

National Association of Manufacturers, 277 Park Ave., New York 10017.
Productivity and Production in Industry, Pp. 14.
Industrial Research and Development, Pp. 14.
The Role of Marketing, Pp. 14.
Wages and Prices in an Industrial Economy, Pp. 10.
Industry's Profits, Pp. 10.

National Board of Fire Underwriters, 85 John St., New York 10038. **Fire Hazards in the Plastics Industry,** Pp. 70.

Shell Chemical Company, 1 Shell Plaza, Houston, TX 77002. **Blow Molding Troubleshooting Guide,** Pp. 18.

Society of the Plastics Industry, Inc., 250 Park Ave., New York 10017. **The Story of the Plastics Industry,** 1966, Pp. 40.

Union Carbide Corp., 270 Park Ave., New York 10017. **Polyethylene Powders for Rotational Molding,** Pp. 33.

U.S. Industrial Chemicals Company, 99 Park Ave., New York 10016.
Coating Fabrics with Microthene® Polyethylene Powder, Pp. 16.
Blow-Up Ratios in Blown Film Extrusion, Pp. 5.
Microthene® Dispersions for Metal Coating, Pp. 6.
Microthene® Microfine Polyolefins — An Introduction, Pp. 12.
Molding Microthene® Polyethylene Powder, Pp. 17.
Petrothene® Polyolefins — A Processing Guide, Pp. 151.
Polyethylene Blow Molding — An Operating Manual, Pp. 72.
Polyethylene Extrusion Coating — An Operating Manual, Pp. 64.
Polyethylene Film Extrusion — An Operating Manual, Pp. 56.
Rotational Molding of Microthene® Polyethylene Powder, Pp. 24.

Sources of Illustrations

The following is a list of those firms whose photographs, diagrams, and information were used in the preparation of this book. Many other firms assisted also. This is not intended to be a complete list of plastics firms in the field nor is it a complete list of those who furnished materials to be considered.

AAA Plastic Equipment Company,
 Ft. Worth, TX
Adamson Division of Wean United,
 Pittsburgh, PA
Allied Chemical Corporation,
 Morristown, NJ
American Packaging Corporation,
 Hudson, OH
American Society for Testing and Materials,
 Philadelphia, PA
Apex Machine Company,
 Ft. Lauderdale, FL
Arco/Polymers, Inc.,
 Pittsburgh, PA
Artisan Industries, Inc.,
 Waltham, MA
Ashdee Division of George Koch Sons, Inc.,
 Evansville, IN
Ashland Chemical Company,
 Columbus, OH
Beloit Corporation,
 Beloit, WI
Berkley and Company, Inc.,
 Spirit Lake, IA
Binks Manufacturing Company,
 Franklin Park, IL
Black Clawson Company,
 Fulton, NY
Boonton Molding Company,
 Boonton, NJ
Borg-Warner Chemicals,
 Washington, WV
Branson Sonic Power Company,
 Danbury, CT
Brodhead-Garrett Company
 Cleveland, OH
Butterworth Manufacturing Company,
 Bethayres, PA
Caig Laboratories, Inc.,
 Westbury, NY
Carver, Fred S., Inc.,
 Menomonee Falls, WI
Cavitron Ultrasonics Division,
Cavitron Corporation,
 Long Island City, NY
Celanese Plastics Company,
 Newark, NJ

Chemplex Company,
 Rolling Meadows, IL
Clifton Hydraulic Press Company,
 Clifton, NJ
Comet Industries, Inc.,
 Elk Grove, IL
Crown Molding Company,
 Arlington, TX
Custom Aircraft Products, Inc.,
 Muncie, IN
Dake Corporation,
 Grand Haven, MI
Dependable Machine Company, Inc.,
 Cedar Grove, NJ
DeVilbiss Company,
 Toledo, OH
DiAcro, Division Houdaille Industries, Inc.,
 Lake City, MN
Dow Corning Corporation,
 Midland, MI
Dri-Print Foils Inc.,
 Rahway, NJ
DuBois, J. Harry,
 Morris Plains, NJ
DuKane Corporation,
 St. Charles, IL
DuPont de Nemours, E. I. and Company,
 Wilmington, DE
Durez Division Hooker Chemical Corporation,
 N. Tonawonda, NY
Dusenbury, John, Company, Inc.,
 Clifton, NJ
Eastman Chemical Products, Inc.,
 Kingsport, TN
Egan Machinery Company,
 Somerville, NJ
Fabricon Automotive Products,
 River Rouge, MI
Farrel Company,
 Rochester, NY
Finish Engineering Company,
 Erie, PA
Ford Motor Company,
 Dearborn, MI
French Oil Mill Machinery Company,
 Piqua, OH

General Motors Corporation,
  Detroit, MI
General Electric Company,
  Plastics Department,
  Pittsfield, MA
Gladen Enterprises, Inc.,
  Bay City, MI
Glas-Craft of California,
  Division Ransburg Corporation,
  Sun Valley, CA
Glassmaster Plastics Company,
  Lexington, SC
B. F. Goodrich Chemical Company,
  Cleveland, OH
Grote Manufacturing Company,
  Madison, IN
HPM Division,
  Koehring Company,
  Mt. Gilead, OH
Hayssen Manufacturing Company,
  Sheboygan, WI
Halmatic, Ltd.,
  Havant, Hampshire,
  England
Hercules, Inc.
  Wilmington, DE
Hendrick Manufacturing Corporation,
  Marblehead, MA
Hull Corporation,
  Hatboro, PA
Improved Machinery Inc.,
  Nashua, NH
Industrial Arts Supply Company,
  Minneapolis, MN
Jaw-Bar Plastics Corporation,
  Birmingham, MI
Kamweld Products Company, Inc.,
  Norwood, MA
Koehring Corporation,
  Springfield, MA
Kohler General, Inc.,
  Sheboygan Falls, WI
Krones, Inc.,
  Franklin, WI
Liberty Machine Company,
  Paterson, NJ
Machinery Division,
  Midland-Ross Corporation,
  New Brunswick, NJ
Marblette Corporation,
  Long Island City, NY
Master Unit Die Products, Inc.,
  Greenville, MI
McClean-Anderson, Inc.,
  Milwaukee, WI
McGraw-Hill Publishing Company,
  New York, NY
McNeil, William R.,
  St. Petersburg, FL
McNeil-Femco-McNeil Corporation,
  Cuyahoga Falls, OH

Minneapolis Star and Tribune Company,
  Minneapolis, MN
Modern Plastic Machinery Corporation,
  Clifton, NJ
NRM Corporation,
  Akron, OH
Nalge Company,
  Division Sybron Corporation,
  Rochester, NY
National Aeronautics and Space Administration,
  Houston, TX
National Association of Manufacturers,
  Washington, D.C.
National Automatic Tool Manufacturing Company,
  Richmond, IN
New Hermes Engraving Machine Corporation,
  New York, NY
Newbury Industries, Inc.,
  Newbury, OH
Nosco Plastics, Inc.,
  Erie, PA
Ogden Sales, Inc.,
  Arlington Heights, IL
Orbit, Inc.,
  Deming, NM
Owens-Corning Fiberglass Corporation,
  Toledo, OH
Packaging Aids Corporation,
  San Francisco, CA
Packaging Industries, Inc.,
  Hyannis, MA
Philco-Ford Corporation,
  Blue Bell, PA
Plastic Coating Equipment Company, Inc.,
  E. Hartford, CT
Plastics Engineering Company,
  Sheboygan, WI
Plastics Engineering Magazine,
  Greenwich, CT
Plastics Machinery and Equipment Magazine,
  Denver, CO
Plastics Technology Magazine,
  New York, NY
Plastics World Magazine,
  Denver, CO
Polymer Corporation,
  Reading, PA
Polymer Machinery Corporation,
  Berlin, CT
Portage Industries Corporation,
  Portage, WI
Premier Vacuum Process,
  Division of Avnet, Inc.,
  Maspeth, NY
Process Equipment Corporation,
  Belding, MI
Progressive Machine Company,
  Paterson, NJ
Pyles Industries, Inc.,
  Wixom, MI

Rainville Company,
  Middlesex, NJ
Ransburg Corporation,
  Indianapolis, IN
Rawal Engravers, Inc.,
  Villa Park, IL
Reed-Prentice Division,
Package Machine Company,
  Longmeadow, MA
Reinforced Plastics/Composites Institute,
  New York, NY
Rennco Inc.,
  Homer, MI
Rohm and Haas Company,
  Philadelphia, PA
Ross, Charles, and Son Company,
  Long Island, NY
Rosedale Plastic Containers, Inc.,
  Buffalo, NY
Rostone Corporation,
  Lafayette, IN
Scott Paper Company,
Foam Division,
  Chester, PA
Seelye Plastics Company,
  Minneapolis, MN
Shoe Form Company, Inc.,
  Auburn, NY
Simplomatic Manufacturing Company,
  Chicago, IL
Solidyne, Inc.,
Sealomatic Division,
  Brooklyn, NY
Thermotron Division,
  Bay Shore, NY
Sonics and Materials, Inc.,
  Danbury, CT
Society of the Plastics Industry, Inc., The
  New York, NY
Sossner Steel Stamps, Inc.,
  Long Island City, NY

Sterling Extruder Corporation,
  Plainfield, NJ
Stewart Bolling and Company,
  Cleveland, OH
Tennant, C., Sons and Company of New York,
  New York, NY
Testing Machines, Inc.,
  Amityville, NY
Thermark Corporation,
  Hammond, IN
Thermolator Company, Division, Van Dorn Company,
  Indianapolis, IN
Thiokol Chemical Corporation,
  Trenton, NJ
Tinius Olsen Testing Machine Company, Inc.,
  Willow Grove, PA
Uniloy Division Hoover Ball and Bearing Company,
  Saline, MI
Union Carbide Corporation,
  New York, NY
Uniroyal, Inc.,
  New York, NY
Ultra Sonic Seal,
  Broomall, PA
U. S. Industrial Chemicals Company,
  New York, NY
Van Nostrand Reinhold Company,
  New York, NY
Vega Enterprises Inc.,
  Decatur, IL
Venus Products, Inc.,
  Kent, WA
Wagner Engraving Company,
  St. Louis, MO
Westinghouse Electric Corporation,
  Hampton, SC
Wheelabrator-Frye, Inc.,
  Mishawaka, IN
Williams-White & Company,
  Moline, IL

Index

Abbreviations, of plastics materials, (table) 382-385
Accumulator, in blow molding, 114
Acrylic, tests on, 35, 37
Acrylic plastic, transparency of, 18
Advertising, in classroom corporation, 375
Adhesives —
  plastics, 24, 292, (tables) 294
  and types of plastics, (table) 312
Air-inhibited resins, 198
Aluminum molds, 124, 125
American Society for Testing and Materials (ASTM), standards handbook, 44
Asbestos, and phenolics, 14
Ashing, polishing plastics through, 290
Assembly line, 341
Atoms, in plastics makeup, 7-8, 50-55
Autoclaves, 52, 198
Automation, and plastics, 345

Baekeland, Dr. Leo H., 2
Bait casting, operation sequence for full mold, (table) 262-263
Bakelite®, 2
Bead foam molding, 246
  operation sequence for, 248, (table) 250-253
Beads, hollow plastic, 25
Belt, endless blow molder, 116
Bema coating machine, 181
Billiard balls, and plastics invention, 1
Biodegradation, 29
Blending, in calendering, 180
Blister forming, 160
Blow molding, 108-131
  operation sequence, 119, (table) 120-123
Blown film, extruding, 92
Bodied adhesives, 292
Bonding —
  chemical, 50

industrial plants for, 346
of plastics, 283-316
Boron filament, in reinforced plastics, 191
Buffing, of plastics, 290
Burning test, 41
Business, starting a manufacturing, 350
Bylaws, of classroom corporation, 353, 367

Calender coating, 182
  industrial equipment for, 184
  materials for, 183
Calendering, 178
  industrial equipment, 182
Capital, in classroom corporation, 358
Captive operation, as manufacturing method, 343
Carbon atom, 52
Carbon backbone, 50
  copolymers, chemistry of, 389
Careers, in plastics chemistry, 58
Casting —
  aluminum in blow molding, 125
  defined, 26
  dip and coating, 255, 256
  of epoxy molds, 82
  full mold plastisol, 261
  operation sequence for full mold bait, (table) 262-263
  operation sequence for polyester resins, (table) 221-223
  operation sequence for resin foam, 241, (table) 242-245
  operation sequence for slush, (table) 264-265
  plastic materials, 217-232
  resin foam, 239
  silicone and urethane molds, 230
  synthetic wood, 229
  vinyl fabric, 185-187

Catalysts, 9, 52, 53, 54
  polyester resin, 198
Cavity vacuum forming, 155
Celluloid®, 1-3
Cellulose nitrate, 1
Cementing, of plastics, 292-295
Ceramic molds, in thermoforming, 175
Chains, types of chemical, 57
Characteristics, of specific plastics, (table) 382-385
Charpy test, impact, 45, 46
Charter, in classroom corporation, 352
Chemical bonding, 50
Chemicals, plastics makeup, 7
Chemistry —
  of copolymers, (table) 389
  of plastics, 50-58
  of polymers with ethylene chain, (table) 388
  of polymers with nitrogen chain, (table) 391
  of polymers with oxygen chain, (table) 390
  of thermoset structures, (table) 392
Chlorine atom, 52
Chopper, fiberglass, 192, 194, 199
Classroom corporation, 350-379
Coaters, in plastics industry, 346
Coating —
  calender, **182**
  and decorating plastics, 317-339
  and dip casting, 256
  and dip casting operation sequence, (table) 258-260, 261
  electrostatic powder, 334
  extrusion, 187-189
  fluidized bed, 328
  knife, 184-185
  methods of, 324-325
  operation sequence for fluidized bed, 330, (table) 332-333
Cobalt naphthenate, 198

411

Cohesives, for plastics, 293
Cold molding, 135
Color —
   pigments, 198
   as plastic property, 19
Compatibility, of plastic types, (table), 302
Composting, and plastics, 28
Compreg lamination, 214
Compression molding, 132-154
   fiberglass, 191, 194, 195
   operation sequence, 142, (table) 144-147
Compression molds, 140
Compression testing, 47
Continuous tube, blow molding, 115
Cooperative ownership, 352
Copolymer plastics, 9
   chemistry of, 56, (table) 389
Copper wire test, 42
Corporate divisions, 347-349
Corporation, forming classroom, 351
Corporations, complete manufacturing, 344
Costs, relative plastic materials, (table) 382-385
Covalent bonding, 50
Cull and runner, in transfer molding, 136
Cushioning, plastics as, 19
Cutting, processes, 284-289

Decorating —
   and coating plastics, 317-339
   by painting, 323-325
   industrial plants for, 346
   of plastics in molds, 336
   by silk screen, 322
   of blow molded products, 124
Destructive tests, 33
   instructions for, 39
Die cutting, of plastics, 286-287
Dielectric sealing, of plastics, 307
Dies —
   blown film, 93
   extruder, 88, 90-91, 104
   for extrusion coating, (table) 189
   for hot foil stamping, 318
   two-color molding, 336, 337

Dip casting, 255, 256
   operation sequence for, (table) 258-260, 261
Dip coating, of plastics, 324
Distribution, of classroom corporation products, 375
Dividend, in classroom corporation, 379
Dope cements, 292
Double-cavity molds, 72, 73
Draft angle, in injection molding, 72
Drape vacuum forming, 155
Drilling, of plastics, 289, (table) 290
Dynamic thermofusion, 255

Economy, and plastics, 4-6
Elasticity, 16
Elastomer plastics, 16
Elastomeric adhesives, 292
Electrical insulation, 12, 13
Electrons, in plastics, 50
Electroplating, 326
Electrostatic force, in thermoplastics, 57
Electrostatic powder coating, 334
Embossing, in vinyl fabric casting, 185, 186
Energy, and plastics, 30, 31
Engle® process 255, 256, 268
Engineering, manufacturing consulting, 344
Engraving, of plastics, 336
Environmental management, and plastics, 28
Epoxy —
   and glass laminates, 193
   tests on, 35, 38
Epoxy molds —
   cast, 127
   casting, 82
Ethylene chain, polymers based on, (table) 388
Ethylene monomer molecule, 50
Expansion, defined, 26
Expansion processes, 233-254
Extruder —
   operation behavior, (table) 101
   parts of, 87-88
   in plastics industry, 346
Extrusion, 87-107
   intermittent blow molding, 117

   operation sequence, 97, (table) 98-101
   of plastic foam, 235
Extrusion blow molding, 108
Extrusion coating, 187-189

Fabric, casting vinyl, 185-187
Fasteners, for plastics, 295
Fastening, methods of plastic, (table) 313
Fabricating, industrial plants for, 347
Fabrication, of plastics, 283-316
   and plastic strength, 11, 12
   in reinforced plastics, 191-201
Fiberglass lamination, operation sequence, 201, (table) 202-207
Fibers, plastic, 24
Filament winding, in fiberglass, 195
Filaments, plastic, 24
Filing, of plastics, 290
Film —
   extruding plastic, 89
   plastic, 22
Film processors, in plastics industry, 346
Finance and control —
   in classroom corporation, 376
   as industrial division, 349
Financing, of classroom corporation, 358
Finishing processes —
   defined, 27
   fabrication and bonding as, 283-316
   industrial plants for, 347
Fiberglass —
   choppers, 199
   laboratory laminating equipment, 198
   manufacturing flowchart, 196
Fixed mold, in blow molding, 114
Flash mold, 140
Flat-back molds, 72
Flexural tests, 49
Flotation, 29
Flow charts, 368, 369, 370
Flow coating, of plastics, 324

Flow number, phenolic, 150
Flow rate, testing plastic, 48
Fluidized bed coating, 328
   operation sequence for, 330, (table) 332-333
Fluorine atom, 52
Fluorocarbon —
   insulation, 13
   tests on, 35, 37
Fluxing, in calendering, 180
Foam —
   backing for vinyl fabric, 186, 187
   expandable polystyrene bead, 246
   plastic, 233-235
Foam casting —
   operation sequence for resin, 241, (table) 242-245
   resin, 239
Foam molding, operation sequence for bead, 248, (table) 250-253
Foam-in-place casting, 240
Foil, hot stamping, 317, 321
Formica®, molds of, 200
Forming, roll, 178-189
Free blowing, in thermoforming, 157
Friction, 16
   and extrusion, 88
Full mold bait casting, operation sequence for, (table) 262-263
Full mold plastisol casting, 261

Gel coat resins, 198
Glass fibers, in reinforced plastics, 191-201
Granules, 24-25
Gravity, testing specific, 42
Gypsum, in cast molds, 127, 129
Gypsum molds, in thermoforming, 172-174

Heat —
   and chemical process, 53, 54
   and plastic insulation, 14
Heat fusion, 255
Heat sealing —
   industrial plants for, 346
   of plastics, 303

Heat staking, plastics fastener, 299
Heiser® process, 255, 256, 272
Hinges, plastic fasteners, 297
History, of plastics, 4
Hollow beads, plastic, 25
Hot foil stamping, 317
   operation sequence for, 321
Hot gas welding, of plastics, 307
Hot plate welding, 312
Hot wire cutting, of plastics, 289
Hyatt, John W., 1-2

Identification tests, for thermoplastics, (table) 386-387
Impact strength, of plastics, 11
Impact testing, 45
Impulse sealing, of plastics, 304
Incineration, and plastic wastes, 28
Incorporation —
   application for, 353
   certificate of, 356
Industrial production —
   organizing for, 349
   of plastics, 340-379
Industrial relations, as division of industry, 348
Industry —
   divisions within, 347-349
   the plastics, 345
Inhibitor —
   chemical, 198
   in chemical process, 52
Injection, of foamed plastics, 236
Injection blow molding, 110
Injection molding, 59-86
   operation sequence, 74, (table) 76-80
   two-color, 336, 337
Injection molds, hand operated, 70
Insulation —
   electrical, 12, 13
   heat, 14
Interchangeability, of parts, 340
In-the-mold decorating, 336
Ivory, plastics replacing, 1
Izod impact test, 45, 46

Job shops, as manufacturing method, 344
Joints, welding, 309

K shell, atomic, 51

Lamination, 190-216
   defined, 26
   operation sequence of fiberglass, 201, 202-207
   thermoplastic operation sequence, 209, 210-213
   of vinyl plastic sheets, (table) 213
Laminators, in plastics industry, 346
Liquids, plastic, 24
Log sheets, sample experimentation, 393-394

Machining —
   industrial plants for, 347
   of plastics, 283
Manual rotocasting, operation sequence for, (table) 266-267
Manufacturers —
   of plastics, 345-347
   of plastics by trade names, 395-398
Manufacturing —
   in classroom corporation, 362
   as industrial division, 348
   starting business, 350
   types of systems, 343
Marketing —
   in classroom corporation, 372
   as industrial division, 348
Mark-offs, in vacuum forming, 156
Mask spray painting, of plastics, 324
Mass production, of plastics, 340-379
MIK peroxide, in polyester resin, 198
Mechanical fasteners, and types of plastics, (table) 312
Mechanical fastening, of plastics, 295
Mechanical stretch forming, 157

Melt index, and laboratory injection molding, 84, 85
Melt index test, 48
Metal coating, of plastics, 326
Metal molds, in thermoforming, 174
Microballons®, casting, 219
Mill, roll forming, 178
Mold makers, 347
Molders, in plastics industry, 346
Molding —
  bead foam, 246
  blow, 108-131
  compression and transfer, 132-154
  of foamed plastics, 236
  full mold casting, 261
  injection, 59-86
  methods of plastic, 26-27
  operation sequence for bead form, 248, (table) 250-253
  operation sequence for single-axis rotational, (table) 274
  operation sequence for static powder, 269, (table) 270-271
  operation sequence for two-axis rotational, (table) 278-281
  single-axis rotational, 272
  static powder, 268
  two-axis rotational, 275
  two-color plastic, 336
Molds —
  cast aluminum, 125
  cast epoxy, 127
  casting, 217
  casting silicone and urethane, 230
  ceramic thermoforming, 175
  compression, 140-141
  decorating plastics in, 336
  fiberglass, 199, 200
  flash, 140
  fully positive, 133, 140
  hand operated, 70
  hardwood, 129
  machined aluminum, 124
  making compression, 149
  making injection, 81-83
  metal thermoforming, 174
  plaster thermoforming, 172
  plastic thermoforming, 174
  resin foam casting, 239, 240
  rising and fixed, 113, 114
  semipositive, 140
  wooden thermoforming, 172
Molecules —
  in plastics, 50
  in plastics makeup, 7-8
Monoforming, 161
Monomer molecule, 50
Monomeric cements, 293

Names, of specific plastics, (table) 382-385
Nitrogen atom, 52
Nondestructive tests, 33
Nylon strands, in reinforced plastics, 191

Odor, testing heated, 40
Operation analysis sheet, sample, 366
Operation process chart, sample, 367
Organization chart, of classroom corporation, 361
Oxygen atom, 52
Ozite®, 11

Packaging —
  in classroom corporation, 372
  thermoforming, 159
Painting, of plastics, 323-325
Parison, defined, 108
Parting line, in injection molding, 72
Partnership, 352
Pellets, plastic, 24, 25
Peroxide, in polyester resin, 198
Phenolic molding, fillers for, 151
Phenolic plastics, tests on, 35, 38
Phenolic resin, 2
Pigments, polyester color, 198
Plaster molds, in thermoforming, 172, 174
Plastics —
  audiovisual sources, 402
  characteristics of specific, (table) 382-385
  chemistry of, 50-58
  costs of types, (table) 382-385
  growth of industry, 4-6
  history of, 1-3, (table) 4
  makeup of, 7-8
Plastisol dip casting, operation sequence for, (table) 258-260, 261
Plating, metal coating, 326
Plenum chamber, in lamination, 194, 195
Plexiglas®, tests on, 35, 37
Plowback, of profit in classroom corporation, 378
Plug forming, 157
Plunger injection molding, 59, 69
  of foamed plastics, 236
Polishes, solvent, 291
Polishing, by buffing and ashing, 290
Polyester —
  as hard plastic, 10
  operation sequence for water-extended resins, (table) 226-228
  resins in lamination, 191-198
  tests on, 35, 38
  water-extended, 219
  water-extended resins, 224
Polyester resins —
  casting, 220
  operation sequence for casting, (table) 221-223
Polyethylene —
  as soft plastic, 11
  tests on, 35, 37
Polymer chain, 52
Polymer chemistry, of specific plastics, (tables) 388-391
Polymerization, 53, 54
Polypropylene —
  as soft plastic, 11
  tests on, 35, 37
Polyvinyl chloride, tests on, 35, 37
Polystyrene bead foam molding, 246
Polystyrene, tests on, 35, 37
Positive molds, 133
Powders, plastic, 24, 25
Pre-Expander, for polystyrene beads, 248
Preforming, in lamination, 194
Preforms, 134
Premix, lamination method, 194

Prepolymer —
  in compression molding, 134
  in lamination, 198
Press, lamination, 191
Presses, compression and transfer, 141, 142, 152
Press fit, plastics fastener, 299
Pressure —
  and chemical process, 53, 54
  in compression molding, (table) 143
  forming, 156
  in injection molding, 73
  in lamination of vinyl sheets, (table) 213
Pricing, in classroom corporation, 373
Processes —
  blow molding, 108-131
  casting plastic materials, 217-232
  coating and decorating, 317-339
  compreg lamination, 214
  compression and transfer molding, 132
  expansion, 233-254
  extrusion, 87
  fabrication and bonding, 283-316
  lamination, 190-216
  injection molding, 59
  molding, 26-27, (table) 27
  roll forming, 178-189
  thermoforming, 155
  thermofusion, 255-282
  thermoplastic lamination, 207
Processors, plastics, 346-347
Production —
  in classroom corporation, 371
  industrial plastics, 340-379
  organizing for, 349
Profile shapes, plastic, 23
Profiles, plastic extrusion, 87
Profit and loss, in classroom corporation, 377
Proprietorship, 351
Prospectus, for classroom corporation, 360
Publications, plastics listing, 399-402

Purchase order, in classroom corporation, 378
Pyrolysis, 31

Quality control, in classroom corporation, 370

Reactive adhesives, 292
Reactor tank, 52
Recycling —
  and extrusion, 105
  of plastics, 29
References, plastics, 381
Reinforced plastics, 191-198
  industrial materials, 196-198
  manufacturers, 346
Release paper, in vinyl fabric casting, 185, 186
Requisition, in classroom corporation, 378
Research and development (R&D) —
  in classroom business, 350-351
  of classroom corporation, 362
  industrial, 347
Resins —
  casting polyester, 220
  laminating, 198
  operation sequence for casting polyester, (table) 221-223
Resin foam casting, 233, 239
  operation sequence for, 241, (table) 242-245
Ribbon extrusion, 95
Rising mold, in blow molding, 113
Rivets, plastic fasteners, 298
Rod —
  extruding plastic, 89
  plastic, 23
Roll forming, 178-189
  defined, 27
Roller coating, of plastics, 325
Rotational molding —
  operation sequence for single-axis, (table) 274
  operation sequence for two-axis, (table) 278-281
  single-axis, 272
  two-axis, 275
Rotary wheel, blow molder, 116

Rotocasting, operation sequence for manual, (table) 266-267
Routing, to cut plastics, 285-286
Rovings, fiberglass strands, 192
Rubber, casting urethane molds, 230

Safety —
  and compression molding, 148
  and extrusion, 97
  in thermoforming, 172
Sales, industrial, 348
Sanding, of plastics, 290
Sanitary landfill, and plastics, 28
Sawing, plastics, 284-286
Scraping, of plastics, 290
Screw extrusion, of plastic foam, 235
Screw injection molding, 60, 66
Screws, plastic fasteners, 296
Sealing, of plastics, 303-307
Shearing, of plastics, 286
Sheeting —
  extruding plastic, 89, 91
  plastic, 22
Sheeting processors, in plastics industry, 346
Shrink packaging, 160
Silicone —
  in casting, 219
  and temperatures, 15
  tests on, 35, 38
Silk screen decorating, 322
Single-axis rotational molding, 272
  operation sequence for, (table) 274
Skin packaging, 160
Slush casting, operation sequence for, (table) 264-265
Snap fit, plastics fasteners, 299
Softening point, testing, 39
Solubility test, 41
Solvents, to polish plastics, 291-292
Solvent cementing, 293, 296
Specific gravity, testing, 42
Speed nuts, plastic fasteners, 296

415

Spin welding, 310
Spray painting, of plastics, 323
Spray-up, lamination process, 193, 194
Spray and wipe painting, 325
Spring clips, plastic fasteners, 296
Staking, heat and ultrasonic, 299, 300
Stamping —
  hot foil, 317
  operation sequence for hot foil, 321
Static molding, operation sequence for, 269, (table) 270-271
Static powder molding, 268
Static thermofusion, 255
Stock certificates, of classroom corporation, 360
Strand extrusion, 95
Strength, impact and plastics, 11
Stretch forming, mechanical, 157
Structural molding, of foamed plastics, 236
Styrofoam®, for insulation, 14
Subcontractors, as manufacturing method, 343
Substrate, in coating plastics, 182
Supported material, in coating plastics, 182
Symbols, for flow charts, 369
Synthetic, plastic as, 7-8
Synthetic wood, 219
  cast, 229

Teflon® —
  as insulation, 13
  tests on, 35, 37
Temperatures —
  in compression molding, 143
  conversion table to metrics, 394
  for extrusion, (table) 102, 103
  in injection molding, 73
  in lamination of vinyl sheets, (table) 213
  and plastics, 15

Tensile testing, 44
Terpolymer plastics, 9
  chemistry of, 56, 57
Tests —
  destructive and nondestructive, 32-49
  to identify thermoplastics, (table) 386-387
Thermal sealing, of plastics, 303
Thermoforming, 155-177
  defined, 26
  foam-in-place, 240
  operation sequence, 171, (table) 168-171
Thermofusion, 255-282
  defined, 26
Thermoplastic lamination, 207-213
Thermoplastics —
  chemical chains of, 57
  defined, 8
  and extrusion temperatures, (table) 102, 103
  grades for laboratory injection molding, (table) 85
  identification tests for, (table) 386-387
  and injection molding, 65, 66
  injection molding temperatures, (table) 80
  processes used on, (table) 27
  properties of, (table) 20
  tests on, (tables) 35, 37
  welding chart, 308
Thermosetting plastics —
  chemical chains of, 57
  chemistry of, (table) 392
  defined, 8
  and injection molding, 65, 66
  methods of fastening, (table) 312
  processes used on, (table) 27
  properties of, (table) 20
  tests on, (tables) 35, 38
Thixotropic materials, 198
Trade names —
  of plastics, 395-398
  of plastic materials, (table) 382-385
Transfer molding, 135-154
Transparency, of plastics, 18

Tubes —
  continuous blow molding, 115
  plastic, 23
Tubing, extruding plastic, 94
Two-axis rotational molding, 275
  operation sequence for, (table) 278-281
Time —
  in injection molding, 74
  in lamination of vinyl sheets, (table) 213

Ultrasonic sealing, of plastics, 306
Ultrasonic staking, plastics fastener, 300
Ultrasonic welding, 312
  characteristics, (table) 301
Urethane, in casting, 219

Vacuum forming, 155
Vacuum metalizing, of plastic surfaces, 326, 327
Vinyl fabric —
  casting, 185-187
  operation sequence, 186
Vinyl plastic sheets, laminating, (table) 213

Water-extended polyester resins, 219, 224
  operation sequence for, (table) 226-228
Webbing, thermoforming defect, 171
Weight, of plastics, 13
Welding —
  industrial plants for, 346
  of plastics, 307-314
  and types of plastics, (table) 312
  ultrasonic characteristics, (table) 301
Welds, of plastic, 315
Wire coating, 94
Wire cutting, hot, 289
Wood —
  cast synthetic, 229
  molds, 129
  synthetic, 219
Wood molds, in thermoforming, 172

MICHIGAN
CHRISTIAN
COLLEGE
LIBRARY
ROCHESTER, MICH.